Platinum and Other Metal Coordination Compounds in Cancer Chemotherapy 2

Platinum and Other Metal Coordination Compounds in Cancer Chemotherapy 2

Edited by

H. M. Pinedo

Free University
Amsterdam, The Netherlands

and

J. H. Schornagel

Netherlands Cancer Institute
Amsterdam, The Netherlands

Plenum Press • New York and London

Library of Congress Cataloging-in-Publication Data

On file

Proceedings of the Seventh International Symposium on Platinum and Other Metal Coordination Compounds in Cancer Chemotherapy, held March 1 – 4, 1995, in Amsterdam, The Netherlands

ISBN 0-306-45287-1

A Division of Plenum Publishing Corporation
233 Spring Street, New York, N. Y. 10013

10 9 8 7 6 5 4 3 2 1

Printed in the United States of America

PREFACE

The 7th International Symposium on Platinum and other metal coordination compounds in Cancer Chemotherapy, ISPCC '95, organized by the European Cancer Centre, was held in Amsterdam, the Netherlands, March 1-4, 1995. As with previous ISPCC meetings, the goal of ISPCC '95 was to bring together clinicians, clinical investigators, scientists, and laboratory workers from many disciplines to promote further collaboration and cooperation in the development of new platinum and other metal coordination compounds as well as in new ways to use 'classical' drugs as cisplatin and carboplatin in the treatment of cancer.

Important aspects addressed by experts in the field included the synthesis and activity of new platinum compounds, the biochemistry and molecular pharmacology as well as the clinical pharmacology of this class of antineoplastic agents, an overview of current clinical studies, one special minisymposium on the mechanisms of cell kill of platinum, and one on resistance against platinum compounds. Finally, the current status of development of nonplatinum metal complexes was discussed.

This volume contains the contributions of the various speakers at ISPCC '95 and provides an up-to-date and comprehensive overview of this important class of anticancer agents, ranging from synthesis and molecular pharmacology on one hand to clinical pharmacology and clinical investigations on the other hand.

The Organizing Committee and Editors wish to express their gratitude to the contributors to this volume, to the various organizations and pharmaceutical companies for their generous sponsoring of ISPCC '95, and to the Plenum Publishing Company for their help in producing this volume.

Herbert M. Pinedo
Jan H. Schornagel

CONTENTS

SYNTHESIS AND ACTIVITY OF PLATINUM COMPOUNDS

CLINICAL PHARMACOLOGY

BIOCHEMISTRY AND MOLECULAR PHARMACOLOGY

CLINICAL STUDIES

NMR SPECTROSCOPY OF PLATINUM DRUGS : FROM DNA TO BODY FLUIDS

Kevin J. Barnham,[1] Susan J. Berners-Price,[2] Zijian Guo,[1]
Piedad del Socorro Murdoch[1] and Peter J. Sadler*[1]

[1]Department of Chemistry, Birkbeck College, University of London
Gordon House and Christopher Ingold Laboratories
29 Gordon Square, London WC1H 0PP, UK.
[2]School of Science, Griffith University
Nathan, Brisbane, Australia 4111

INTRODUCTION

In this paper we illustrate how NMR spectroscopy can provide new insights into the mechanism of action of platinum anticancer drugs, including the detection of intermediates in DNA platination reactions and metabolites in body fluids. The use of modern inverse detection methods and pulsed field gradients in particular now allows work to be carried out on intact biological samples at concentrations of physiological relevance.

We address the problems of the detection of aqua-chloro intermediates during reactions of cisplatin with DNA, the possible involvement of ring-opened species in the mechanism of action of carboplatin, the role of L-methionine adducts in activation and deactivation of platinum drugs, and show that both substitution and reduction are likely to be important reactions for Pt(IV) anticancer complexes.

USEFUL NUCLEI

Table 1 lists some of the nuclei which are useful for the study of platinum drugs.[1,2] ^{1}H is the most sensitive (only radioactive ^{3}H has a higher magnetic moment), and together with ^{13}C and ^{15}N are invaluable for investigating the structure and dynamics of platinum complexes and their ligand exchange and redox (Pt(II)/Pt(IV)) reactions with biomolecules. We will discuss only solution studies here although solid-state NMR spectroscopy can also be applied to platinum compounds.

Table 1. Properties of some of the nuclei useful in the study of platinum anticancer complexes.

Nucleus	Nuclear spin	Abundance	Receptivity[1]
^{1}H	1/2	99.98	5682
^{13}C	1/2	1.11	1
^{14}N	1	99.63	5.7
^{15}N	1/2	0.37	0.02
^{31}P	1/2	100	377
^{195}Pt	1/2	33.8	19.1

[1] Receptivity = Sensitivity x natural abundance

^{14}N NMR spectroscopy can be useful for ammine and amine complexes, but it is a quadrupolar nucleus and quadrupolar relaxation, which dominates when the environment of ^{14}N has a low symmetry, can lead to very broad lines and a consequent reduction in sensitivity. On the other hand, one advantage of short relaxation times is that rapid pulsing can be employed and a large number of transients can be acquired in a short time. Thus it is possible to follow reactions of cisplatin in blood plasma and cell culture media at millimolar drug concentrations and to detect ammine release.[3]

^{195}Pt NMR spectroscopy would often be the method of choice. It has a very large chemical shift range (10,000 ppm). Pt(II) is readily distinguishable from Pt(IV), and ligand substitutions usually produce predictable chemical shift changes. However, the sensitivity of detection is relatively low (concentration limit for direct detection of ^{195}Pt is ca. 10 mM) which precludes detection of Pt in physiological fluids. The receptivity can be improved by a factor of three (Figure 1) by isotopic enrichment of the 33.8% natural abundance of ^{195}Pt to > 95% , although this is expensive. Also the linewidths of ^{195}Pt resonances can be very large, especially with ^{14}N ligands and when the complex has a low symmetry (i.e. worse for Pt(II) than Pt(IV)). The latter is due to relaxation *via* chemical shift anisotropy, which is also the cause of broadening of ^{195}Pt satellites of ^{1}H, ^{13}C or ^{15}N NMR resonances from ligands coordinated to Pt. Since this relaxation is proportional to B_o^2, the satellites are often broadened beyond detection at high observation fields (e.g. > 7 T).

The low receptivity of ^{15}N limits its usefulness for directly detected ^{15}N NMR studies of Pt ammine and amine complexes. The sensitivity of detection can be improved by ^{15}N isotopic enrichment combined with enhancement by polarisation transfer from ^{1}H (e.g. ^{15}N-{^{1}H} DEPT and INEPT pulse sequences). These have the additional advantage of allowing more rapid pulsing as the repitition time of the pulse sequence is governed by the ^{1}H rather than the longer ^{15}N spin-lattice relaxation time (T_1). For example, ^{15}N-{^{1}H} DEPT NMR methods have allowed detection of rapidly changing intermediates in the reaction of ^{15}N-cisplatin with glutathione[4] and also ammine release following reaction with intracellular components in intact red blood cells at concentrations as low as 1 mM.[5] The maximum enhancement in ^{15}N signal intensity achievable *via* polarization transfer is only 9.8 (γ_H/γ_N) which means that inverse (^{1}H-detected) ^{15}N methods are usually preferred due to the superior enhancement for ^{15}N-detection (*vide infra*). However, direct ^{15}N-{^{1}H} DEPT/INEPT methods can be of value, for example in situations where ^{1}H NMR resonances are very broad.

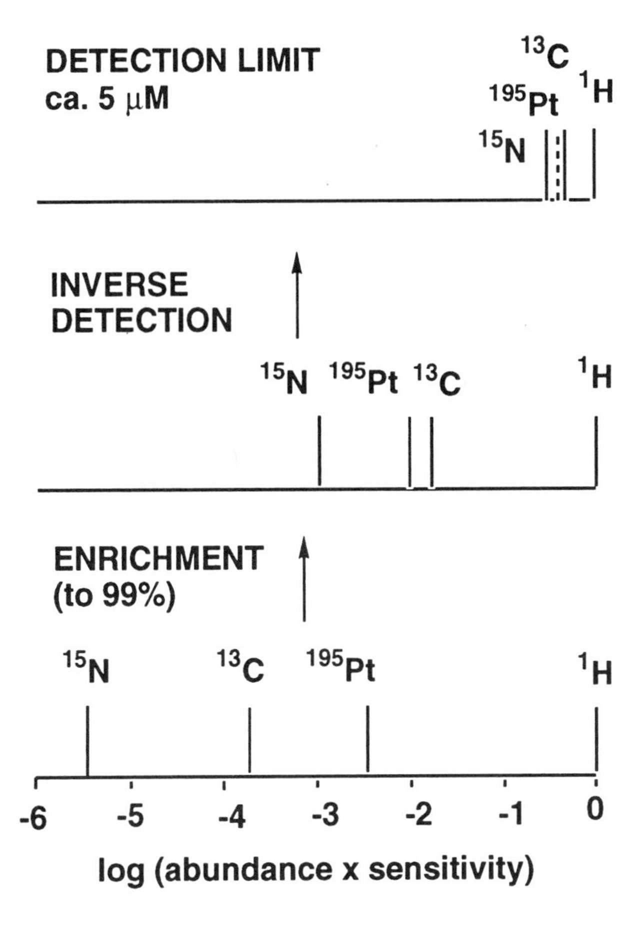

Figure 1. The theoretical increase in receptivity (abundance x sensitivity) obtainable by isotope enrichment and inverse ^{1}H detection of ^{13}C, ^{15}N and ^{195}Pt. In practice inverse ^{1}H-{^{195}Pt} detection is limited by the linewidths of the ^{195}Pt satellites which are often broadened beyond detection.

INVERSE DETECTION METHODS

By the use of inverse detection methods (i.e. ^{1}H-detected ^{13}C or ^{15}N), the sensitivity of ^{13}C and ^{15}N can be greatly improved (e.g. for ^{15}N by a maximum of $(|\gamma_H|/|\gamma_N|)^{5/2} = 306$), such that signals can often be detected at concentrations as low as 5 μM, Figure 1. However ^{1}H-detected inverse methods are restricted to ^{13}C and ^{15}N atoms which have measurable spin-spin couplings to ^{1}H (i.e. ammine, primary and secondary amines, not tertiary amines). In practice the best are the large one-bond couplings (ca. 73 Hz for $^{15}NH_3$). Besides the high sensitivity, inverse detection also brings a simplification of complicated spectra since it is possible to detect only those protons which are directly attached to the labelled ^{13}C or ^{15}N atoms in the sample. This is especially important for investigations of ^{1}H NMR spectra of body fluids or cell culture media which consist of thousands of overlapping resonances. These can be completely filtered out.

Although ^{1}H NMR resonances can be detected from NH protons with ^{14}N present in natural abundance (99.6%), they are often broad because of the quadrupolar relaxation of ^{14}N

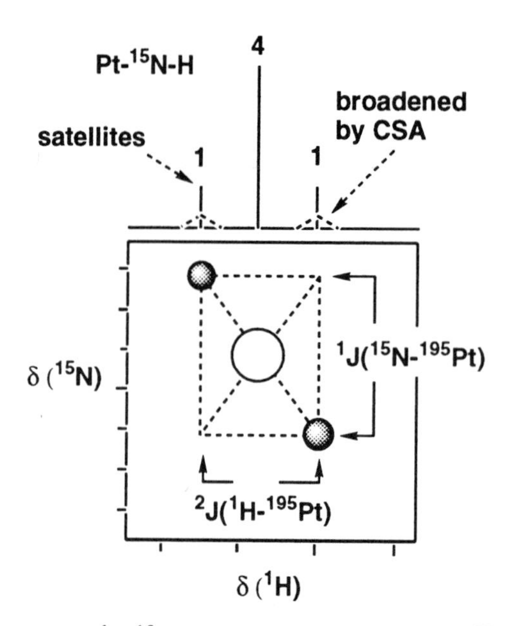

Figure 2. General appearance of a 2D [^{1}H, ^{15}N] HMQC or HSQC spectrum. The ^{195}Pt satellites are usually more intense for symmetrical Pt species (Pt(IV) rather than Pt(II)).

(I = 1). Also it is necessary to work in H_2O as opposed to D_2O since NH protons in platinum ammine and amine complexes usually exchange with D within minutes. There are some exceptions and studies of such exchange rates may themselves give valuable information about unusual NH interactions in complexes. Introduction of ^{15}N by synthetic labelling usually gives rise to a sharp ^{1}H NMR doublet for a Pt-NH group in H_2O. The resonances move progressively to lower field on changing from Pt-NH_3, to Pt-NH_2 to Pt-NH, and those for Pt(IV) are at lower field compared to those of Pt(II) (*vide infra*).

[^{1}H, ^{15}N] NMR SPECTROSCOPY

To obtain spectra from Pt-^{15}NH systems we use heteronuclear single (or multiple) quantum coherence (HSQC and HMQC) pulse sequences. By acquiring only the first increment in a two-dimensional experiment, a 1D ^{1}H spectrum is obtained containing only resonances from protons bonded to ^{15}N; resonances for CH and OH (including water) are eliminated. In practice the water resonance is so intense that it is usually necessary to use additional solvent suppression techniques (e.g. presaturation). Alternatively, the use of sequences in which coherence selection is achieved by the use of pulsed-field gradients allows greatly improved water suppression so that, with care, it is possible to detect peaks very close to water and at very low concentrations (down to ca. 5 μM). If ^{15}N decoupling is also employed (e.g. the GARP method), then each distinct type of Pt-NH_3 resonance appears as a singlet, sometimes together with the broadened ^{195}Pt satellites. In a 2D [^{1}H, ^{15}N] spectrum, the ^{195}Pt satellites appear on a diagonal, Figure 2.

The combined detection of ^{1}H and ^{15}N in a 2D inverse NMR experiment is especially powerful since the ^{15}N NMR chemical shift is diagnostic of the trans ligand,[6,7] as indeed is the

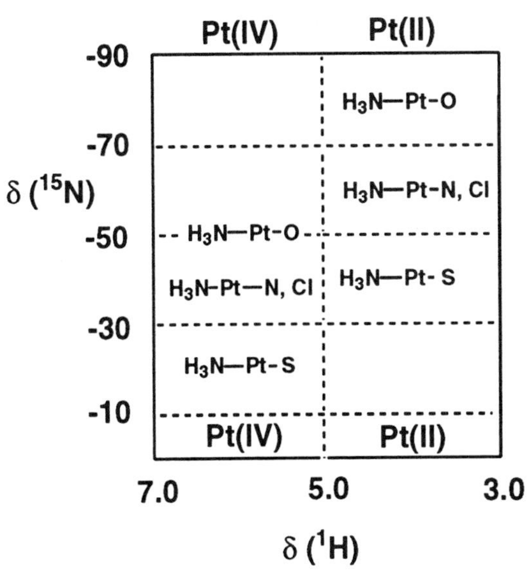

Figure 3. Variation of ^{1}H and ^{15}N NMR chemical shifts with the *trans* ligand in Pt-NH_3 complexes. A similar picture is obtained for amine complexes but the shifts are offset in both dimensions, e.g. for ethylenediamine ^{1}H peaks are shifted to low field by ca 2 ppm , and ^{15}N shifts to low field by ca 40 ppm. Thus not only can Pt(II) be distinguished from Pt(IV), but also Pt-NH_3 from Pt-NH_2 and Pt-NH.

one-bond coupling constant $^1J(^1H-^{15}N)$. Pt(II) and Pt(IV) ammine/amine complexes can be distinguished by the combination of ^{1}H and ^{15}N shifts, Figure 3.

As it takes only a few minutes to acquire a 1D ^{15}N-edited ^{1}H spectrum, reaction pathways can be followed by observing time-dependent changes in the intensity of the Pt-NH ^{1}H resonances. 2D [^{1}H,^{15}N] spectra need be acquired only at selected time intervals to aid assignment of Pt-N-H species *via* their ^{15}N chemical shifts.

The following sections provide some examples of the applications of [^{1}H, ^{15}N] and related techniques in the study of platinum anticancer drugs.

REACTIONS OF Pt(IV) COMPLEXES WITH GLUTATHIONE

In general Pt(II) and Pt(IV) can readily be distinguished by [^{1}H, ^{15}N] NMR spectroscopy, and in some complexes exchange of NH protons with solvent is much faster for Pt(IV) than for Pt(II) complexes. This can be related to the low pK_a values of NH_3 ligands on Pt(IV). For example the pK_a of $[Pt(en)_3]^{4+}$ is 5.5 and $[Pt(NH_3)_6]^{4+}$ 7.9, as discussed by Erickson *et al.*[8] This means that some of the neutral Pt(IV) ammine and amine complexes which are being tested for anticancer activity may be negatively-charged at neutral pH. NH exchange can be slowed down by lowering the pH, making it possible to study these complexes by [^{1}H, ^{15}N] NMR.

We have investigated reactions of the Pt(IV) complex *cis, trans, cis*-$[PtCl_2(OH)_2(NH_3)_2]$ with the tripeptide glutathione (γ-Glu-L-Cys-Gly). Not only do we detect [^{1}H, ^{15}N] NMR peaks for the reduction product cisplatin and for Pt(II) thiolate complexes, but also for a range of substituted Pt(IV) species, Figure 4. These include equatorial and axial

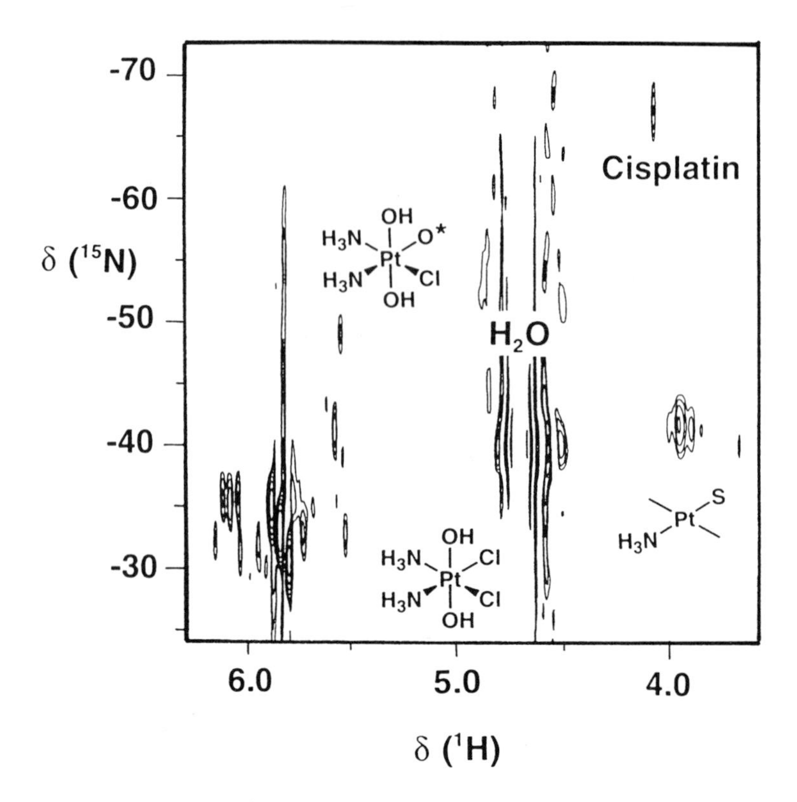

Figure 4. [^{1}H, ^{15}N] NMR spectrum of the Pt(IV) complex *cis, trans, cis*-[$PtCl_2(OH)_2(NH_3)_2$] after reaction with the tripeptide glutathione (1:1, 5 mM, pH 3.7, 310 K) for 3 h. Some tentatitive assignments are indicated. The most intense cross-peaks are for the starting complex (with streaking in the ^{15}N dimension due to relaxation effects) together with its ^{195}Pt satellites. Several products also have cross-peaks in the Pt(IV) region of the spectrum. The peak near -50/5.5 ppm is probably due to a Pt(IV) complex containing an NH_3 ligand *trans* to oxygen (O*). On the right hand side, cross-peaks for cisplatin and Pt(II)-NH_3 complexes with sulfur as the *trans* ligand can be seen.

substitutions. It is possible that other functional groups on GSH could be acting as ligands for Pt(IV) e.g. carboxylate groups.

[^{1}H, ^{15}N] NMR spectroscopy can provide insights into photoactivation reactions of Pt(IV) complexes and the possibility that some of the products observed in Figure 4 are light-induced cannot be ruled out yet. Although exposure to light was low, no attempt was made to totally exclude it.

CISPLATIN HYDROLYSIS AND ACTIVATION

Since the hydrolysis products of cisplatin *cis*-[$PtCl(H_2O)(NH_3)_2$]$^+$ and *cis*-[$Pt(H_2O)_2(NH_3)_2$]$^{2+}$ are potential intermediates in its reactions, we characterized them by [^{1}H, ^{15}N] NMR.[9] By observation of ^{1}H and ^{15}N NMR chemical shifts as a function of pH, their pK_a values were determined to be 6.41, and 5.37 and 7.21, respectively. Such titrations can be carried out at low concentration (millimolar) and any hydroxo-bridged species which form during the course of the reaction can be readily detected. In principle, the pK_a values of aqua ligands on any Pt(II) ammine or (primary or secondary) amine complex can be determined by the same method. Such information is valuable in understanding structure-activity relationships because of the reactivity of bound aqua ligands but inertness of hydroxo ligands.

^{15}N-Cisplatin in water at 310 K for 40 h (at equilibrium) gave rise to [^{1}H, ^{15}N] resonances for unreacted cisplatin, the monoaqua and diaqua adducts in a ratio of 0.64:0.35:0.01, respectively, from which an equilibrium constant of 2.72 for the first stage of cisplatin hydrolysis was calculated.[9]

NH HYDROGEN-BONDING INTERACTIONS IN NUCLEOTIDE ADDUCTS

The course of the reaction between ^{15}N-cisplatin and the mononucleotide 5′-guanosine monophosphate (5′-GMP) can be followed by [^{1}H, ^{15}N] NMR. The short-lived aqua-chloro intermediate is detectable in the early stages, followed by the formation of the mono- and bis-GMP adducts.[10] The large low field shift of the NH_3 ^{1}H NMR resonance for *cis*-$[Pt(GMP)_2(NH_3)_2]^{2+}$ was notable. The pH dependence of this peak mirrored that of the ^{31}P phosphate resonance suggesting an interaction between the 5′-phosphate and the Pt-NH protons.[11] Such H-bonding has been proposed previously, and often invoked to explain the preferential kinetic reactivity of guanines with a 5′-phosphate group. Indeed this large low-field shift was not seen with 3′-GMP or G itself.

A similar behavior was observed with $[Pt(en)]^{2+}$ adducts of GMP and AMP. The low field shifts of AMP complexes were even larger on account of the slow head-to-tail isomerization of coordinated AMP. Similarly Pt-NH-5′-phosphate H-bonding is detectable for pGpG adducts but not for GpG adducts.[12]

NMR data[13] on $[Pt(en)(5'\text{-GMP})_2]^{2+}$ suggest that such H-bonding can occur in solution both for dianionic 5′-phosphate groups and for monoanionic groups (as they are in DNA). We crystallized this complex at pH 2 where the phosphate is monoanionic and in the X-ray structure there is clearly PtNH-5′-phosphate H-bonding. This is also present in solution because a ^{31}P-{^{1}H} nuclear Overhauser effect is observed. The X-ray structure emphasizes the potential role which water molecules play in the H-bonding network. There are electrostatically-bound axial water molecules 3.3 Å from Pt which are H-bonded to C6O and Pt-NH. We conclude that there is a rearrangement of the H-bonding network when the charge on phosphate is changed from minus one to minus two.

PATHWAYS OF DNA PLATINATION REACTIONS

We have investigated the detailed kinetics of the reaction of ^{15}N-cisplatin with the decamer oligonucleotide d(ACATGGTACA) and with the duplex containing the complementary strand.[14] This has allowed us to determine directly for the first time the lifetime of the aqua-chloro intermediate (8 min), and to observe the formation of monofunctional and bifunctional adducts in a single experiment (Figure 5). Our kinetic data (Figure 6) are in close agreement with those determined or estimated by Bancroft *et al.*[15] from ^{195}Pt NMR studies of

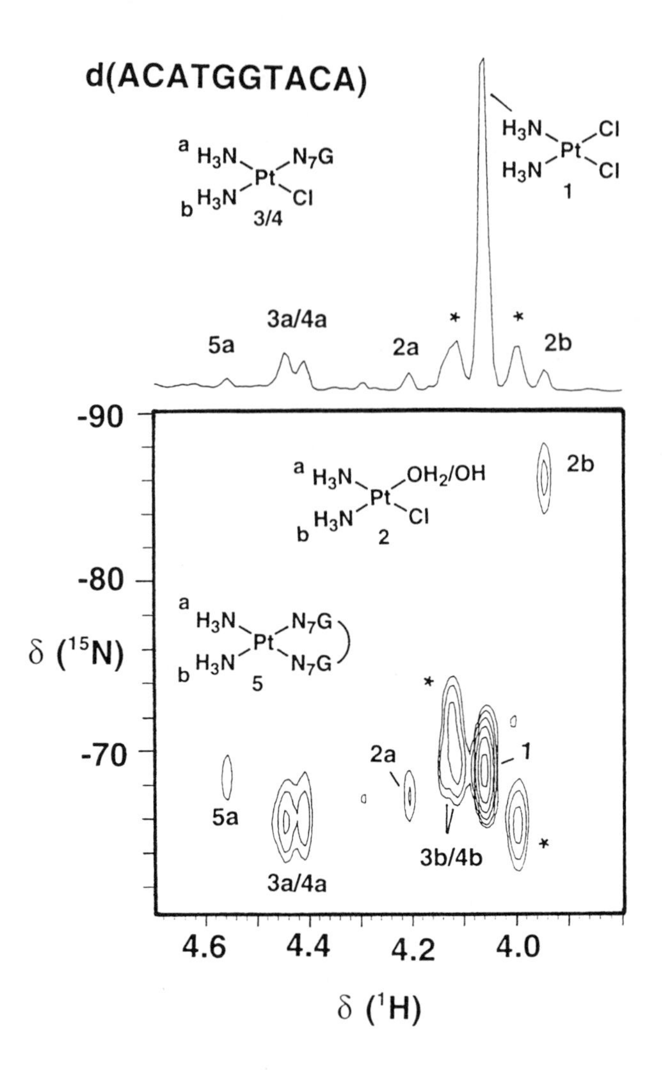

Figure 5. [^{1}H, ^{15}N] 2D NMR spectrum of ^{15}N-cisplatin after reaction with the single strand d(ACATGGTACA) (1:1 mol ratio, 1.3 mM pH 7.1, 310 K) for 1.5 h

cisplatin and monoaqua cisplatin reactions with chicken erythrocyte DNA fragments (ca. 40 base pairs) at higher concentration (ca. 15 mM and using enriched ^{195}Pt).

One of the monofunctional adducts is present during the course of the reaction to a greater extent than the other, in line with the findings of Chottard, Kozelka and co-workers[16] that there is a preferential rate of ring closure (platinated 3′-G ring-closing faster than platinated 5′-G). Other products were observed during the later stages of the reaction that may be due to N7/N1 crosslinks (pK_a of N1 being lowered by platination of N7).

The [^{1}H, ^{15}N] NMR shifts of duplex platinated with cisplatin to give a bifunctional GG adduct are sensitive to duplex melting. Curiously, a large low field shift of one of the Pt-NH_3 resonances is observed just after the duplex melts, which suggests that a platinated single strand still possesses secondary structure.

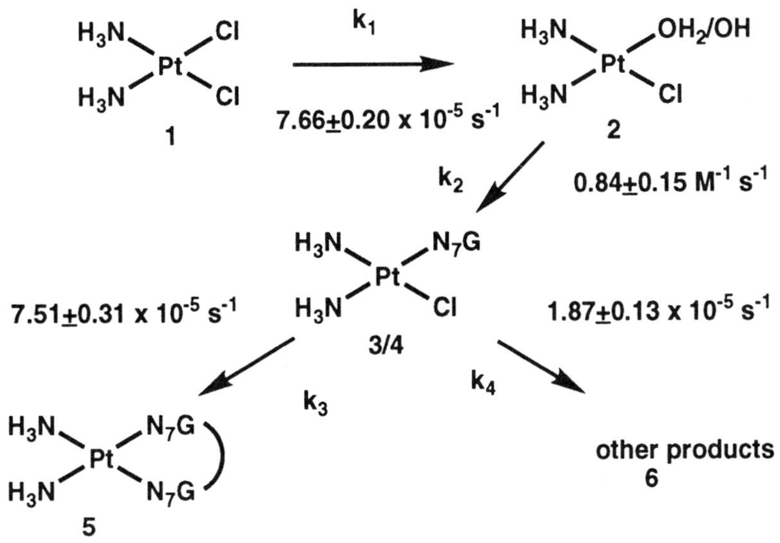

Figure 6. Summary of kinetic parameters for platination of the single strand d(ACATGGTACA) with cisplatin at 310 K, pH 7 (ref 14). No distinction was made between the two monofunctional adducts to simplify the kinetic fit. Note that both the initial (k_1) and mono- to bi-functional (k_3) stages are controlled by the hydrolysis rate.

METHIONINE METABOLITES

Pt(II) being a "soft" metal ion is known to have a very high affinity for "soft" ligand atoms such as sulfur. Most of the rescue agents being tested for the removal of Pt from the body are sulfur-containing ligands, e.g. glutathione, diethyldithiocarbamate, and N-acetyl-L-cysteine. The amino acid and thioether L-methionine is thought to play a role in the metabolism of cisplatin and the complex [Pt(L-Met)$_2$] was isolated from the urine of patients over 10 years ago.[17] Then it was described as having the *trans* geometry. The complex is readily produced during chemical reactions of cisplatin with L-Met and by the use of ^{15}N-labelling we have characterized it in detail by ^{1}H and ^{15}N NMR spectroscopy. In solution [Pt(L-Met)$_2$] exists as a mixture of *cis* and *trans* isomers, the former predominating by ca. 87:13, and each of these exists as a further 3 diastereomers due to the presence of two chiral coordinated S atoms.[18,19] The *cis* and *trans* isomers can be separated by HPLC but the inversion of S is too rapid to allow separation of the diastereomers.

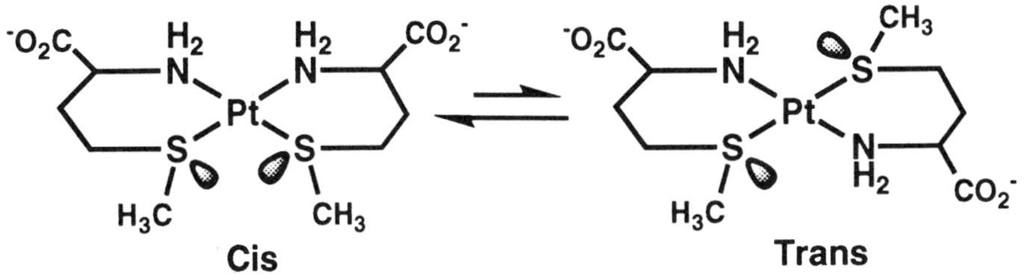

Remarkably the rates of *cis-trans* isomerization of [Pt(L-Met)$_2$] are very slow with a half-life of ca. 22.4 hours for the conversion of *cis* into *trans* and 3.2 h for *trans* to *cis* at 310 K. The electronic distributions are different in the two isomers. Not only does the *cis* isomer have a more extensive hydrophobic face (retained longer on a reverse-phase HPLC column) but it has a higher partial charge on Pt (+1.74) compared to the trans isomer (+1.15). Both of these factors could have a major influence on their passage through membranes and on their excretion rates.

At physiological pH (7.4) these isomers appear to be unreactive, although in the presence of chloride at lower pH, one of the chelate rings can open up.

There is considerable interest in the potential for methionine adducts of cisplatin as the cause of side-effects, e.g. nephrotoxicity, and cytotoxicity. Recently data for male Wistar rats have been reported by Deegan *et al.*[20] We confirmed by [^{1}H, ^{15}N] NMR spectroscopy using ^{15}N-labelled cisplatin and L-methionine that under the conditions they described (overnight incubations of mixtures of cisplatin and excess L-methionine at 310 K) that they were investigating the properties of [Pt(L-Met-S,N)$_2$]. Interestingly they report that this species lacks cisplatin-associated renal toxicity but is significantly cytotoxic towards C6 glioma cells.

Basinger *et al.*[21] have reported that simultaneous iv administration of L-methionine with cisplatin results in a significant reduction of nephrotoxicity in rats nearing Walker 256 carcinosarcoma, without any apparent effect on antineoplastic activity. However, L-methionine administration does not prevent accumulation of Pt in the kidney. It would be interesting to discover how much direct chemical reaction of L-Met with cisplatin occurs under the conditions of these administrations and [^{1}H, ^{15}N] NMR work could play a valuable role in such studies.

METHIONINE SULFUR DISPLACEMENT BY NUCLEOTIDES

Monodentate S-bound L-Met is a much more reactive ligand on Pt(II) than *S,N*-chelated L-Met and can be substituted by guanine N7. This could provide novel pathway for DNA platination.

For example when [Pt(dien)Cl]$^+$ reacted with L-Met in the presence of GMP, initially little GMP coordinated to Pt, only L-Met, but in the later stages of the reaction coordinated L-Met was displaced by N7 of GMP.[22] Van Boom and Reedijk have observed a similar reaction involving intramolecular displacement of thioether S by N7 of guanine in guanosyl-L-homocysteine.[23] Moreover we found that displacement of S by N7 on [Pt(dien)]$^{2+}$ was selective for GMP vs AMP, CMP or TMP or the amino acid L-His. It is notable that thioethers such as L-Met react with Pt(II) amines faster than thiols such as glutathione,[24] and reactions of thiols tend to be irreversible. We have also carried out reactions of cisplatin with 5′-GMP in the presence of L-methionine.[25] These are complicated by the labilization of ammonia ligands due to the strong trans influence of sulfur. Again we observe reversible binding of monodentate S-bound L-Met during the course of the reaction, and the products include [Pt(NH$_3$)(GMP-*N7*)(L-Met-*S,N*)]$^-$ where N7 is *trans* to S. If analogous adducts can form on DNA, then interesting new possibilities arise for their stabilization, recognition and repair, especially if the sulfur ligand is not L-Met itself but a Met-containing peptide or protein.

Monodentate S-bound N-acetyl-L-methionine in the complexes [Pt(en)(NAcMet-*S*)Cl] and [Pt(en)(NAcMet-*S*)$_2$] can readily be displaced by N7 of GMP or GpG.[26] This is

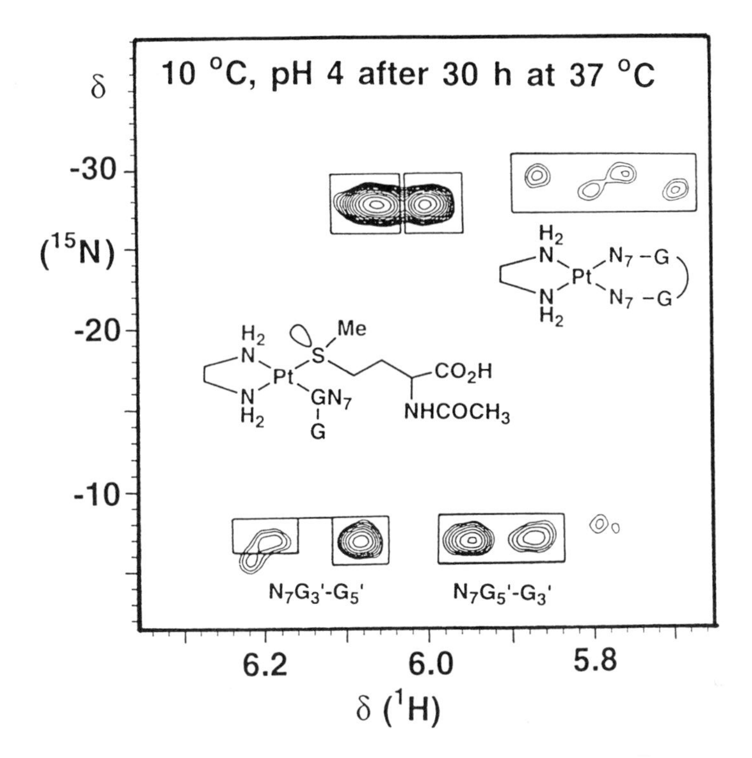

Figure 7. [^{1}H, ^{15}N] NMR spectrum recorded at 283 K of a solution containing [Pt(en)(NAcMet-*S*)Cl] and GpG in a 1:1 mol ratio after reaction at 310 K for 30 h. Tentative assignments are indicated; peaks for H_2N-Pt *trans* to N have ^{15}N shifts near -30 ppm and *trans* to sulfur near -7 ppm. At higher temperatures the doublets for the monofunctional adducts (lower part of spectrum) coalesce into singlets consistent with rapid S inversion.

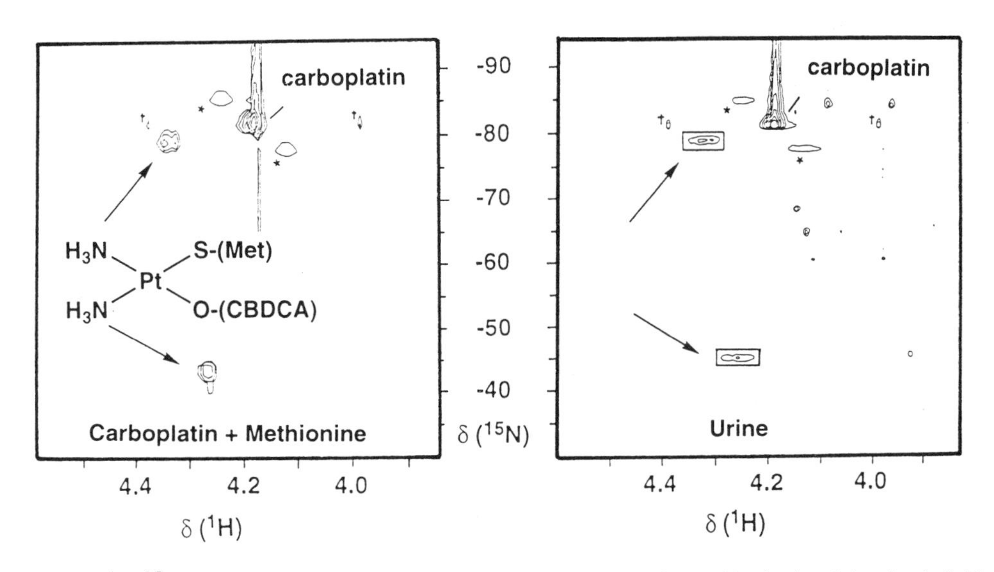

Figure 8. [^{1}H, ^{15}N] NMR spectrum of a solution containing carboplatin and L-methionine in a 1:1 mol ratio 3.5 h after mixing (left), and of urine collected from mice treated with carboplatin (right). (adapted from ref 28)

illustrated for the former complex on reaction with GpG in Figure 7. Two intermediates are detected which can be tentatively assigned to monofunctional adducts with either the 3′-G or the 5′-G bound to Pt. It is interesting that they exhibit slow inversion of coordinated S at low temperature e.g. 283 K (Figure 7), and also have a wide range of ^{1}H Pt-NH chemical shifts suggesting that hydrogen-bonding or other interactions differ significantly in these complexes. We are investigating further the stabilization of monofunctional intermediates with peptide derivatives.

RING-OPENED CARBOPLATIN ADDUCTS

Some ring-opened carboplatin complexes are surprisingly stable. The half-life of [Pt(CBDCA-*O*)$(NH_3)_2$(L-Met-*S*)], for example, is ca. 1 day at 310 K. The stability may be attributable to intramolecular H-bonding involving the uncoordinated amino and carboxyl groups of methionine and the monodentate CBDCA and ammine ligands.

A ring-opened carboplatin adduct can be detected during reactions of carboplatin with 5′-GMP.[27] This has unusual ^{1}H NMR chemical shifts, with each proton on the four-membered cyclobutane ring being magnetically non-equivalent and shifted to high field. Modelling showed that this is attributable to close hydrophobic contact with the purine ring of 5′-GMP bound by N7. Such hydrophobic contacts could be important for the stabilization of monofunctional adducts on DNA. Ring-opening by hydrolysis or chloride attack on CBDCA is very slow and the mechanism of reaction of carboplatin with 5′-GMP appears to involve direct attack.

H_3N Pt O O — GMP → H_3N Pt O O N_7-GMP

Carboplatin   Ring-opened

CHARACTERISATION OF METABOLITES IN URINE

[^{1}H, ^{15}N] NMR spectroscopy can be used to detect a wide range of metabolites in animal urine after administration of ^{15}N-labelled Pt complexes.[28] An example is shown in Figure 8. Apart from the peak for unreacted carboplatin, the most intense peaks in the spectrum (boxed) are a pair assignable to ammine ligands *trans* to oxygen and sulfur. Other peaks are visible in the Pt-NH_3 trans to S region and also in the trans to N, Cl region. The major peaks have shifts very similar to those of the ring-opened complex [Pt(CBDCA-*O*)$(NH_3)_2$(L-Met-*S*)](Figure 8). Also notable is the presence of peaks for other metabolites, one of which (-45.5/3.89 ppm) may be a glutathione conjugate.

H_3N Pt O O — L-Met → H_3N Pt O O S-Met — chelation → H_3N Pt N S + CBDCA

Carboplatin   Ring-opened

In ^1H NMR spectra of urine from patients treated with carboplatin, peaks for both intact drug and free CBDCA ligand can be assigned.[29] In initial investigations[30] of the [^1H, ^{15}N] NMR spectra of the urine from mice treated with ^{15}N-cisplatin, about 20 different types of Pt-NH_3 species were detected, including at least four with sulfur as the *trans* ligand (thioethers and thiols), and when *cis, trans, cis*-[$PtCl_2(OAc)_2(NH_3)_2$] was administered, no peaks for Pt(IV) ammine metabolites were observed, only peaks for Pt(II) complexes. The combination of NMR with chromatography should be a powerful one for further work on the indentification of novel Pt metabolites.

DETECTION OF PLATINATION SITES ON PROTEINS

[^1H, ^{15}N] NMR spectroscopy can be used to study sites of platination by Pt(II) ammines and amines even on proteins as large as serum transferrin (80 kDa). Transferrin is a single-chain glycoprotein which has two similar binding sites for Fe(III) ions situated in interdomain clefts in the N-terminal half (N-lobe) and C-terminal half of the molecule. Diferric transferrin is taken up by cells via receptor-mediated endocytosis and we are therefore interested in the possibility that this mechanism could be used to deliver Pt to tumour cells since they are known to overexpress such receptors.

A combined ^1H, ^{15}N and ^{13}C approach can be used for studying reactions of ^{15}N-cisplatin with diferric transferrin labelled at the εCH_3 groups of the Met residues with ^{13}C.[31,32] With the use of inverse detection techniques, it is possible to work at transferrin concentrations of ca. 500 μM or lower. It was clear that there is one major platination site, Met 256 in the N-lobe. This is one of two Met residues which is freely solvent-accessible, based on the X-ray crystal structure of FeC-HTF. The other is Met 499 in the C-lobe. In all there are nine Met residues in HTF, five in the N-lobe and four in the C-lobe, most of which are partially or completely buried.

CONCLUSIONS

Although, in principle, the method of choice for determining the oxidation state and types of bound ligands in platinum complexes is ^{195}Pt NMR spectroscopy, it suffers from the disadvantages of lack of sensitivity (detection limit ca 10 mM) and line broadening in unsymmetrical complexes (relaxation via chemical shift anisotropy). More sensitive, providing there is an N-H bond in the platinum complex of interest, is [^1H, ^{15}N] NMR with a detection limit ca. 5 μM using ^{15}N-labelled complexes. Here the ^{15}N resonances are detected inversely via ^1H observation. Since the ^{15}N chemical shifts are diagnostic of both the nature of the ligand *trans* to ^{15}N and of Pt(II) vs Pt(IV), this method can also be used to identify Pt coordination spheres.

Work can now be carried out using this technique at physiologically-relevant concentrations of platinum drugs in biological fluids (e.g. urine), and even with macromolecules (e.g. the blood plasma protein transferrin, 80 kDa). This technique has allowed the first direct detection to be made of the aqua-chloro intermediate during reactions of cisplatin with DNA. Studies such as these promise to provide new insights into the metabolism of platinum complexes and into factors responsible for their recognition of specific base sequences on DNA, and hence should be invaluable in the drug design process.

Acknowledgements

We thank the Medical Research Council, the Association for International Cancer Research, Australian NH&MRC, Royal Society, Wolfson Foundation, British Council, and EC (HCM and COST) for their support for our work. We are grateful to the NMR Centres at Mill Hill, Birkbeck College, and Queen Mary and Westfield College for the provision of NMR facilities, and to our collaborators who have contributed to various aspects of the work described here.

REFERENCES

1. F.M. Macdonald and P.J. Sadler, ^{195}Pt NMR Spectroscopy: applications to the study of anticancer drugs, *in*: "Biochemical mechanisms of platinum antitumour drugs", D.C.H. McBrien and T.R. Slater, eds., IRL Press Ltd., Oxford, 361 (1986).
2. P.J. Sadler, NMR Studies of metallodrugs, in: "NMR Spectroscopy in Drug Research", Alfred Benzon Symposium 26, J.W. Jaroszewski, K. Schaumburg and H. Kofod, eds., Munksgaard, Copenhagen, 252 (1988).
3. R.E. Norman and P.J. Sadler, ^{14}N NMR studies of amine release from platinum anticancer drugs: models and human blood plasma, *Inorg. Chem.*, 27: 3583 (1988).
4. S.J. Berners-Price and P.J. Kuchel, The reaction of *cis*- and *trans*-$[PtCl_2(NH_3)_2]$ with reduced glutathione studied by ^{1}H, ^{13}C, ^{195}Pt and ^{15}N-{^{1}H} DEPT NMR, *J. Inorg. Biochem.* 38: 305 (1990).
5. S.J. Berners-Price and P.J. Kuchel, The reaction of *cis*- and *trans*-$[PtCl_2(NH_3)_2]$ with reduced glutathione inside human red blood cells studied by ^{1}H and ^{15}N-{^{1}H} DEPT NMR, *J. Inorg. Biochem.* 38: 327 (1990).
6. T.G. Appleton, J.R. Hall, S.F. Ralph, ^{15}N and ^{195}Pt NMR spectra of platinum ammine complexes: *trans*- and *cis*-influence series based on ^{195}Pt-^{15}N coupling constants and ^{15}N chemical shifts, *Inorg. Chem.* 24: 4685 (1985).
7. I.M. Ismail and P.J. Sadler, ^{195}Pt and ^{15}N NMR Studies of Anti-Tumor Complexes, *JACS Symp. Ser.* 209: 171 (1983).
8. L.E. Erickson, D.J. Cook, G.D. Evans, J.E. Sarneski, P.J. Okarma, A.D. Sabatelli, Multinuclear NMR studies of platinum(IV) complexes 1. Acid dissociation, anation and intramolecular catalysis of bipyridyl proton exchange in [aqua(bipyridine)(1,3,5-triaminocyclohexane)platinum(IV)]$^{4+}$, *Inorg. Chem.* 29:1958 (1990).
9. S.J. Berners Price, T.A. Frenkiel, U. Frey, J.D. Ranford and P.J. Sadler, Hydrolysis products of cisplatin: pK_a determinations via [^{1}H, ^{15}N] NMR spectroscopy, *JCS Chem. Comm.* 789 (1992).

10. S.J. Berners-Price, T.A. Frenkiel, J.D. Ranford, P.J. Sadler, Nuclear magnetic resonance studies of N-H bonds in platinum anticancer complexes: detection of reaction intermediates and hydrogen bonding in guanosine 5′-monophosphate adducts of $[PtCl_2(NH_3)_2]$, *JCS Dalton Trans.*2137 (1992).
11. S.J. Berners-Price, U. Frey, J.D. Ranford and P.J. Sadler, Stereospecific hydrogen-bonding in mononucleotide adducts of platinum anticancer complexes in aqueous solution, *J. Am. Chem. Soc.* 115: 8649 (1993).
12. S.J. Berners-Price, J.D. Ranford and P.J. Sadler, [^{1}H, ^{15}N] Investigations of Pt-NH hydrogen bonding in d(GpG), and d(TpGpG)-*N7,N7* adducts of $[Pt(en)]^{2+}$ in aqueous solution, *Inorg. Chem.* 33: 5842 (1994).
13. K.J. Barnham, C.J. Bauer, M.I. Djuran, M.A. Mazid, T. Rau and P.J. Sadler, Outer-sphere macrochelation in [Pd(en)(5′-GMP-*N7*)$_2$]·9H$_2$O and [Pt(en)(5′-GMP-*N7*)$_2$]·9H$_2$O : X-ray crystallography and NMR spectroscopy in solution. *Inorg. Chem.* 34: 2826 (1995).
14. K.J. Barnham, S.J. Berners-Price, U. Frey, T.A. Frenkiel and P.J. Sadler, Platination pathways for reactions of cisplatin with GG single-stranded and double-stranded decanucleotides, *Angew. Chem.* (1995) in press.
15. D.P. Bancroft, C.A. Lepre and S.J. Lippard, ^{195}Pt NMR kinetic and mechanistic studies of *cis*- and *trans*-diamminedichloroplatinum(II) binding to DNA, *J. Am. Chem. Soc.* 112: 6860 (1990).
16. F. Gonnet, J. Kozelka, J.-C. Chottard, Cross-linking of adjacent guanine residues in an oligonucleotide by *cis*-$[Pt(NH_3)_2(H_2O)_2]^{2+}$: kinetic analysis of the two-step reaction, *Angew. Chem. Int. Ed. Engl.* 31: 1483 (1992).
17. C.M. Riley, L.A. Sternson, A.J. Repta, S.A. Slyter, Monitoring the reactions of cisplatin with nucleotides and methionine by reversed-phase high-performance liquid chromatography using cationic and anionic pairing ions, *Anal. Biochem.* 130: 203 (1983).
18. R.E. Norman, J.D. Ranford and P.J. Sadler, Studies of platinum(II) methionine complexes: metabolites of cisplatin, *Inorg. Chem.* 31: 877 (1992).
19. P. del S. Murdoch, J.D. Ranford, P.J. Sadler and S.J. Berners-Price, *Cis-trans* isomerization of [Pt(L-methionine)$_2$]: metabolite of the anticancer drug cisplatin, *Inorg. Chem.* 32: 2249 (1993).
20. P.M. Deegan, I.S. Pratt and M.P. Ryan, The nephrotoxicity, cytotoxicity and renal handling of a cisplatin-methionine complex in male Wistar rats, *Toxicology.* 89: 1 (1994).
21. M.A. Basinger, M.M. Jones and M.A. Holscher, L-Methionine antagonism of *cis*-platinum nephrotoxicity, *Toxicol. Appl. Pharmacol.* 103: 1 (1990).
22. K.J. Barnham, M.I. Djuran, P. del S. Murdoch and P.J. Sadler, Intermolecular displacement of S-bound L-methionine on platinum(II) by guanosine 5′-monophosphate: implications for the mechanism of action of anticancer drugs, *JCS Chem. Commun.* 721 (1994).
23. S.S.G.E. van Boom and J. Reedijk, Unprecedented migration of $[Pt(dien)]^{2+}$ (dien = 1,5-diamino-3-azapentane from sulfur to guanosine-*N7* in guanosyl-L-homocysteine (sgh), *JCS Chem. Commun.* 1397 (1993).

24. M.I. Djuran, E.L.M. Lempers and J. Reedijk, Reactivity of chloro(diethylenetriamine)platinum(II) and aqua(diethylenetriamine)platinum(II) ions with glutathione, S-methylglutathione, and guanosine 5′-monophosphate in relation to the antitumor activity and toxicity of platinum complexes, *Inorg. Chem.* 30, 2648 (1991).
25. K.J. Barnham, M.I. Djuran, P. del S. Murdoch and P.J. Sadler, manuscript in preparation.
26. K.J. Barnham, Z. Guo and P.J. Sadler, manuscript in preparation.
27. U. Frey, J.D. Ranford and P.J. Sadler, Ring-opening reactions of the anticancer drug carboplatin: NMR characterization of *cis*-[Pt(NH_3)$_2$(CBDCA-*O*)(5′-GMP-*N7*)] in solution, *Inorg. Chem.* 32: 1333 (1993).
28. K.J. Barnham, U. Frey, P. del S. Murdoch, J.D. Ranford and P.J. Sadler, [Pt(CBDCA-*O*)(NH_3)$_2$(L-methionine-*S*)]: ring-opened adduct of the anticancer drug carboplatin ("Paraplatin"). Detection of a similar complex in urine by NMR spectroscopy, *J. Am. Chem. Soc.* 116: 11175 (1994).
29. J.D. Ranford, P.J. Sadler, K. Balmanno and D.R. Newell, ^{1}H NMR Studies of human urine :urinary elimination of the anticancer drug carboplatin, *Magn. Res. Chem.* 29: S125 (1991).
30. K.J. Barnham, D.R. Newell, P.J. Sadler, unpublished results.
31. E.J. Beatty, M.C. Cox, T.A. Frenkiel, G. Kubal, A.B. Mason, P.J. Sadler and R.C. Woodworth, Ga^{3+}-induced structural changes in human serum transferrin: [^{1}H, ^{13}C] NMR studies of methionine residues in the N-lobe, *in:* Metal Ions in Biology and Medicine, Eds. Ph. Collery, N.A. Littlefield, J.C. Etienne, John Libby Eurotext, Paris, pp315-320 (1994).
32. K.J. Barnham, M.C. Cox, P.J. Sadler, J.D. Hoeschele, T.A. Frenkiel, A.B. Mason and R.C. Woodworth, manuscript in preparation.

TRANS-DIAMMINEDICHLOROPLATINUM (II) IS NOT AN ANTITUMOR DRUG: WHY?

Rozenn Dalbiès, Marc Boudvillain and Marc Leng

Centre de Biophysique Moleculaire, CNRS
Rue Charles Sadron
45071 ORLEANS Cedex 2, FRANCE

INTRODUCTION

The clinical activity of the antitumor drug *cis*-diamminedichloroplatinum(II) (*cis*-DDP) is thought to be related to its ability to form bifunctional lesions in DNA. *trans*-Diamminedichloroplatinum(II) (*trans*-DDP), the stereoisomer of *cis*-DDP, is clinically ineffective, although it forms bifunctional cross-links in the *in vitro* reaction with DNA[1-4]. A question still unsolved is why only *cis*-DDP is an antitumor drug.

The binding of *cis*-DDP to DNA induces large distortions in the double helix, which are generally considered as related to the clinical activity of this compound. These distortions have been characterized by several techniques[1-4]. Less work has been devoted to *trans*-DDP adducts in DNA. As concerns the distortions induced by the (G1,G3)-intrastrand cross-links, probably one of the major lesions in *trans*-DDP modified DNA[5-7], the conclusions of two recent studies[8,9] are in disagreement. This led us to reinvestigate some properties of oligonucleotides containing a single *trans*-{$Pt(NH_3)_2$[d(GXG)-*N7*-G,*N7*-G]} intrastrand cross-link (X stands for an adenine (A), thymine (T) or cytosine (C) residue). This paper deals with the stability of these adducts within single- and double-stranded oligonucleotides.

The bifunctional lesions in DNA of *cis*- and *trans*-DDP are considered as stable in conditions close to physiological conditions. However, Comess et al.[10] have reported the rearrangement of the *trans*-{$Pt(NH_3)_2$[d(CGCG)-*N7*-G,*N7*-G]} cross-link into the thermodynamically favoured 1,4-*trans*-{$Pt(NH_3)_2$[d(CGCG)-*N3*-C,*N7*-G]} cross-link within a single-stranded oligonucleotide. They considered that the rearrangement was specific of the d(CGCG) sequence. We show that the linkage isomerization reaction occurs in the two sequences d(CGAG) and d(CGTG). As concerns double-stranded oligonucleotides, preliminary results have shown that *trans*-{$Pt(NH_3)_2$[d(GXG)-*N7*-G,*N7*-G]} intrastrand cross-links rearrange into interstrand cross-links within 22-mer duplexes of central sequence d(PyGXGPy).d(PuCYCPu), where Y is one of the four common base residues, Py and Pu complementary pyrimidine and purine residues[11]. We extend these results to other sequences flanking the adducts. With regard to the significant amount of cross-links between two G residues in native DNA after modification with *trans*-DDP[7], we searched whether *trans*-$\{Pt(NH_3)_2[d(GXG)\text{-}N7\text{-}G,N7\text{-}G]\}$ intrastrand cross-links were stable in some sequences in *trans*-DDP-modified DNA. We find that, at low level of platination, (G1,G3), (G1,G4)- and (C1,G4)-intrastrand cross-links, if formed, are minor adducts in *trans*-DDP modified DNA restriction fragments.

RESULTS

Oligonucleotides of various sequences but containing a single central d(GXG) sequence were reacted with *trans*-DDP under acidic conditions (pH 3.1-3.6)[8-11]. After incubation during 24 h at 37 °C , and then treatment in 10 mM thiourea (10 min at 37 °C) to remove monofunctional adducts[6,10], the oligonucleotides containing a (G1,G3)-intrastrand cross-link were purified by FPLC[8,11]. The stability of the (G1,G3)-intrastrand cross-links has been studied as a function of the conformation (single-stranded versus double-stranded oligonucleotides).

I) Single-stranded oligonucleotides

Instability of the intrastrand cross-link. The single-stranded dodecamer 5'-d(CCTCGAGTCTCC) containing a single *trans*-$\{Pt(NH_3)_2[d(CGAG)\text{-}N7\text{-}G,N7\text{-}G]\}$ intrastrand cross-link was incubated in 10 mM $NaClO_4$, 5 mM phosphate buffer, pH 7 and at 60 °C. At various times, aliquots were withdrawn and analyzed by three different techniques: 1) enzymatic hydrolysis by T4 DNA polymerase followed by gel electrophoresis. 2) C18 reverse phase HPLC. 3) 24% polyacrylamide gel electrophoresis under denaturing conditions (figure 1).

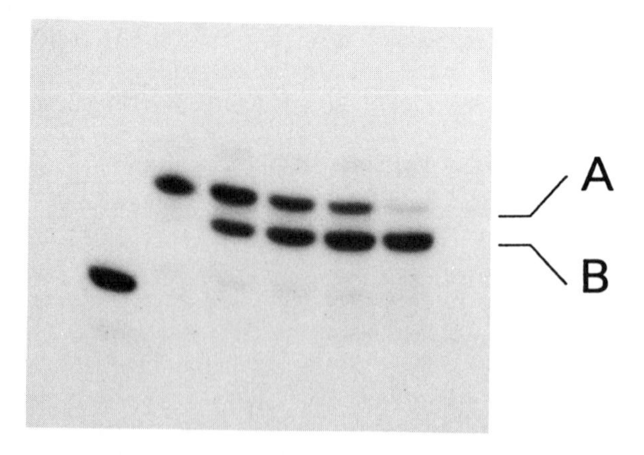

Figure 1: Instability of *trans*-$\{Pt(NH_3)_2[d(CGAG)\text{-}N7\text{-}G,N7\text{-}G]\}$ cross-link in d(CCTCGAGTCTCC) followed by gel electrophoresis under denaturing conditions. Incubation times are indicated in hours above the lines. Lane U refers to the unplatinated oligonucleotide. The bands corresponding to (C1,G4)- and (G1,G3)-intrastrand cross-links are indicated by A and B respectively, on the right side of the figure.

The three experiments revealed the formation of a new compound, while the proportion of the initial platinated oligonucleotide decreased until the two species reached an equilibrium at a 94/ 6 ratio in favour of the new compound. The $t_{1/2}$ for the disappearance of the (G1,G3)-cross-link is about 4 h.

The new product was identified as the oligonucleotide containing the 1,4-*trans*-{Pt(NH_3)$_2$[d(CGAG)-*N3*-C,*N7*-G]} intrastrand cross-link, by 1) reaction with DMS; the 5' G residue was reactive in the oligonucleotide containing the 1,4-*trans*-{Pt(NH_3)$_2$[d(CGAG)-*N3*-C,*N7*-G]} intrastrand cross-link[10,12]. 2) C18 reverse phase HPLC analysis of the digests after hydrolysis by endonuclease P1 and alkaline phosphatase[10,12-14].

These results resembled those reported by Comess et al.[10] and it was likely that we were dealing with a similar linkage isomerization reaction until the two compounds reached equilibrium (K_{eq}= 16 ± 3 in the 30-80 °C temperature range).

Parameters influencing the rearrangement. *Sequence.* Similar experiments were done with an oligonucleotide containing a single *trans*-{Pt(NH_3)$_2$[d(CGTG)-*N7*-G,*N7*-G]} intrastrand cross-link, and the results resembled those just described. The $t_{1/2}$ for the disappearance of the (G1,G3)-intrastrand cross-link is about 8 h at 60 °C (Keq ≈ 6). In first approximation, we conclude that the rearrangement is not drastically dependent on the nature of the intervening base residue (C[10], A or T). Oligonucleotides containing a single *trans*-{Pt(NH_3)$_2$[d(GXG)-*N7*-G,*N7*-G]} intrastrand cross-link within d(AGAGA), d(GGTGT) or d(TGAGC) sequences, respectively, were incubated in the conditions described above. In first approximation, no new products were detected. The replacement of the 5' C residue by a A, G or T residue prevents the linkage isomerization reaction. Moreover, no rearrangement occured when the C residue was adjacent to the adduct on its 3' side.
Temperature. The stability of the *trans*-{Pt(NH_3)$_2$[d(CGAG)-*N7*-G,*N7*-G]} intrastrand cross-link within d(CCTCGAGTCTCC) was studied as a function of temperature in the range 30-80 °C. From the values of the rate constants, the activation parameters were obtained by following the procedure of Comess et al.[10] with minor modifications. We found ΔH^{*} = 79 ± 5 kJ.mol^{-1} and ΔS^{*} = - 91 ± 16 J.mol^{-1}.K^{-1} while Comess et al.[10], in the case of d(TCTACGCGTTCT), found ΔH^{*} = 91 ± 2 kJ.mol^{-1} and ΔS^{*} = - 58 ± 8 J.mol^{-1}.K^{-1}.
Concentration, salt and pH. The oligonucleotide concentration (1-100 μM), the ionic strength (10-400 mM), the nature of the salt (NaCl or $NaClO_4$) and the pH (in the range 5-9) did not change significantly the linkage isomerization rate within d(CCTCGAGTCTCC).
Oligonucleotide length. Shortening the length of the oligonucleotide up to four nucleotides, as in d(pCGAG), did not prevent the rearrangement. The $t_{1/2}$ values were in the range 4-12 h, at 60°C within the dodecamer, the hexamer d(TCGAGT) and the 5'-end-phosphorylated tetramer. As concerns the 5'-end-dephosphorylated d(CGAG), the $t_{1/2}$ was significantly higher (about 50 h) showing that the phosphate group, if not essential for the reaction, favored it.

II) Double-stranded oligonucleotides

Rearrangement of *trans*-{Pt(NH_3)$_2$[d(GXG)-N7-G,N7-G]} intrastrand cross-links into interstrand cross-links. The stability of the *trans*-{Pt(NH_3)$_2$[d(GXG)-*N7*-G,*N7*-G]} intrastrand cross-links was studied within several 22-mer duplexes of central sequence d(GXG).d(CYC). (for sake of clarity, the duplexes will be abbreviated by their central sequence). The oligonucleotide containing a single *trans*-{Pt(NH_3)$_2$[d(GXG)-*N7*-G,*N7*-G]} intrastrand cross-link was 5'-end labeled, paired with its complementary strand and incubated in 200 mM $NaClO_4$, 5 mM phosphate buffer, pH 7 and at 37 °C. Aliquots were withdrawn at various times and analyzed by gel electrophoresis under denaturing conditions. As a function of time, the intensity of the initial band decreased, while a new band appeared, which migrated much more slowly. It corresponded to the duplex containing an interstrand cross-link. When Y is a purine residue, two interstrand cross-links are formed, in a ratio 80/20. The example of the platinated duplex d(AGAGA).d(TCTCT) is presented figure 2 (the two chelated Gs are indicated by the symbol *).

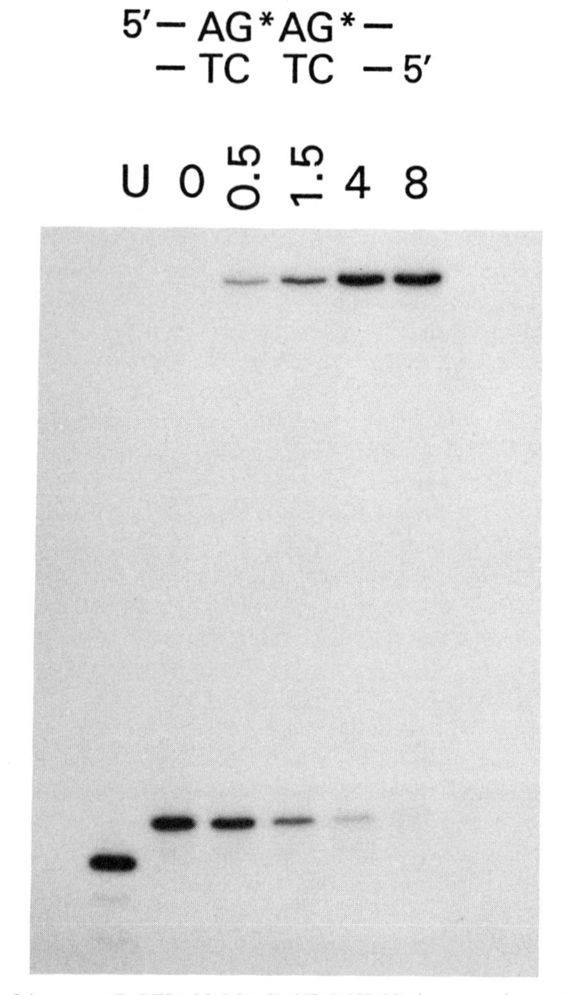

Figure 2: Instability of the *trans*-{$Pt(NH_3)_2$[d(CGAG)-*N7*-G,*N7*-G]} intrastrand cross-link within the duplex d(AGAGA).d(TCTCT). Autoradiogram of a denaturing 24% polyacrylamide gel. Incubation times are indicated in hours above the lines. Lane U refers to the unplatinated oligonucleotide.

The nature and location of the base residues involved in the interstrand cross-link were determined by two sets of experiments: Maxam-Gilbert footprinting experiments, and reverse-phase HPLC analysis of the products after enzymatic digestion by endonuclease P1 and alkaline phosphatase. The analysis of the results revealed that the platinum was bound to the 5' G residue of the upper strand and its complementary C residue (when two interstrand cross-links were formed, only the major one was characterized). A schematic representation of the rearrangement reaction is given in figure 3.

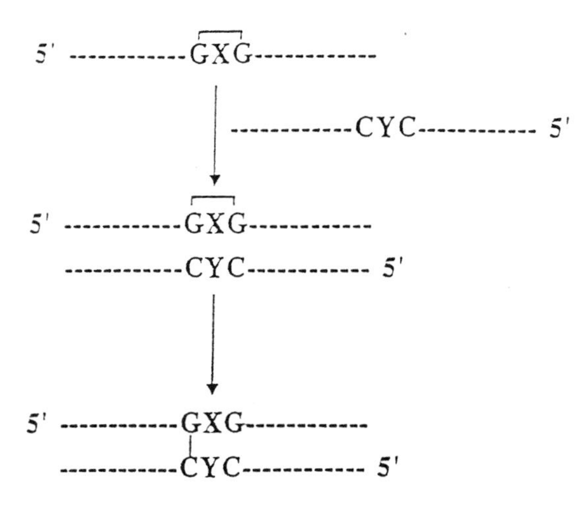

Figure 3: Schematic representation of the linkage isomerization reaction within duplexes. The platinated oligonucleotides were paired with their complementary strand and incubated in 0.2 M NaClO4, 5 mM phosphate buffer, pH 7, at 37 °C.

Parameters influencing the rearrangement. *Sequence.* To determine the influence of the sequence on the linkage isomerization reaction rate, several 22-mer duplexes of central sequence d(GXG).d(CYC) containing a single (G1,G3)-intrastrand cross-link were studied as described above. The rearrangement of the intrastrand cross-link into an interstrand cross-link occurred whatever the nature of X and Y. It was still observed when the intervening X residue was replaced by a propylene bridge. The half-lives of the intrastrand cross-link within the 22-mer duplexes d(TGXGT).d(ACYCA) are summarized in table 1. These results suggest that the intervening X residue does not play a major role in the linkage isomerization reaction. The nature of Y is of more importance: the reaction rate was about 5 times lower for Y= Pu than for Y= Py. Moreover, the linkage isomerization reaction occurred whatever the chemical nature of base pairs adjacent to the adduct (on its 5' or 3' side); however, when one of the adjacent residue was mismatched, no interstrand cross-links were formed.

Table 1. Half-lives (in hours) of the *trans*-{$Pt(NH_3)_2$[d(GXG)-*N7*-G,*N7*-G]} intrastrand cross-links within various d(TGXGT).d(ACYCA) duplexes[1], at 37 °C in 200 mM $NaClO_4$, 5 mM phosphate buffer, pH 7.

Nature of the Y residue	Nature of the X residue		
	Py	A	propylene
Py	2.5 ± 0.5 h	5.5 ± 0.5 h	1.5 ± 0.25 h
Pu	13 ± 1 h	24 ± 2 h	

[1] d(TGXGT).d(ACYCA) is the abbreviation for 22-mer duplexes having for central sequences d(TGXGT).d(ACYCA).

Concentration and salt. The reaction rate was independent of the duplex concentration, which excluded an interduplex reaction. The reaction rate was the same in $NaClO_4$ or NaCl, in the salt concentration range 50-500 mM.
Temperature. The dependence of the rate constant upon temperature was studied with the two duplexes d(TGAGT).d(ACTCA) and d(AGAGA).d(TCTCT). The kinetic activation parameters were deduced from the usual Eyring formula[10,15] and found to be in the range $\Delta H^{\neq} = 82 \pm 5$ $kJ.mol^{-1}$ and $\Delta S^{\neq} = -67 \pm 15$ $J.mol^{-1}.K^{-1}$.

III) Native DNA after reaction with *trans*-DDP

To know whether the (G1,G3)-intrastrand cross-links are stable in some sequences, the location of these sites was searched in *trans*-DDP modified double-stranded DNA by means of digestion with T4 DNA polymerase. It is known that the lesions induced in DNA by *cis*-DDP block the 3'-5' exonuclease activity of T4 DNA polymerase[16-19] whereas the *trans*-DDP-interstrand cross-links do not block it[20]. We have shown that, in single- and double-stranded oligonucleotides, (G1,G3)-, (G1,G4)- and (C1,G4)-intrastrand cross-links of *trans*-DDP stopped the exonucleolytic progression of the enzyme[21,22].

Mapping experiments were carried out on several DNA restriction fragments after reaction with *cis*- or *trans*-DDP during 24 h at 37 °C at an input molar drug/ nucleotide ratio equal to 0.5 %. Monofunctional adducts were removed by incubation of *trans*-DDP-modified DNA in 10 mM thiourea during 10 min at 37 °C. As revealed by atomic absorption analysis, about 80 % of the total bound platinum residues were removed. After enzymatic hydrolysis with T4 DNA polymerase, the samples were analyzed by gel electrophoresis. The exonucleolytic progression of the enzyme was stopped at d(GpG) and d(ApG) sites in *cis*-DDP-modified DNA. Very few stops were found in *trans*-DDP-modified DNA, which led us to conclude that (G1,G3)-, (G1,G4)- and (C1,G4) intrastrand cross-links were not the prominent lesions formed in these conditions. On the other hand, about 15 % of the total bound platinum residues (before treatment with thiourea) were involved in interstrand cross-links, as revealed by gel electrophoresis under denaturing conditions[23].

DISCUSSION

This work was devoted to the study of the stability of *trans*-{$Pt(NH_3)_2$[d(GXG)-*N7*-G,*N7*-G]} intrastrand cross-links within single-stranded and double-stranded oligonucleotides.

Initially, Comess et al. have found that the *trans*-{$Pt(NH_3)_2$[d(CGCG)-*N7*-G,*N7*-G]} intrastrand cross-link rearranges into the 1,4-*trans*-{$Pt(NH_3)_2$[d(CGCG)-*N3*-C,*N7*-G]} intrastrand cross-link within a single-stranded oligonucleotide while *trans*-{$Pt(NH_3)_2$[d(CGAG)-*N7*-G,*N7*-G]} intrastrand cross-link is stable[9,10]. In disagreement with this result, we find that the nature of the intervening base (A, T or C) does not play a major role in this reaction. The rearrangement rates and the equilibrium constants are not very different within the d(CGCG)[10], d(CGAG) and d(CGTG) sequences. The residues flanking the d(CGXG) sequence do not strongly influence the rearrangement, since it occurs in oligonucleotides as short as the d(CGAG) tetramer. The linkage isomerization reaction can be considered as direction- and sequence-specific in that it does not occur when the C residue is adjacent to the adduct on its 3' side or when the 5' C residue is replaced by a A, T or G residue.

Recently, we have shown that the *trans*-{$Pt(NH_3)_2$[d(GCG)-*N7*-G,*N7*-G]} intrastrand cross-links rearrange to interstrand cross-links within 22-mer duplexes of central sequence d(PyGXGPy).d(PuGYGPu)[11]. The interstrand cross-link was located between the 5' G and its complementary C residues. The kinetics of reaction were investigated in several duplexes as a function of the chemical nature of the (X, Y) base pair and of the base residues adjacent to the adduct (Py or Pu). We find that the rearrangement occurs whatever the nature of the (X, Y) base pair and of the adjacent base pairs.

The activation parameters of the double helix promoted reaction were determined for two platinated duplexes. The values of activation entropy can support a reaction proceeding through solvent associated intermediate[24]. On the other hand, there were no evidences for a predominent solvolysis which implies the formation of transient monofunctionnal adducts; trapping of these intermediate species was not observed in NaCl solutions. We propose that the rearrangement results from a direct nucleophilic attack of the platinum(II) centre by the C residue complementary to the platinated 5' G residue. Such a mechanism requires that the oligonucleotides adopt a conformation allowing the reactive N atom of the entering C residue to attack the platinum(II) centre along a line perpendicular to the platinum square-plane. The conformational change induced in the double-helix by the (G1,G3)-intrastrand cross-link is still unknown. A NMR study of the d(CCTCGAGTCTCC).d(GGAGACTCGAGG) duplex has been reported[9]. Nevertheless, the sample was heated and probably the 1,3-intrastrand cross-links were in part transformed into interstrand cross-links. In a 22-mer duplex, chemical probes suggest a local denaturation over 4 base pairs including the 3 base pairs at the level of the adduct and the 5' base pair adjacent to the adduct[8]. This local denaturation could allow the C residue to be in the position required for a direct nucleophilic substitution.

Recently, we have shown that in the duplexes of central sequence d(GXG).d(CYC) containing a single monofunctional adduct *trans*-$\{Pt(NH_3)_2(dG)Cl\}^+$, the closure of the monofunctional adducts to bifunctional cross-links is slow ($t_{1/2} > 15$ h) and yields mainly interstrand cross-links[11,22]. Intrastrand cross-links were hardly detected. In *trans*-DDP-modified DNA, interstrand cross-links are formed between complementary G and C residues, and represent 10-20% of the total platinum bound to DNA[20]. On the other hand, a substantial part of bifunctional cross-links are formed between two G residues[7] which suggested that (G1,G3)-intrastrand cross-links could be stable in some sequences. To answer this point, several *trans*-DDP-modified DNA restriction fragments (r_i= 0.5 %) were digested by T4 DNA polymerase. Very few stops were detected which lead us to exclude (G1,G3)-, (G1,G4)- and (C1,G4)-intrastrand cross-links as the major adducts.

It has been reported[6] that in *trans*-DDP-modified DNA, the closure of the monofunctional adducts to bifunctional cross-links is slow ($t_{1/2} \simeq 24$ h) and depends upon the drug-to-nucleotide ratio, in the range 1-10 %. On the other hand, Bancroft et al.[25] find short half-lives for the monofunctional adducts (3.9 h) in small DNA fragments (40 base pair average length) modified at a drug-to-nucleotide ratio equal to 7 %. All these results suggest that the rate of the cross-linking reaction and the nature of the bifunctional adducts depend upon the level of DNA platination. At low level of platination, we find that after 24 h of incubation at 37 °C the prominent lesions are the monofunctional adducts ($\simeq$ 80 %) while the bifunctional adducts are mainly interstrand cross-links. *In vivo*, the drug-to-nucleotide ratio is low; the interstrand cross-linking reaction being slow[20], monofunctional adducts can be trapped by compounds such as glutathione[6,25]. It is tempting to argue that *trans*-DDP has no antitumor activity because it does not form bifunctional adducts which are supposed to be the lesions related to the antitumor activity of *cis*-DDP.

It has been already shown that *cis*-$\{Pt(NH_3)_2(dG)(Am)\}^{n+}$ monofunctional adducts (Am is an heterocyclic amine) are unstable within double-stranded oligonucleotides and can rearrange to interstrand cross-links after releasing of Am[14,26,27]. The linkage isomerization reaction of the *trans*-$\{Pt(NH_3)_2[d(GXG)$-*N7*-G,*N7*-G]} intrastrand cross-links in single- and double-stranded oligonucleotides are other examples of the complexity of the platinum-DNA interactions. In addition, both rearrangements to interstrand cross-links (in *cis*-$\{Pt(NH_3)_2(Am)Cl\}^{n+}$- and *trans*-DDP-modified DNA) also occur in DNA.RNA duplexes[11,14].

They are promising tools in the context of the antisense strategy[28,29] to irreversibly cross-link antisense oligonucleotides to their targets.

Acknowledgments: We are indebted to Dr. D. Payet, A. Schwartz and C. Aussourd for their assistance and helpful suggestions. This work was supported in part by la Ligue contre le Cancer, l'Association pour la Recherche sur le Cancer and U.E. (projects CHRX-CT92-0016, D1-92-002 and CHRX-CT94-0482).

REFERENCES

1. A. Eastman, The formation, isolation and characterization of DNA adducts produced by anticancer platinum complexes, *Pharmacol. Ther.*, 34, 155-166 (1987).
2. S. L. Bruhn, J. H. Toney, and S. J. Lippard, Biological processing of DNA modified by platinum compounds, *in* "Progress in Inorganic Chemistry: Bioinorganic Chemistry", S. J. Lippard, ed., Wiley, New York (1990).
3. J. Reedijk, The relevance of hydrogen bonding in the mechanism of action of platinum antitumor compounds, *Inorg. Chim. Acta*, 198: 873 (1992).
4. M. Sip and M. Leng, DNA, cis-platinum and intercalators: catalytic activity of the DNA double helix, *in* "Nucleic Acids and Molecular Biology" (vol. 7), F. Eckstein and D. M. J. Lilley, eds., Springer, Berlin (1993).
5. A. L. Pinto and S. J. Lippard, Sequence-dependent termination of *in vitro* DNA synthesis by *cis*- and *trans*-diamminedichloroplatinum(II), *Proc. Natl. Acad. Sci. U. S. A.*, 82: 4616 (1985).
6. A. Eastman and M. A. Barry, Interactions of *trans*-diamminedichloroplatinum(II) with DNA: formation of monofunctional adducts and their reaction with glutathione, *Biochemistry*, 26: 3303 (1987).
7. A. Eastman, M. M. Jennerwein and D. L. Nagel, Characterization of bifunctional adducts produced in DNA by *trans*-diamminedichloroplatinum(II), *Chem. Biol. Interact.*, 67: 71 (1988).
8. M. F. Anin and M. Leng, Distortions induced in double-stranded oligonucleotides by the binding of *cis*- and *trans*-diamminedichloroplatinum(II) to the d(GTG) sequence, *Nucleic Acids Res.*, 18: 4395 (1990).
9. C. A. Lepre, L. Chassot, C. E. Costello, and S. J. Lippard, Synthesis and characterization of *trans*-$[Pt(NH_3)_2Cl_2]$ adducts of d(CCTCGAGTCTCC). d(GGAGACTCGAGG), *Biochemistry*, 29: 811 (1990).
10. K. M. Comess, C. E. Costello, and S. J. Lippard, Identification and characterization of a novel linkage isomerization in the reaction of *trans*-diamminedichloroplatinum(II) with 5'-d(TCTACGCGTTCT), *Biochemistry*, 29: 2102 (1990).
11. R. Dalbiès, D. Payet, and M. Leng, DNA double helix promotes a linkage isomerization reaction in *trans*-diamminedichloroplatinum(II) modified DNA, *Proc. Natl. Acad. Sci. U. S. A.*, 91: 8147 (1994).
12. A. Eastman, Characterization of the adducts produced in DNA by *cis*-diamminedichloroplatinum(II) and *cis*-dichloro(ethylenediamine) platinum(II), *Biochemistry*, 22: 3927 (1983).
13. A. M. Fichtinger-Shepman, J. L. Van der Veer, P. H. Lohman, and J. Reedijk, Adducts of the antitumor drug *cis*-diamminedichloroplatinum(II) with DNA: formation, identification and quantitation, *Biochemistry*, 24: 707 (1985).
14. D. Payet, F. Gaucheron, M. Sip, and M. Leng, Instability of the monofunctional adducts in *cis*-$[Pt(NH_3)_2(N7\text{-}N\text{-methyl-2-diazapyrenium})Cl]^{2+}$-modified DNA:

rates of cross-linking reactions in *cis*-platinum-modified DNA, *Nucleic Acids Res.*, 21: 5846 (1993).
15. R. G. Wilkins, The deduction of mechanisms, *in* "Kinetics and Mechanism of Reactions of Transition Metal Complexes" (2nd ed.), VCH, Weinhem, Germany (1991).
16. J. M. Malinge, A. Schwartz, and M. Leng, Characterization of the ternary complexes formed in the reaction of *cis*-diamminedichloroplatinum(II), ethidium bromide and nucleic acids, *Nucleic Acids Res.*, 15: 1779 (1987).
17. R. Rahmouni, A. Schwartz, and M. Leng, Importance of DNA sequence in the reaction of d(ApG) and d(GpA) with *cis*-diamminedichloroplatinum(II), *in* "Platinum and Other Metal Coordination Compounds in Cancer Chemotherapy", S. B. Howell, ed., Plenum Press, New-York (1988).
18. N. J. Rampino and V. A. Bohr, Rapid gene-specific repair of cisplatin lesions at the human DUG/DHFR locus comprising the divergent upstream gene and dihydrofolate reductase gene during early G_1 phase of the cell cycle assayed by using the exonucleolytic activity of T4 DNA polymerase, *Proc. Natl. Acad. Sci. U. S. A.*, 91: 10977 (1994).
19. Y. Zou, B. Van Houten, and N. Farrell, Sequence specificity of DNA-DNA interstrand cross-links formation by cisplatin and dinuclear platinum complexes, *Biochemistry*, 33: 5404 (1994).
20. V. Brabec and M. Leng, DNA interstrand cross-links of *trans*-diamminedichloroplatinum(II) are preferentially formed between guanine and complementary cytosine residues, *Proc. Natl. Acad. Sci. U. S. A.*, 90: 5345 (1993).
21. R. Dalbiès, M. Boudvillain, and M. Leng, Linkage isomerization reaction of intrastrand cross-links in *trans*-diamminedichloroplatinum(II)-modified single-stranded oligonucleotides, *Nucleic Acids Res.* (in press).
22. M. Boudvillain, R. Dalbiès, and M. Leng, Rearrangement of the 1,3-intrastrand cross-links into interstrand cross-links in transplatinum-modified DNA (submitted for publication).
23. M. A. Lemaire, A. Schwartz, A. R. Rahmouni, and M. Leng, Interstrand cross-links are preferentially formed at the d(GC) sites in the reaction between *cis*-diamminedichloroplatinum(II) and DNA, *Proc. Natl. Acad. Sci. U. S. A.* 88: 1982 (1991).
24. U. Belluco, Substitution reactions, *in* "Organometallic and coordination Chemistry of Platinum", Academic Press, London (1974).
25. D. P. Bancroft, C. A. Lepre, and S. J. Lippard, ^{195}Pt NMR kinetic and mechanistic studies of *cis*- and *trans*-diamminedichloroplatinum(II) binding to DNA, *J. Am. Chem. Soc.*, 112: 6860 (1990).
26. F. Gaucheron, J. M. Malinge, A. J. Blacker, J. M. Lehn, and M. Leng, Possible Catalytic activity of DNA in the reaction between the antitumor drug *cis*-diamminedichloroplatinum(II) and the intercalator N-methyl-2,7-diazapyrenium, *Proc. Natl. Acad. Sci. U. S. A.*, 88: 3516 (1991).
27. D. Payet and M. Leng, DNA, cisplatinum and heterocyclic amines: catalytic activity of the DNA double helix, *in* "Structural Biology: The State of the Art", R. H. Sarma and M. H. Sarma, eds, Adenine, Guilderland, NY (1994).
28. C. Hélène and J. J. Toulmé, Specific regulation of gene expression by antisense, sense and anti-gene nucleic acids, *Biochim. Biophys. Acta*, 1049: 99 (1990).
29. W. S. Marshall and M. H. Caruthers, Phosphorothioate DNA as a potential therapeutic drug, *Science*, 259: 1564 (1993).

IMINOETHERS AS CARRIER LIGANDS: A NOVEL *TRANS*-PLATINUM COMPLEX POSSESSING *IN VITRO* AND *IN VIVO* ANTITUMOUR ACTIVITY.

Mauro Coluccia[1] Maria A. Mariggiò,[1] Angela Boccarelli,[1] Francesco Loseto,[1] Nicola Cardellicchio,[3] Paola Caputo,[2] Francesco P. Intini,[2] Concetta Pacifico,[2] and Giovanni Natile[2*]

[1]Dipartimento di Scienze Biomediche e Oncologia Umana and
[2]Dipartimento Farmaco-Chimico
Università di Bari
70125 Bari, Italy
[3] CNR, Istituto Sperimentale Talassografico
74100 Taranto, Italy.

INTRODUCTION

The platinum-based anticancer drug *cis*-$[PtCl_2(NH_3)_2]$ (*cis*-DDP) is one of the most effective drugs available for the treatment of human tumours,[1] whereas its *trans* isomer (*trans*-DDP) is inactive.[2-6] The clinical efficacy of *cis*-DDP is, however, limited by tumour cell resistance present either in the onset of treatment (intrinsic) or after an initial response (acquired).[7] *Cis*-DDP resistance appears to be mediated by factors which reduce platinum-DNA adduct formation, e.g. reduced intracellular accumulation, increased inactivation by intracellular thiols, increased repair[8] or tolerance of platinum-DNA adducts.[9]

A possible solution to the problem of drug resistance could come from the use of platinum complexes which bind to DNA in a manner distinct from that of *cis*-DDP. Suitable for this purpose are platinum complexes with *trans* geometry which have been shown to bind to DNA in a way different from that of *cis*-DDP.[10-14] Therefore, notwithstanding the structure-activity relationships of platinum complexes contemplate that the *trans* isomers are inactive as antitumour agent,[2-6] a few groups have pursued the idea of activating this geometry by a proper choice of the carrier ligands.[15-19]

The substitution of iminoethers for ammines in diamminedichloroplatinum(II) complexes has lead to a new series of active *trans* platinum compounds.[18] Some features of the biological activity of these complexes are here described.

RESULTS

Iminoether complexes are similar to DDP species in having two chlorine and two nitrogen donor ligands, and the nitrogen atoms carry a proton suitable for hydrogen bonding.[21]

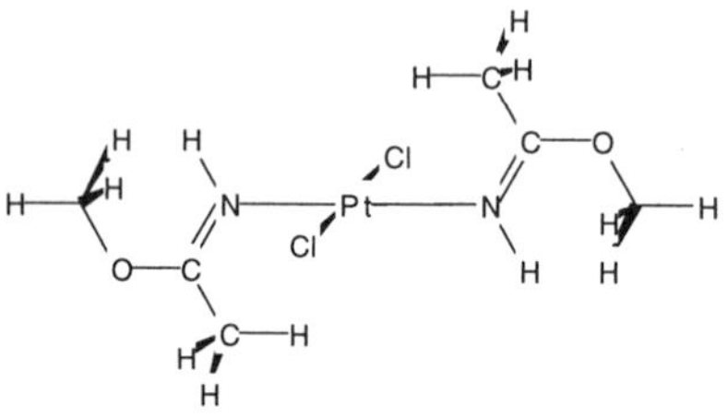

trans-EE 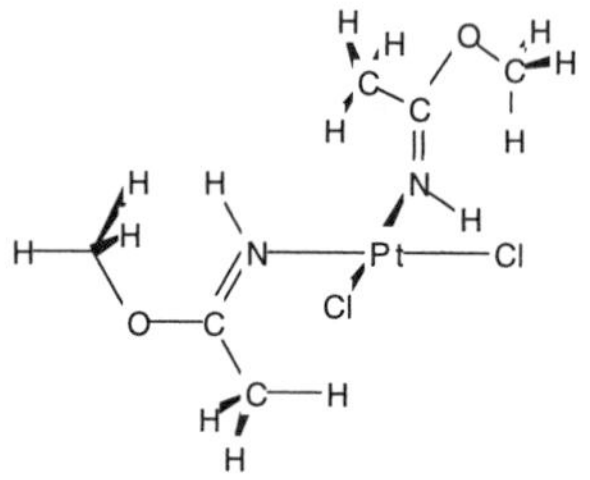 *cis-EE*

Like for *cis-* and *trans*-DDP, the nitrogen ligands are non-leaving groups in water solution at physiological pH and are presumably retained in adduct formation with biological substrates. The hydrolysis involves exclusively the chloride ligands and leads initially to the formation of a monoaqua species, then further transformations lead to a mixture of several species. However the geometry of the complex (*cis* or *trans*) and the configuration of the iminoether ligands (*E*) are always retained.

Both the *cis* and *trans* isomers of the iminoether complex react slower with DNA than *cis* and *trans* isomers of the ammine species, but they reach approximately the same level of binding after 24 h reaction time. Therefore it appears that the imino ether ligands, probably because of their steric hindrance, slow down the reaction with a sterically demanding nucleophile such as a nucleobase inserted in a DNA duplex. The different geometry of the complexes (*cis* or *trans*) does not appear to influence significantly the rate of DNA platination. Figure 1.

The results obtained with the fluorescent terbium probe indicate that, independently from the type of ligand (ammine or imino ether) *cis* complexes disrupt the duplex more than the *trans* complexes. Therefore, *trans-EE*, similarly to trans-DDP, does not induce on DNA the local conformational alterations which are considered typical of the antitumour-active platinum complexes. More detailed structural studies in progress support this conclusion. Figure 2.

The ability of platinum-iminoether complexes to form inter-strand cross-links (ICLs) was evaluated by heat denaturation/renaturation experiments on purified calf

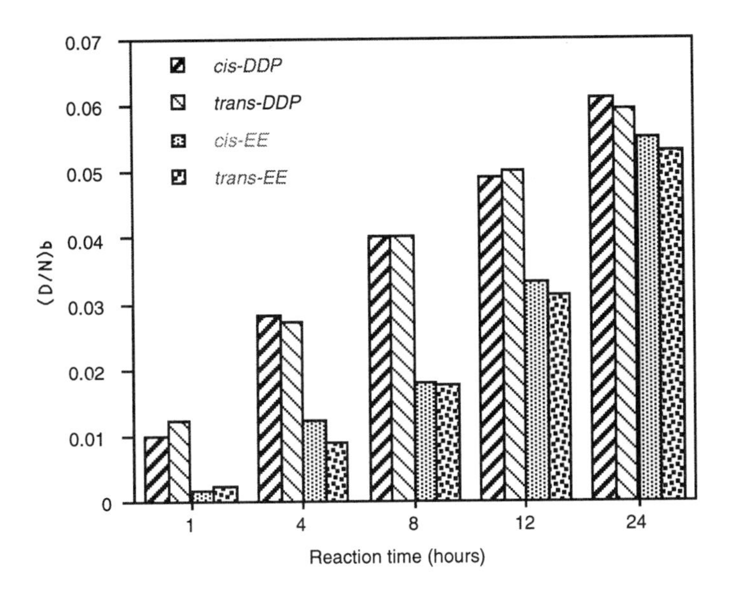

Figure 1. Kinetics of binding of *cis*-DDP, *trans*-DDP, *cis-EE* and *trans-EE* to calf thymus DNA in 2 mM Tris-HCl, pH 7.4, 37 °C, at 0.08 drug/nucleotide formal ratio. At fixed time intervals, unbound platinum was removed by centrifugation through Sephadex G-50 columns; the DNA concentration and the amount of platinum bound/nucleotide, $(D/N)_b$, were measured by UV and flameless atomic absorption spectroscopy, respectively.

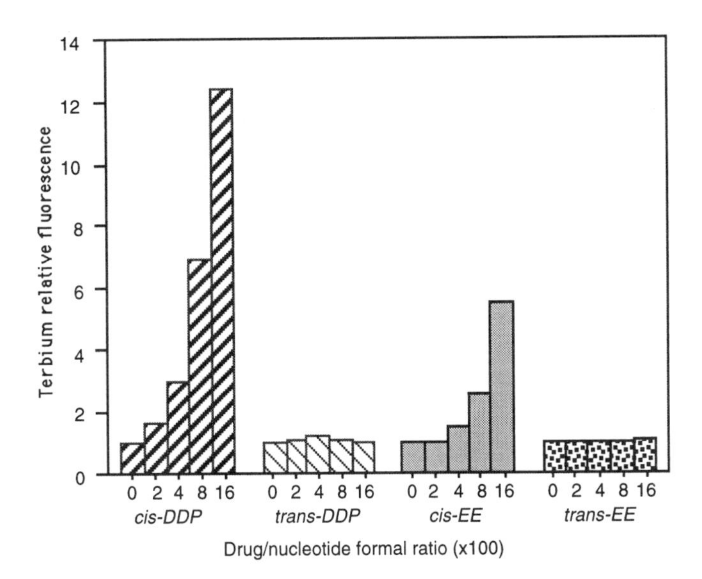

Figure 2. Change in terbium fluorescence produced on calf thymus DNA after 6 h reaction in 2 mM Tris-HCl, pH 7.3, with *cis*-DDP, *trans*-DDP, *cis-EE* and *trans-EE* at drug/nucleotide formal ratios ranging from 0.02 to 0.16. Fluorescence of untreated control was set at unity. Each value represents the mean of three independent experiments.

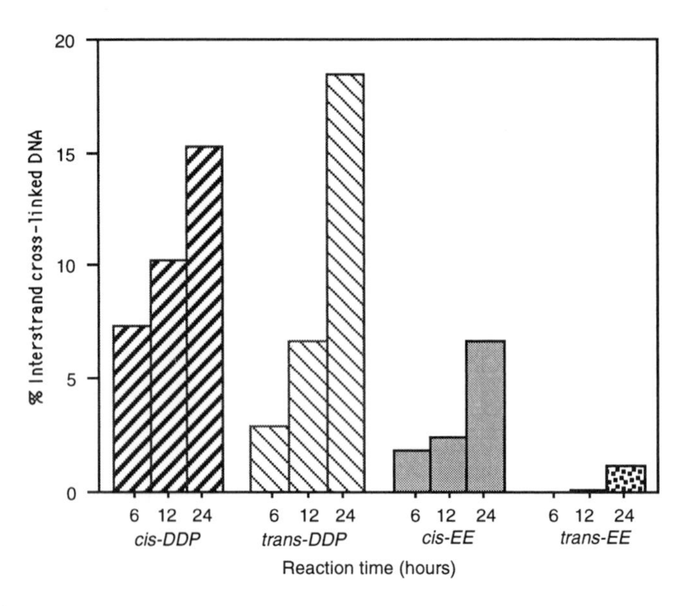

Figure 3. Kinetics of interstrand cross-link formation in calf thymus DNA treated with *cis*-DDP, *trans*-DDP, *cis-EE* and *trans-EE* in 2 mM Tris-HCl, pH 7.4, 37 °C, at $2x10^{-4}$ drug/nucleotide formal ratio. The percentage of cross-linked DNA was calculated from the relative fluorescence of ethidium bromide remaining after heat denaturation of DNA (90 °C, 10 min). Each value represents the mean of three independent experiments.

thymus DNA (Figure 3). The results demonstrate that *trans-EE* has a greatly reduced ability to form DNA interstrand cross-links and there appears to be a striking difference between the behaviours of *trans-EE* and *trans*-DDP.

The effects on DNA synthesis and cell growth of P388 cells exposed to different concentrations of platinum complex for 5 h was also investigated in parallel to the DNA interstrand cross-link formation (Table 1).

Trans-EE is most effective in inhibiting the DNA synthesis and cell proliferation but does not form detectable interstrand cross-links on cellular DNA. In contrast, *cis*-DDP, *trans*-DDP and *cis-EE* complexes form DNA interstrand cross-links which are proportional to the drug concentration and correlate with the inhibition of DNA synthesis and cell proliferation.

The observation made for cell toxicity and inhibition of DNA syntheses, that the potency of *cis* and *trans* isomers is reversed as a consequence of the substitution of the iminoethers for ammines, has also been confirmed for mutagenicity and antitumour activity.

Trans-EE is definitely more mutagenic than *cis-EE* towards the TA 100 strain of *Salmonella typhimurium* (Figure 4) while the opposite trend is observed for *trans*- and *cis*-DDP.[22] Both *cis*- and *trans-EE*, and *cis*- and *trans*-DDP, are inactive towards the TA 98 strain which undergoes mutation *via* a frame-shift mechanism.

The *in vivo* antitumour activity was also significantly greater for *trans-EE* than for *cis-EE* (P388 leukaemia; %T/C = 170 and 144, respectively) and, unlike *cis-EE*, *trans-EE*

Tab. 1. *In vitro* effects of platinum complexes on DNA synthesis, interstrand cross-link formation and cell proliferation of P388 cells.

Complex	Concentration (mM)	% inhibition of DNA synthesis	% ICL formation	% Inhibition of cell proliferation
cis-DDP	3.75	29±4	4.6±0.9	83±0.7
	7.5	40±12	7.1±1.1	91±0.6
	15	54±12	16.2±1.4	97±0.6
cis-EE	15	15±2	ND	78±4
	30	27±4	2.7±0.2	86±2
	60	43±9	7.6±0.5	92±2
trans-DDP	15	41±6	3.9±0.2	76±1.5
	30	67±9	9±0.9	91±3
	60	89±5	17.4±1.1	98±1
trans-EE	3.75	55±3	ND	80±5
	7.5	86±1	ND	98±2
	15	96±1	ND	100±1

P388 cells (500,000/mL) were incubated for 5 h with platinum complexes in RPMI 1640 medium. At the end of incubation time, the DNA was extracted and the %ICL formation was evaluated on 20 μg aliquots of DNA from control and treated cells. The inhibition of DNA synthesis was measured in parallel as changes in [methyl-^{3}H]thymidine incorporation after pulse-labeling for the last 20 min of incubation. The inhibition of cell growth was evaluated after 48 h of post-incubation in drug-free medium (trypan blue exclusion test). Data are expressed as mean ± SD of three independent experiments. ND = not detectable.

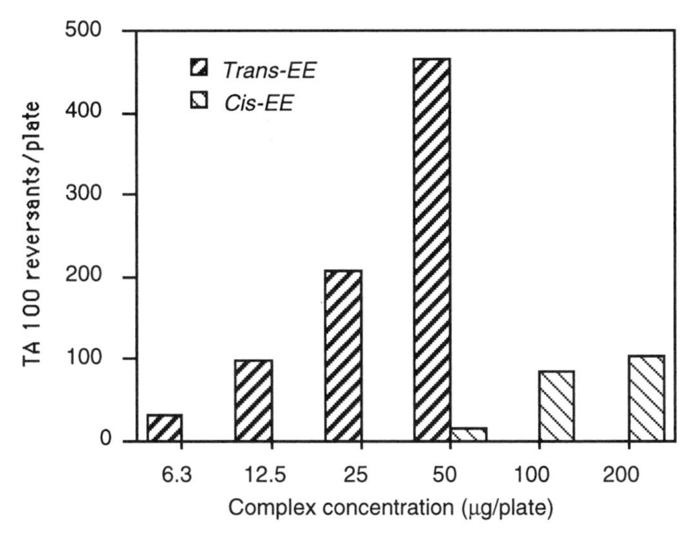

Figure 4. Mutagenic activity of *trans-EE* and *cis-EE* towards TA 100 strain of Salmonella typhimurium. Revertant colonies are expressed as mean value of three independent experiments. The spontaneous revertants (116 ± 15) have been subtracted.

was effective also on a cisplatin-resistant subline of P388 leukemia (%T/C = 133).[18] *Trans-EE* also showed potent activity towards subcutaneous Lewis lung carcinoma.[20]

DISCUSSION

Two other classes of active *trans*-platinum complexes have been reported, they have structural formulae *trans*-$[PtCl_2(L)(L')]$, where L = L' are planar ligands such as pyridine or thiazole,[15-17] and *trans*-$[PtCl_2(OH)_2(NH_3)\{NH_2(C_6H_{11})\}]$,[19] the latter is a platinum(IV) compound derived from a platinum(II) species which probably behaves as a classical *trans* platinum complex.

A lower range in cytotoxicity was observed for the above compounds compared to *cis*-DDP (this is not our case). Moreover, in common with our findings, they appear to react more slowly with GSH than *trans*-DDP itself.[20]

By analogy with classical alkylating agents, the cytotoxicity of platinum drugs has been attributed to the formation of DNA-DNA cross-links. Both *cis*-DDP and *trans*-DDP are capable of forming DNA intrastrand and interstrand cross-links and controversy remains as to explain the contrasting antitumour effects of the two congeners.[10] *Trans*-DDP is stereochemically incapable of forming the 1,2-intrastrand d(GpG) or d(ApG) cross-links (the major adduct formed by *cis*-DDP)[10,22,23] suggesting that the differences in antitumour activity may result from the different nature of distortions induced in DNA by the various intrastrand cross-liks. Others have suggested that the inactivity of *trans*-DDP results from a high proportion of monofunctional adducts on DNA (which may rapidly react with glutathione before they can be converted to more toxic bifunctional adducts)[23,14] and/or from preferential recognition and repair of the induced DNA lesions.[25] More recently, the possible importance of interstrand cross-links and their repair in determining the cytotoxicity of *cis*-DDP, has been highlighted;[8,26] while *cis*-DDP preferentially forms ISCs between guanine residues,[27] *trans*-DDP has been shown to preferentially form ISCs between guanine and complementary cytosine residues.[14]

The cytotoxic activity of *trans*-iminoether complexes does not appear to correlate with measured ICLs which are among the critical molecular lesions able to inactivate the DNA as a template for replication. However, before drawing any conclusion it is necessary to check if *trans-EE* does not form DNA cross-links which are thermally unstable and that is why such cross-links are not detected. It has been demonstrated that the ICLs induced by some non metal-based anticancer drugs such as acridines,[28] anthracyclines[29] and nitrogen mustards[30,31] may be destroyed either in severe alkaline conditions (alkaline elution method) or at high temperature (heat denaturation/renaturation assay).[31-34] This behaviour could also apply to the iminoether compounds if one of the drug-DNA interaction does not involve directly the metal atom but is mediated by an iminoether ligand. For instance metathesis of the alkoxide group of the iminoether ligand by the amine group of a nucleobase could create a cross-link unstable at high pH or high temperature. Such a type of interaction with DNA would place the platinum-iminoether complexes at the junction between *cis*-DDP and non-metallic cytotoxic agents which also are known to react bifunctionally with DNA.[31-34] This could represent the novelty of this new class of drugs. Work is in progress to clarify this point.

In conclusion, as a consequence of the substitution of iminoethers for ammines, several biological properties between *cis* and *trans* geometry such as cell toxicity, inhibition of DNA synthesis, mutagenicity and antitumour activity are reversed. The new carrier ligands reduce the activity of the *cis* isomer, which is basically a *cis*-DDP like drug, while providing a different reaction pathway for the *trans* isomer. The different mechanism of action of *trans-EE* could be the premise for a possible clinical use of this drug in combination with *cis*-DDP.

ACKNOWLEDGEMENTS

This work was supported by the contributions of CNR (A.C.R.O. Project), MURST (Contribution 40%) and EC (Contract C11-CT92-0016 and COST Chemistry project D1/02/92).

REFERENCES

1. P.J. Loehrer and L.H. Einhorn, Cisplatin, *Ann. Inter. Med.* 100: 704-713 (1984).
2. T.A. Connors, M.J. Cleare, and K.R. Harrap, Structure-activity relationships of the antitumour platinum coordination complexes, *Cancer Treat. Rep.* 63: 1499-1502 (1979).
3. A. Eastman, The formation, isolation, and characterization of DNA adducts produced by anticancer platinum complexes, *Pharmacol. Ther.* 34: 155-166 (1987).
4. J. Reedijk, The mechanism of action of platinum antitumor drugs, *Pure Appl. Chem.* 59: 181-192 (1987).
5. S.L. Bruhn, J.H. Toney, and S.J. Lippard, Biological processing of DNA modified by platinum anticancer drugs, *Prog. Inorg. Chem.* 38: 477-516 (1990).
6. M. Sip, A. Schwartz, F. Vovelle, M. Ptak, and M. Leng, Distortion induced in DNA by *cis*-platinum interstrand adducts, *Biochemistry*, 31: 2508-2513 (1992).
7. R.F. Ozols and R.C. Young, Chemotherapy of ovarian cancer, *Semin. Oncol.* 11: 251-263 (1984).
8. W. Zhen, C.J. Link, P.M. O'Connor, E. Reed, R. Parker, S.B. Howell, and V. Bohr, Increased gene-specific repair of cisplatin interstrand crosslinks in cisplatin-resistant human ovarian cancer cell lines, *mol. Cell. Biol.* 12: 3689-3698 (1992).

9. L.R. Kelland, New platinum antitumor complexes, *Crit. Rev. Oncol/Hematol.* 15: 191-219 (1993).
10. K.M. Comess and S.J. Lippard, Molecular aspects of platinum-DNA interactions, in: "Molecular Aspects of Anticancer Drug-DNA Interactions," Vol. 1, 134-168. S. Neilde and M. Waring, ed., Macmillan Press Ltd., London (1993).
11. A. Eastman, M.M. Jennerwein, and D.L. Nagel, Characterization of bifunctional adducts produced in DNA by *trans*-diamminedichloroplatinum(II), *Chem. Biol. Interact.* 67: 71-80 (1988).
12. A.P. Pinto and S.J. Lippard, Sequence-dependent termination of *in vitro* DNA synthesis by *cis*- and *trans*-diamminedichloroplatinum(II), *Proc. Natl. Acad. Sci. USA*, 82: 4616-4619 (1985).
13. C.A. Lepre, K.G. Strothkamp, and S.J. Lippard, Synthesis and ^{1}H NMR spectroscopic characterization of *trans*-[Pt$(NH_3)_2${d(AGGCCT) N7-A(1), N7-G(3)}], *Biochemistry*, 26: 5651-5657 (1987).
14. V. Brabec and M. Leng, DNA interstrand cross-links of *trans*-diamminedichloroplatinum(II) are preferentially formed between guanine and complementary cytosine residues, *Proc. Natl. Acad. Sci. USA*, 90: 5345-5349 (1993).
15. N. Farrell, T.T.B. Ha, J.-P. Souchard, F.L. Wimmer, S. Cros, and N.P. Johnsin, Cytostatic *trans*-platinum(II) complexes. *J. Med. Chem.* 32: 2240-2241 (1989).
16. M. Von Beusichem and N. Farrell, Activation of the trans geometry in platinum antitumor complexes. Synthesis, characterization and biological activity of complexes with planar ligands pyridine, N-methylimidazole, thiazole and quinoline. The crystal and molecular structure of *trans*-dichloro bis(thiazole) platinum(II). *Inorg. Chem.* 31: 634-639 (1992).
17. N. Farrell, L.R. Kelland, J.D. Roberts, and M. Von Beusichem, Activation of the trans geometry in platinum antitumor complexes: a survey of the cytotoxicity of *trans* compounds containing planar ligands in murine L1210 and human tumor panels and studies on their mechanism of action. *Cancer Res.* 52: 5065-5072 (1992).
18. M. Coluccia, A. Nassi, F. Loseto, A. Boccarelli, M.A. Mariggiò, D. Giordano, F.P. Intini, P.A. Caputo, and G. Natile, A *trans*-platinum complex showing higher antitumor activity than the cis congeners. *J. Med. Chem.* 36: 510-512 (1993).
19. L.R. Kelland, C.F.J. Barnard, K.J. Mellish, M. Jones, P.M. Goddard, M. Valenti, A. Bryant, B.A. Murrer, and K.R. Harrap, A novel *trans*-platinum coordination complex possessing *in vitro* and *in vivo* antitumor activity. *Cancer Research*, 54: 5618-5622 (1994).
20. M. Coluccia, M.A. Mariggiò, A. Boccarelli, F. Loseto, N. Cardellicchio, P.A. Caputo, F.P. Intini, and G. Natile, Unpublished results.
21. R. Cini, P.A. Caputo, F.P. Intini, and G.Natile, Mechanistic and stereochemical investigation of iminoethers formed by alcoholysis of coordinated nitriles: X-ray

crystal structure of *cis*- and *trans*-[bis(1-imino-1-methoxyethane)dichloroplatinum(II)], *Inorg. Chem.*, in press.
22. M. Coluccia, M. Correale, F.P. Fanizzi, D. Giordano, L. Maresca, M.A. Mariggiò, G. Natile, and M. Tamaro, Mutagenic activity of some platinum complexes: chemical properties and biological activity, *Toxicol. Environm. Chem.*, 8: 1-8 (1984).
23. A. Eastman and M.A. Barry, Interaction of *trans*-diammine-dichloroplatinum(II) with DNA: formation of monofunctional adducts and their reaction with glutathione. *Biochemistry*, 26: 3303-3307 (1987).
24. G.L. Cohen, J.A. Ledner, W.R. Bauer, H.M. Ushay, C. Caravana, and S.J. Lippard, Sequence dependent binding of *cis*-dichlorodiammineplatinum(II) to DNA, *J. Am. Chem. Soc.* 102: 2487-2488 (1980).
25. R.B. Ciccarelli, M.J. Solomon, A. Varshavsky, and S.J. Lippard, *In vivo* effects of *cis*- and *trans*-diamminedichloroplatinum(II) on SV40 chromosomes: differential repair, DNA-protein cross-linking, and inhibition of replication. *Biochemistry*, 24, 7533-7540 (1985).
26. S.W. Johnson, R.P. Perez, A.K. Godwin, A.T. Yeung, L.M. Handel, R.F. Ozols, and T.C. Hamilton, Role of platinum-DNA adduct formation and removal in cisplatin resistance in human ovarian cancer cell lines. *Biochem. Pharmacol.* 47: 687-697 (1994).
27. M.A. Lemaire, A. Schwartz, A.R. Rahmouni, and M. Leng, Interstrand crosslinks are preferentially formed at the d(GC) sites in the reaction between cis-diamminedichloroplatinum(II) and DNA. *Proc. Natl. Acad. Sci. USA*, 88: 1982-1985 (1991).
28. J. Konopa, J.W. Pawlak, and K. Pawlak, The mode of action of cytotoxic and antitumor 1-nitroacridines. III. *In vivo* interstrand cross-linking of DNA of mammalian or bacterial cells by 1-nitroacridines. *Chem. Biol. Interact.* 43: 175-197 (1983).
39. A. Skladanowski and J. Konopa, Interstrand DNA crosslinking induced by anthracyclines in tumor cells. *Biochemical Pharmacology*, 47: 2269-2278 (1994).
30. K. Kohn and C.L. Spears, Stabilization of nitrogen-mustard alkylations and interstrand crosslinks in DNA by alkali. *Biochim. Biophys. Acta*, 145: 734-741 (1967).
31. D.C. Gruenert and J.E. Cleaver, Sensitivity of mitomycin C and nitrogen mustard crosslinks to extreme alkaline conditions. *Biochem. Biophys. Res. Commun.* 123: 549-554 (1984).
32. J. Konopa, Adriamycin and Daunomycin induce interstrand crosslinks in HeLa S3 cells. *Biochem. Biophys. Res. Commun.* 110: 819-826 (1983).
33. P.G. Parson, Dependence on treatment time of melphalan resistance and DNA cross-linking in human melanoma cell lines. *Cancer Res.* 44: 2773-2778 (1984).
34. E.P. Geiduschek, "Reversible" DNA. *Proc. Natl. Acad. Sci. USA*, 47: 950-955 (1961).

MOLECULAR DYNAMICS SIMULATIONS OF CISPLATIN ADDUCTS WITH DINUCLEOTIDES

Jiří Kozelka

Laboratoire de Chimie et Biochimie Pharmacologiques et Toxicologiques, URA 400
CNRS, Université René Descartes, 45 rue des Saints-Pères, 75270 Paris 06, France

INTRODUCTION

Whereas the molecular basis for the antitumor activity of cisplatin is not yet established, there are numerous experimental results implicating DNA binding in the antitumor mechanism. Whatever happens between platinum binding to DNA and the cell death, the hypothesis that the structural distortion plays a role in the processing of the Pt-DNA adducts does not seem unlikely, and provides the pharmacological motivation for structural studies on these adducts.

While studying the double-stranded decanucleotide d(GCCG*G*ATCGC)-d(GCGATCCGGC), crosslinked at the G* guanines with *cis*-$Pt(NH_3)_2^{2+}$, we were confronted with the problem of multiple minima, a classical difficulty associated with the modeling of flexible molecules such as oligonucleotides.[1] Correlation of NMR data and model structures supported the conclusion that the platinated decamer exists in solution as an equilibrium mixture of several conformations rapidly interconverting on the NMR time-scale. Rather than by means of static models, such equilibria would be, of course, more appropriately described by molecular dynamics (MD) simulations.

Since procedures for MD simulations of nucleic acids are far from being well-established[2], we have cautiously chosen two dinucleotide adducts, *cis*-$[Pt(NH_3)_2\{d(GpG)\}]^+$ (**1**) and *cis*-$[Pt(NH_3)_2\{r(GpG)\}]^+$ (**2**), for our initial MD study. The H8 resonances in the NMR spectra of these two complexes have inverse order, and this inversion has been ascribed to inverse helicities of the base arrangements.[3] Apart from testing different MD protocols, the aim of this work was therefore to identify the molecular interaction(s) responsible for the opposed helicities of these similar compounds.

Platinum and Other Metal Coordination Compounds in Cancer Chemotherapy 2
Edited by H.M. Pinedo and J.H. Schornagel, Plenum Press, New York, 1996

RESULTS AND DISCUSSION

The two main parameters defining the conformation of the dinucleotide complexes studied here are the torsion angles α and β about the Pt-N7 bonds (Figure 1). If for the (hypothetical) conformation with both guanines in the coordination plane, as shown in Figure 1, we define $\alpha=\beta=0$, then $\alpha+\beta$ provides a measure of the helical sense, positive $\alpha+\beta$ values indicating right-handed helicity.

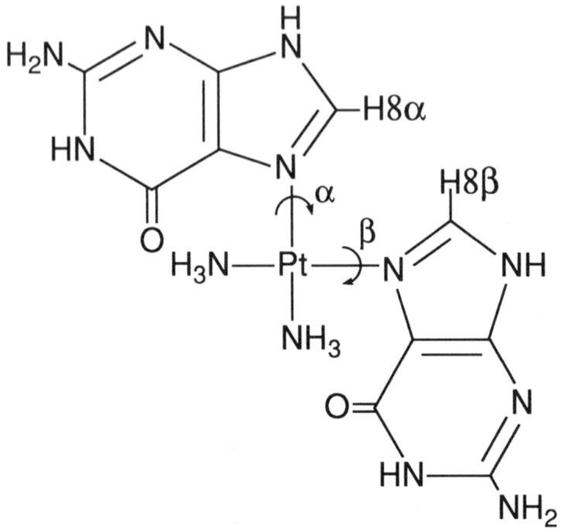

Figure 1. The complex *cis*-$[Pt(NH_3)_2Gua_2]^{2+}$ in the conformation for which we define $\alpha=0°$; $\beta=0°$. Both guanine planes are co-planar with the platinum coordination plane. α and β are positive, if the corresponding guanine, observed in the N7-Pt-NH_3 direction, turns counterclockwise.

In the conformational space defined by the two torsion angles α and β, four low-energy zones exist (Figure 2): two comprising head-to-head (HH) conformations, and two involving head-to-tail (HT) conformations.[3] The NOESY spectra of both **1** and **2** show a strong correlation between the two H8 protons, indicating a short (<4 Å) interatomic distance. HT structures can therefore be ruled out, since they feature a long (4-6 Å) H8-H8 distance.

We have carried out MD simulations of 630 ps at 350 K, starting from conformations belonging to both HH1 and HH2 zones. A correlation between NMR and CD spectra on one hand, and the geometrical features of the conformations occurring during the simulations on the other hand, allowed us to conclude that the solution structures of both **1** and **2** belong to the HH1 zone. Within this zone, however, **1** and **2** prefer (as revealed by the H8 shifts and CD spectra) different energy-minima, **1** the L1 minimum (left-handed helical conformation) and **2** the R2 minimum (right-handed helical conformation). The remaining question is: Why do **1** and **2** prefer opposed helicities?

Figure 3 shows the evolution of three geometrical parameters of **2** with time during a 630 ps simulation at 350 K. These parameters include the helicity, measured as $\alpha+\beta$, the nonbonded distance separating the hydrogen of the 2'-OH group of the 5'-nucleotide with one

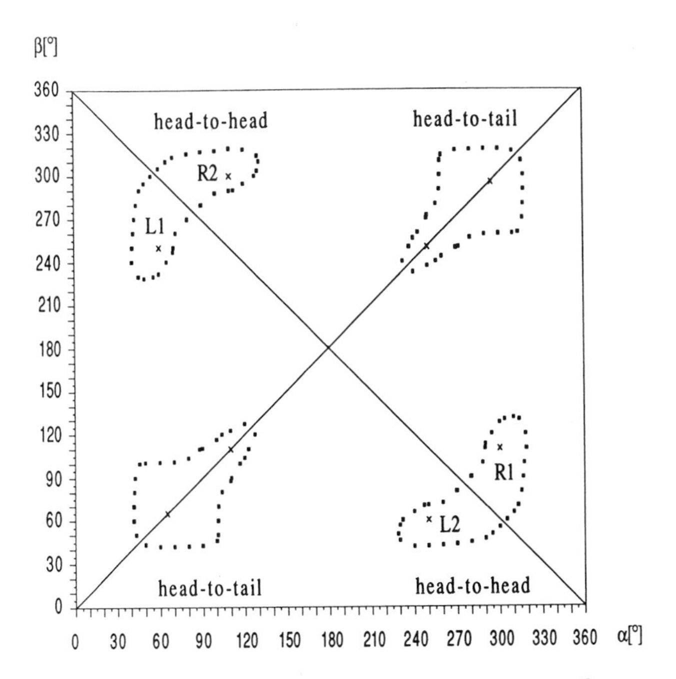

Figure 2. Calculated nonbonded energy map for the complex *cis*-$[Pt(NH_3)_2Gua_2]^{2+}$. Dashed curves encircle low-energy zones. The eight nearly equienergetic minima (marked with a cross) are 15.7 kJ/mol lower in energy than the level curves.The structures corresponding to the minima L1 and L2 are left-handed helicoidal, those corresponding to R1 and R2 are right-handed helicoidal. (Reproduced from Ref. 3 with permission)

terminal oxygen of the phophodiester group, O_A, and the torsion angle χ_2 about the glycosodic bond of the 3'-nucleotide. These three parameters are clearly correlated and delimit two distinct conformational domains, {1}, in which the helicity is predominantly left-handed, the OH...OP separation oscillates around 2.7 Å, and the χ_2 value is in the *syn* range, and {2}, with preponderantly right-handed helicity, a mean OH...OP distance of 4.2 Å, and the χ_2 angle in the *anti* range. Inspection of the α and β angles shows that {1} corresponds to the L1 conformation of the *cis*-$[Pt(NH_3)_2Gua_2]^{2+}$ complex, and {2} to R2. Although the number of interconversions between the two domains (7 during the 630 ps run) is not sufficient to permit a statistical analysis, the simulation suggests that both domains are roughly equally populated, that is, the free energies of the two domains seem comparable. A similar figure results for the simulation of **1** (not shown). Obviously, our force-field is not accurate enough to allow the identification of the (apparently small) energy differences between the conformations {1} and {2}, which cause **1** to prefer {1} and **2** to prefer {2}.

The main neglect in our force-field is the absence of explicit solvent. Let us try to predict how the inclusion of a water bath would influence the relative energy of the two conformations {1} and {2} of **2**. Representative energy-minimized models of both conformations are shown in Figure 4. For conformation {1}, we observe that the 2'-OH group of the 5'-nucleotide forms a three-center hydrogen bond with two oxygen atoms of the phosphodiester group; both

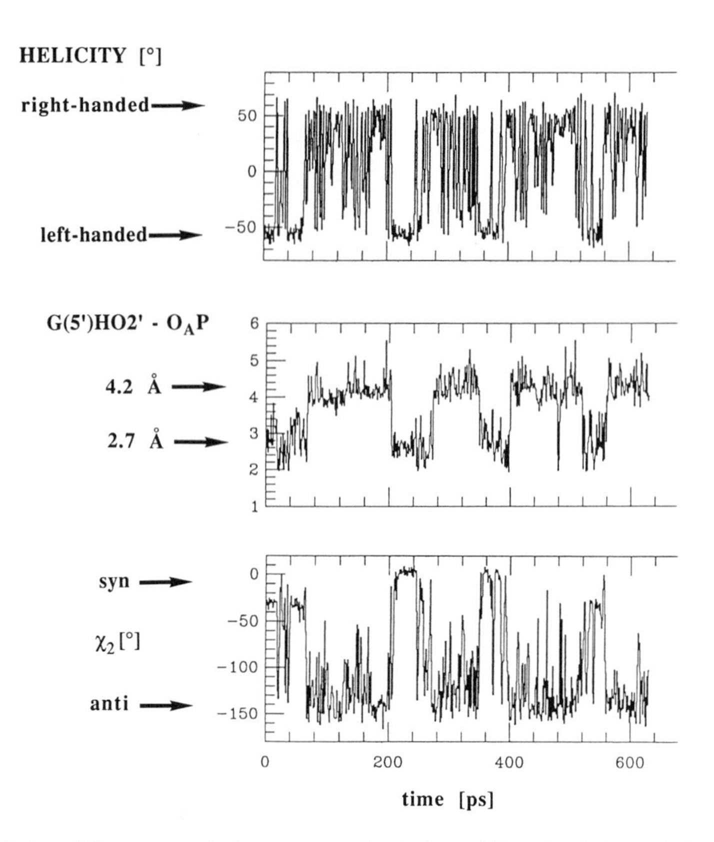

Figure 3. Evolution of three geometrical parameters of **2** during a 630 ps simulation at 350 K. The helicity is measured as the sum $\alpha + \beta$ (see Figure 1 for definition).

H...O separations being ~2.7 Å. For conformation {2}, the shortest H...O separation is ~4.2 Å, as we have seen in Figure 3; this separation precludes a direct hydrogen bond but allows for water-mediated hydrogen bonding with nearly ideal geometry. Such hydrogen bonding was not possible in the MD simulations which were run "*in vacuo*" (the water molecule displayed in Figure 4 (right) was added subsequently in order to visualize this possibility of indirect hydrogen bonding). We believe that inclusion of the solvent will favor the right-handed conformation {2}, since the latter will be stabilized by the water bridge. It is this water-mediated hydrogen bonding to which we attribute the reversal of helicity from left-handed in **1** to right-handed in **2**.

This work clearly demonstrates the complementarity of molecular modeling on one hand, and experimental spectroscopic techniques on the other hand. The initially obscure structural difference between **1** and **2** was discovered by NMR. Molecular modeling provided, in a first turn, a plausible explanation, attributing the inverse order of H8 resonances to opposed helical sense. Subsequently, using MD simulations, we have shown that the 2'-OH...OH_2...OP hydrogen bonding is a likely driving force for the reversal of helicity. The conformations {1} and {2} that we assign as the preponderant conformations to the complexes **1** and **2**, respectively, present distinct features which call, in turn, for a check by modern 2D-NMR techniques. For instance, the *syn* range of χ_2 for conformation {1} and the *anti* range of χ_2 for conformation {2} can be verified by measuring the appropriate H8-H1' distances using NOESY spectroscopy. We are therefore currently re-investigating both **1** and **2** by 2D-NMR.

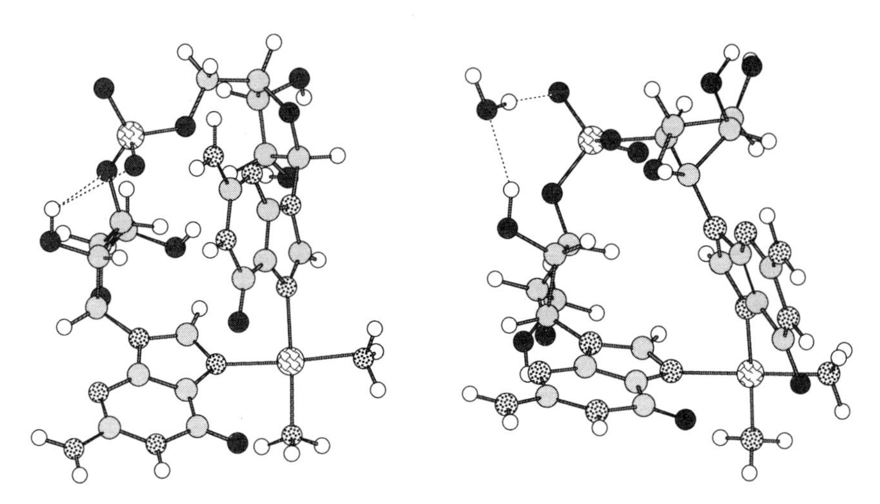

Figure 4. Molecular models representing energy-minimized conformations of complex **2** belonging to the conformational domains {1} (left) and {2} (right). The water molecule displayed in the conformation {2} was not present in the MD simulations but was added before the energy-minimization using the routine ADD of AMBER.

EXPERIMENTAL

The MD simulations were carried out using the MORCAD[4] package coupled to the MOLDYN module of AMBER[5], using a Silicon Graphics Crimson Entry workstation. The AMBER database was appended as described in Ref. 1, except for the atomic charges of the core complex *cis*-$[Pt(NH_3)_2Gua_2]^{2+}$, which were determined by *ab initio* calculations[6] (supplementary material). A distance-dependent dielectric coefficient of 4r was used for energy-minimizations and MD simulations. No cutoff was used for the calculation of nonbonded interactions. 1-4 nonbonded energy was scaled by a factor of 0.5.

The initial structures were energy-minimized using the program ORAL[7] with a norm of the energy gradient of 0.01 as convergence criterion. The simulation protocol comprised a heating period of 14 ps, 1 ps of randomization and 2x4 ps of equilibration, and 630 ps of production. The time step used was 1 fs for production and 0.5 fs for the preceeding periods. During the heating, randomization, and the first equilibration period, the atoms were constrained to their initial positions with a harmonic force constant of 0.1 kcal.mol^{-1}.Å^{-2}. The SHAKE routine was applied to all bonds involving hydrogen atoms. The temperature was held constant by rescaling the velocities if the deviation exceeded 1 K, with a damping factor of 0.4 s.

Acknowledgement

The MD simulations presented here were carried out by two students, Pascale Augé and Frédéric Allain. Technical assistance by Drs. J. A. H. Cognet, J. P. Girault and M. Le Bret, and helpful discussions with Dr. J.-C. Chottard are gratefully acknowledged.

Supplementary material available. Atomic charges for **1** and **2** used in energy-minimizations and MD simulations are available from the author on request.

References

1. F. Herman, J. Kozelka, V. Stoven, E. Guittet, J. P. Girault, T. Huynh-Dinh, J. Igolen, and J.-C. Chottard, A d(GpG)-platinated decanucleotide duplex is kinked, *Eur. J. Biochem. 194*:119 (1990).

2. J. Kozelka, Molecular Modeling of Transition Metal Complexes with Nucleic Acids and their Constituents, *in:* "Metal Ions in Biological Systems", Vol. 33, H. Sigel and A. Sigel, eds., Marcel Dekker, New York, 1995.

3. J. Kozelka, M.-H. Fouchet, and J.-C. Chottard, H8 chemical shifts in oligonucleotides cross-linked at a GpG sequence by *cis*-$Pt(NH_3)_2^{2+}$: a clue to the adduct structure, *Eur. J. Biochem. 205:* 895 (1992).

4. M. Le Bret, J. Gabarro-Arpa, J. C. Gilbert, and C. Lemaréchal, MORCAD, an object-oriented molecular modelling package running on IBM RS/6000 and SGI 4Dxxx workstations, *J. Chim. Phys. 88:* 2489 (1991).

5. U. C. Singh, P. Weiner, J. C. Caldwell, P. A. Kollman, D. A. Case, G. L. Seibel, and P. Bash, AMBER 3.0, University of California, San Francisco, 1986.

6. J. Kozelka, R. Savinelli, and G. Berthier, unpublished results.

7. K. Zimmermann, ORAL: All purpose molecular mechanics simulator and energy minimizer, *J. Comput. Chem. 12:* 310 (1991).

ORALLY ACTIVE DACH-Pt (IV) COMPOUNDS

Yoshinori Kidani,[1#] Ryoichi Kizu,[2] Motoichi Miyazaki,[2] Masahide Noji,[1] Akio Matsuzawa,[3] Yasutaka Takeda,[3] Nachio Akiyama[3] and Masazumi Eriguchi[3]

1. Faculty of Pharmaceutical Sciences, Nagoya City University, 3-1 Tanabe-dori, Mizuho-ku, Nagoya 467, Japan
2. Faculty of Pharmaceutical Sciences, Kanazawa University, 13-1 Takara-machi, Kanazawa 920, Japan
3. The Institute of Medical Science, The University of Tokyo, 4-61-Shirokanedai, Minato-ku, Tokyo 108, Japan

INTRODUCTION

Since the discovery of the antitumor activity of cisplatin by Rosenberg, et al.[1] in 1969, many attempts have been made to prepare highly antitumor active Pt complexes without severe toxicities. The authors discovered a simple preparative separation method of 1,2-diaminocyclohexane (dach) geometrical isomers, cis and trans[2], by Nickel complex formation and trans isomers were resolved into d and *l* optical isomers by the conventional method.

The authors synthesized dichloro 1,2-dach isomer Pt(II) complexes and tested them against L1210[3]. They were highly active and the authors had selected trans-*l*-dach (1R, 2R-dach) Pt(II) complexes the most active compound to develop.

Efficacy: trans-*l*-dach > trans-d-dach > cis-dach

Toxicity: trans-d-dach > trans-*l*-dach > cis-dach

The authors synthesized various trans-*l*-dach Pt(II) complexes. [Pt(oxalato)(1R, 2R-dach)], *l*-OHP was synthesized and is one of the superior Pt(II) complexes to be developed[4]. Prof.Dr. G. Mathé of ICIG, France developed *l*-OHP clinically, cooperated with Laboratoire Roger Bellon and found the superior efficacy and toxicity of *l*-OHP. He reported preclinical study and Phase I and II studies[5,6,7]. *l*-OHP showed efficacy against melanoma, ovarian and testicular cancers, lung cancer, stomach and colorectal cancers. Toxicities were nausea and vomiting, and peripheral sensory neuropathy.

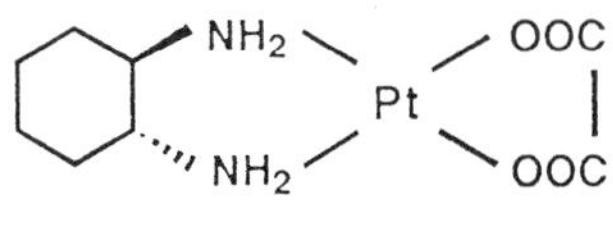

ℓ -OHP, Oxaliplatin

Characteristics of *l*-OHP was no nephrotoxicity, no cardiotoxicity, no cross-resistance[8], non-mutagenicity[8] and low myelosuppression. Being named as "Oxaliplatin" by L. Roger Bellon, it is now being developed clinically by DEBIOPHARM S.A., Switzerland, as a promising next generation dach Pt(II) complex.

[1#] Present address: 3-13-11 Kataseyama, Fujisawa 251, Japan

Platinum and Other Metal Coordination Compounds in Cancer Chemotherapy 2
Edited by H.M. Pinedo and J.H. Schornagel, Plenum Press, New York, 1996

Characteristics of dach Pt(II) and Pt(IV) complexes was that their antitumor activity showed no cross-resistance against cisplatin resistant tumors. 1,2-Diaminocyclohexane complexes did not show cross-resistance against diammine Pt complexes, irrespective of the dach isomers and the valencies, Pt(II)[8] and Pt(IV)[9].

ATTEMPT TO PREPARE ORALLY ACTIVE DACH-Pt(IV) COMPLEXES[10]

In order to enhance "Quality of Life" of cancer patients, preparation of orally active Pt complexes was believed to be very important and significant. It will be beneficial to treat out-patients without hospitalization, and also to treat terminal patients peacefully at hospices and homes without pains. Therefore, the authors attempted the development of oral antitumor dach Pt(IV) complexes. We synthesized [$PtCl_2$ (ox)(1R,2R-dach)], *l*-OHP.Cl[9] and it showed very higher antitumor activity against L1210, ip, but its partition coefficient was not high enough. Orally active dach Pt(IV) complexes should be lipophilic to penetrate through the intestinal membranes, and they must be stable in the stomach at low pH.

Synthesis of [Pt(IV)(trans-OCO-$C_mH_{2m+1})_2$(oxalato)(1R,2R-dach)]
m = 1 - 7, n = m + 1

The authors synthesized various (trans-bis-carboxylato) (oxalato) (1R,2R-dach) platinum(IV) by the reaction between [Pt(IV)(trans-dihydroxo)(ox)(1R,2R-dach)] and various carboxylate anhydrides to afford respective [Pt(IV)(trans-carboxylato)$_2$(ox)(1R,2R-dach)], Cn-OHP.

OH
NH_2 Pt OOC
NH_2 OOC
OH
$(C_4H_9CO)_2O$
$OCOC_4H_9$
NH_2 Pt OOC
NH_2 OOC
$OCOC_4H_9$

C5-OHP

Antitumor Activity of C5-OHP Against L1210, ip

Antitumor activity of [Pt(IV)(trans-valerato)$_2$(ox)(1R,2R-dach)], C5-OHP against L1210, ip was higher than those of [Pt(IV)(trans-carboxylato)$_2$(malonato)(1R,2R-dach)], *l*-MHP.
Antitumor activity of JM216 was lower than that of C5-OHP and JM221 showed almost no activity. Dose-response curves and structure-activity relationships between the antitumor activity, T/C% and the carbon numbers of the carboxylates in the trans-position indicated that the antitumor activity of C5-OHP was the most active compound. It showed very higher activity against L1210/DDP with collateral sensitivity, quite similarly to that of *l*-OHP[8]. (Table 1)

Partition Coefficients (1-Octanol/Water) and Stabilities in Strong Acid Solutions

Partition coefficients (1-octanol/Water) of C5-OHP was 8.6, which was larger than that of JM216, 1.8, measured in my Laboratory, and smaller than that of JM221, 140.
Stability, half-life, $t_{1/2}$(h) of C5-OHP measured in 1N HCl was 5.5h. That of JM216 was 3.6h and that of JM221 was 8.6h. Oral compounds will stay usually in the stomach no more than 5 hours and C5-OHP was considered to be enough stable. (Table 2)

Antitumor Activity of C5-OHP Against L1210, per os.

Oral antitumor activity of C5-OHP was tested against L1210. Twenty mg/mouse, ten times of T/C%, ip, suspended in olive oil, were administered to the starved CDF_1 mice by the stomach catheter, every day, Q01D x 05, or every other day, Q02D x 05. Both schedules gave T/C% 148

Table 1. Antitumor activity of C5-OHP, ℓ-OHP, JM216 and JM221 against L1210 and L1210/DDP.

	Dose, mg/kg					
	100	50	25	12.5	6.25	3.12
	T/C %					
ℓ-OHP L1210			T81	308 (4/6)	253 (1/6)	185
L1210/DDP					278 (6/6)	275 (6/6)
C5-OHP L1210	309 (2/6)	297 (3/6)	258 (1/6)			
L1210/DDP		322 (4/5)	375 (5/5)			
JM216 L1210	161 (1/6)	222 (1/6)	151	126	119	
JM221 L1210		109	103	142	119	145

L1210(10^5 cells/mouse), one group = 6 or 5 CDF_1 mice, ip-ip, administered on days 1, 5 and 9. Numbers in parenthesis are cured mice. T:toxic.

and C5-OHP was found to be an orally active candidate, but the optimal dose was pretty high. (Table 3)

PALUSIBLE METABOLIC PATHWAYS OF C5-OHP[11]

When C5-OHP was administered orally, it must pass through the strong acidic stomach and then it should penetrate through the intestinal membranes. C5-OHP was very stable and less than 1% of C5-OHP might be chlorinated to [$PtCl_2$(ox) (1R,2R-dach)], *l*-OHP.Cl in the stomach.

Chlorination of C5-OHP may take place by two steps to produce mono-chloro and di-chloro compounds that will penetrate through the intestinal membranes. These Pt(IV) complexes were considered to be activated by the reduction to dach Pt(II) complexes in the tumor sites. For the reduction, presence of halogeno ligands is necessary. Therefore, the authors synthesized various trans-mono-Cl-carboxylate dach Pt(IV) intermediates.

Synthesis of Trans-mono-halogeno-carboxylate Intermediates.

The authors prepared an intermediate of C5-OHP complex, [Pt(IV)Cl($OCOC_4H_9$)(oxalato)(1R,2R-dach)], C-5OHP.Cl by the mono-chlorination. C5-OHP was incubated in MeOH-1N HCl for 24h at 37°C and the reaction mixture was purified and isolated by means of HPLC. C4-OHP.Cl and C6-OHP.Cl were prepared similarly.

Table 2. Partition coefficients(1-octanol/water) and Half-life, $t_{1/2}$(h) of trans, cis,cis-[Pt(IV)(OCO-$C_mH_{2m+1})_2$(oxalato)(1R,2R-dach)], m=3, 4 and 5, JM216 and JM221.

Pt complexes	Partition coefficients	Half-live, $t_{1/2}$(h)	
		0.05N HCl	1N HCl
ℓ-OHP	2.3×10^{-2}	2.9	<0.1
C4-OHP	1.0	>50	4.4
C5-OHP	8.6	>50	5.5
C6-OHP	1.0×10^{2}	>50	7.5
JM216	1.8 (0.1*)	5.9	3.6
JM221	1.4×10^{2} (40*)	>50	8.6

* Giandomenico, C.M., et al.(1991), "Platinum and Other Metal Coordination Compounds in Cancer Chemotherapy", Ed. Howell, S.B., Plenum Press, New York, pp 93-100.

$$\text{Partition Coefficient} = \frac{C_i - C_w}{C_w} \times \frac{V_w}{V_o}$$

C_i: initial concentration in aqueous phase
C_w: concentration in aqueous phase
V_o: volume of 1-octanol, V_w: volume of water

Table 3. Antitumor activity of C5-OHP against L1210, po.

	Dose, mg/mouse	Schedule	T/C%
C5-OHP	20	Q01D x 05	148
	20	Q02D x 05	148
	15	Q01D x 05	135
	15	Q02D x 05	145
	10	Q01D x 05	133
	10	Q02D x 05	143

C5-OHP was suspended in olive oil and administered by the stomach catheter, to the starved mice. one group = 5 CDF_1 mice.

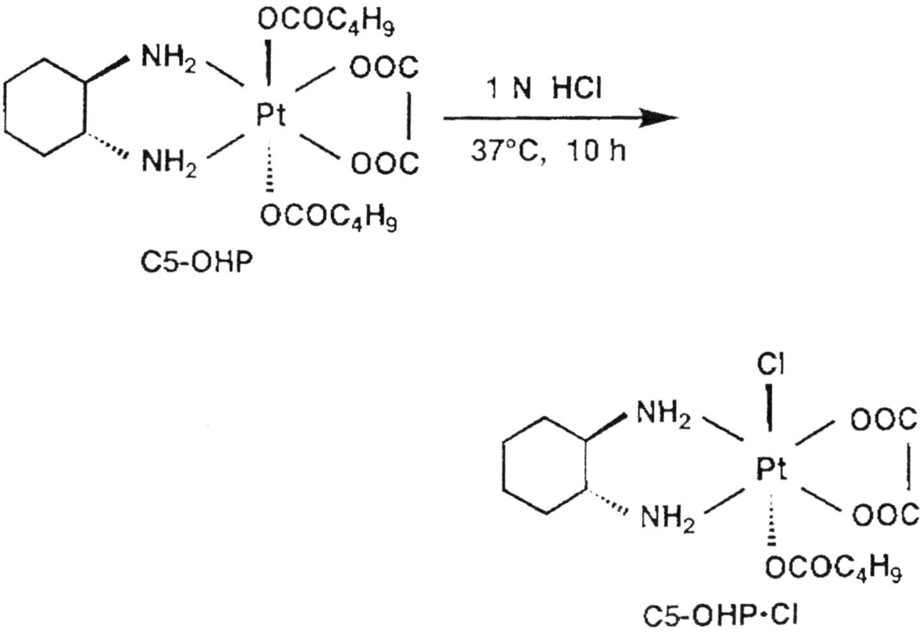

OCOC4H9
NH2
NH2
Pt
OOC
OOC
OCOC4H9
1 N HCl
37°C, 10 h
C5-OHP
Cl
NH2
NH2
Pt
OOC
OOC
OCOC4H9
C5-OHP•Cl

In order to prepare in a large quantity, a new preparative method has been devised. [Pt(IV)(trans-Hal_2)(ox)(1R,2R-dach)], *l*-OHP.Hal (Hal:Cl, Br and I) in MeOH was reacted with one mole of silver butyrate, C4-COOAg, silver valerate, C5-COOAg or silver hexanoate, C6-COOAg and the mixture was stirred for 48h at room temperature in the dark. MeOH was then evaporated to dryness and the residues were purified through chromatography to afford respective intermediates. This method is widely applicable to the preparation of various trans-mono-halogeno intermediates.

Br
NH2
NH2
Pt
OOC
OOC
Br
AgOCOC4H9
room temp.
48 h
l-OHP•Br
Br
NH2
NH2
Pt
OOC
OOC
OCOC4H9
C5-OHP•Br

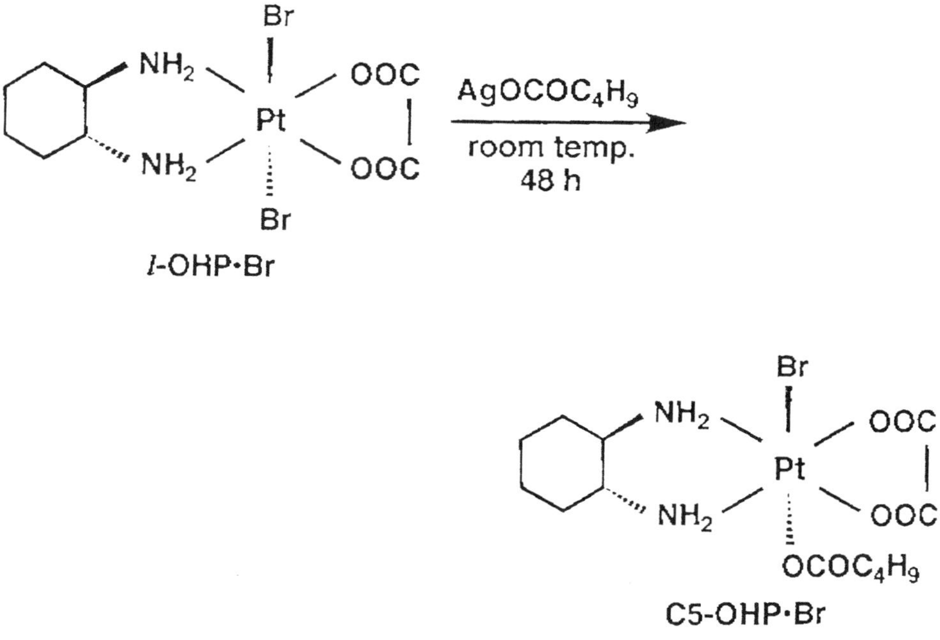

Table 4. Antitumor activities of [Pt(IV)Hal(carboxylato)(oxalato)(1R,2R-dach)] against L1210, ip. Hal:Cl and Br.

	Dose, mg/kg				
	200	100	50	25	12.5
			T/C %		
C4-OHP.Cl		63	195	145	130
C5-OHP.Cl		88	>283 (4/5)	168	
C6-OHP.Cl		80	>280 (1/5)		
C5-OHP.Br			107	>395 (4/5)	
C6-OHP.Br		69	96	>373 (4/5)	
C4-OHP*		160 (1/6)	271 (1/6)	221	156
C5-OHP*	112	309 (2/6)	297 (3/6)	258 (1/6)	
isoC5-OHP*	169 (1/6)	296 (3/6)	223		
C6-OHP*	166	286 (3/6)	214 (1/6)	188 (1/6)	
C5-OHP (L1210/DDP)		>700 (5/5)	>700 (5/5)	69	

L1210 and L1210/DDP(10^5 cells/mouse, one group = 5 CDF_1 mice, (*one group = 6 mice), administered on days 1, 5 and 9. *Cancer Chemotherapy Center, Japanese Foundation for Cancer Research, Tokyo, Japan. Numbers in parenthesis are cured mice.

Antitumor Activity of Intermediates Against L1210, ip.

C4-OHP.Cl, C6-OHP.Cl, C5-OHP.Br and C6-OHP.Br showed higher antitumor activity than C5-OHP against L1210, ip. Optimal doses of the intermediates became smaller by the mono-halogenation. C5-OHP showed collateral sensitivity against L1210/DDP, similarly to that of *l*-OHP. isoC5-OHP showed higher antitumor activity. Optimal doses of C5-OHP and C6-OHP were 100 mg/kg, but those of C5-OHP.Cl and C6-OHP.Cl were 50 mg/kg and those of C5-OHP.Br and C6-OHP.Br were 25 mg/kg. (Table 4)

Partition Coefficients (1-Octanol/Water) and Stabilities, Half-Life, $t_{1/2}$(H) of the Trans-mono-Cl Intermediates in 1N HCl.

Partition coefficients of C4-OHP.Cl, C5-OHP.Cl and C6-OHP.Cl were 0.056, 0.24 and 0.84, respectively. They became smaller than those of C4-OHP, C5-OHP and 6C-OHP. Partition

Table 5. Partition coefficients(1-octanol/water) and stability, $t_{1/2}$(h) in HCl solutions.

	Partition Coefficients	Stability, $t_{1/2}$(h)	
		0.05N HCl	1N HCl
C4-OHP.Cl	5.6×10^{-2}	>50	6.1
C5-OHP.Cl	2.4×10^{-1}	>50	7.4
C6-OHP.Cl	8.4×10^{-1}	>50	8.9
C4-OHP	1.0	>50	4.4
C5-OHP	8.6	>50	5.5
C6-OHP	1.0×10^{2}	>50	7.1

coefficients of Cn-OHP increased about one log unit as an increase of carbon numbers of the trans-carboxylates for the mono-chloro-derivatives, an increase from 4 to 6 of the number of carbons of the trans-carboxylato ligand leads to a one fourth decrease of the partition coefficients.

Half-lives of C4-OHP.Cl, C5-OHP.Cl and C6-OHP.Cl in 1N HCl were about 1.2 times more stable than the trans-bis-carboxylate compounds, C4-OHP, C5-OHP and C6-OHP. Partition coefficients of trans-mono-Cl intermediates became much smaller but the stability in 1N HCl became larger. (Table 5)

Table 6. Antitumor activities of [Pt(IV)Hal(carboxylato)(oxalato)(1R,2R-dach)] against L1210, ip-po, Hal:Cl and Br.

	Schedule	Dose, mg/mouse 20	15	10	5
		T/C %			
C5-OHP.Cl	Q02D x 05		60	83	
C6-OHP.Cl	Q02D x 05		98	>153	
C5-OHP.Br	Q02D x 05		73	>145	>158
C6-OHP.Br	Q02D x 05		88	>153	>168
C5-OHP	Q01D x 05	148	135	133	
	Q02D x 05	148	145		
C6-OHP	Q01D x 05		120		
	Q02D x 05		115		
isoC5-OHP	Q02D x 05		>168		

L1210(10^5 cells/mouse), one group = 5 CDF_1 mice, administered every day, Q01D x 05, or every other day, Q02D x 05. Pt complexes were suspended in olive oil and given by the stomach catheter.

Table 7. Antitumor activity of C5-OHP, together with ascorbic acid, against L1210, ip-po,

C5-OHP	Dose, 15 mg/mouse	T/C% 133
Ascorbate	Dose, 5 mg/mouse	T/C% 99
C5-OHP + Ascorbate	Dose, 15 mg/kg + 5 mg/mouse	T/C 141

L1210(10^5 cells/mouse), one group = 5 CDF_1 mice. Pt complex was suspended in olive oil and administered by the stomach catheter, every other day.

Antitumor Activity of the Trans-mono-halogeno Intermediates Against L1210, ip-po.

Pt compounds, suspended in olive oil, were administered to the starved CDF_1 mice by the stomach catheter. Antitumor activity of C5-OHP.Cl was not high, but we must test at the lower doses. C6-OHP.Cl showed T/C% >153 and C5-OHP.Br and C6-OHP.Br showed much higher activity. Antitumor activity by oral administration showed that the Q02D x 05 system showed better activity at doses of 20 mg/mouse, 10 times of optimal dose, 100 mg/kg, ip. (Table 6)

The antitumor activity of C5-OHP became higher by the simultaneous administration of ascorbic acid. (Table 7)

Reduction of Trans-mono-chloro Intermediate with ascorbate.

Reduction of C4-OHP.Cl with ascorbate was much faster than that of C4-OHP. Similarly, C5-OHP.Cl and C6-OHP.Cl became much faster than C5-OHP and C6-OHP. In the presence of 5 mM ascorbate, half-life, $t_{1/2}$(h) of C5-OHP.Cl was nearly equal to that of JM216 and $t_{1/2}$(h) of C6-OHP.Cl was nearly equal to that of JM221. The reducibility of the compounds may be correlated with the activity. The antitumor activity of C5-OHP became higher by the simultaneous administration of ascorbate. This gives a good support to the proposition that the active species are the reduced Pt(II) compounds. (Table 8)

Table 8. Reduction of trans-mono-chloro intermediates with ascorbic acid, pH 7.5 (HEPES Buffer), at 37°C.

	Half-life, $t_{1/2}$ (h)	
	5 mM	100 mM
C4-OHP.Cl	0.64	<0.1
C5-OHP.Cl	0.83	<0.1
C6-OHP.Cl	1.19	<0.1
C4-OHP	>50	2.56
C5-OHP	>50	2.77
C6-OHP	>50	3.65
JM216	0.76	0.06
Jm221	1.38	0.14

CONCLUSION

Antitumor active new trans-mono-halogeno-carboxylate intermediates showed some promise in the development of oral antitumor dach-Pt(IV) compounds.

1. Optimal doses of antitumor activities, in vivo, ip of trans-mono-halogeno dach-Pt(IV) intermediates became smaller than those of trans-bis-carboxylate dach-Pt(IV) compounds.
2. Partition coefficients (1-octanol/water) of C5-OHP.Cl and C6-OHP.Cl were much smaller than those of trans-bis-carboxylate dach-Pt(IV) compounds.
3. The half-lives, $t_{1/2}$(h) of the intermediates in 1N HCl became longer than those of trans-bis-carboxylate dach-Pt(IV) compounds.
4. Oral antitumor activities of trans-mono-halogeno dach-Pt(IV) intermediates became much higher than that of C5-OHP. Antitumor activities of C5-OHP.Br and C6-OHP.Br showed higher values.
5. Trans-mono-halogenated dach-Pt(IV) compounds were reduced much faster than C5-OHP.

   In conclusion, C5-OHP.Cl and C6-OHP.Cl, as well as C5-OHP.Br and C6-OHP.Br will be promising candidates for the oral antitumor dach-Pt(IV) compounds.

   Finally, trans-mono-halogeno intermediate is a "Pro-drug", and the active species is "*l*-OHP".

ACKNOWLEDGEMENTS

The authors' thanks are due to Dr. R. -Y. Mauvernay, President of DEBIOPHARM S.A., Switzerland for his financial support and to Tanaka Kikinzoku Kogyo for a gift of Pt chemicals.

REFERENCES

1. Rosenberg, B., Trosko, J.E. and Mansour, V.H., Nature, 222, 385 (1969).
2. Saito, R. and Kidani, Y., Chem. Lett., 1976, 123.
3. Kidani, Y., Inagaki, K. and Tsukagoshi, S., Gann, 67, 921 (1976).
4. Kidani, Y., Noji, M. and Tsukagoshi, S., Gann, 69, 263 (1978).
5. Mathé, G., Kidani, Y., Noji, M., Maral, R., Bourut, C. and Chenue, E., Cancer Lett., 27, 135-143 (1985).
6. Mathé, G., Kidani, Y., Triana, K., Brienza, S., Riboud, P., Goldschmidt, E., Excein, E., Despax, R. and Misset, J.L., Biomed. & Pharmacother., 40, 372-376 (1986).
7. Mathé, G., Kidani, Y., Sekiguchi, M., Eriguchi, M., Gedj, G., Peytavin, G., Misset, J.L., Brienza, S., De Vassals, F., Chenu, E. and Bourt, C., Biomed. & Pharmacother., 43, 237-250 (1989).
8. Tashiro, T., Kawada, Y., Sakurai, Y. and Kidani, Y., Biomed. & Pharmacother., 43, 251-260 (1989).
9. Noji, M., Sumi, M., Ohmori, T., Mizuno, M., Suzuki, K., Tashiro, T. and Kidani, Y., J. Chem. Soc. Jpn., 1988, 675 (Japanese).
10. Kidani, Y., Komoda, Y., Tashiro, T., Kizu, R., Miyazaki, M., Abstract, #213, p.125, 8th NCI-EORTC Symposium on new drugs in cancer chemotherapy, March 1994, Amsterdam.
11. Kidani, Y., Kizu, R., Miyazaki, M. and Noji, M., Abstract, S4-21, p.71, The 30th International Conference on Coordination Chemistry, July 1994, Kyoto, Japan.

PHARMACOKINETIC - PHARMACODYNAMIC RELATIONSHIPS

Pierre CANAL and Etienne CHATELUT

Pharmacology Laboratory, Centre Claudius Regaud, 20-24 rue du Pont Saint-Pierre, 31052 Toulouse Cedex, France.

INTRODUCTION

Research of relationships between pharmacokinetics and pharmacodynamics in clinical oncology should result in a better use of anticancer drugs. The goal of these approaches is to improve cancer chemotherapy by optimising drug therapy in individual patients. After some recalls about general principles of pharmacokinetics and pharmacodynamics, the different models used for analysing pharmacokinetics-pharmacodynamics relationships will be defined. Before considering the different methods of adaptive dosing of anticancer drugs, the different reasons of variability in pharmacokinetics will be described briefly.

PHARMACOKINETIC AND PHARMACODYNAMIC TERMINOLOGY

Pharmacokinetics is traditionally described as the mathematical description of the in vivo fate of a drug. Most anticancer drugs are administered by iv route. Under these conditions, pharmacokinetics describes only distribution, metabolism and elimination of the drug.

Central to this description is the plasma concentration versus time profile from which primary pharmacokinetic parameters are derived : peak plasma concentration, area under the plasma concentration versus time curve (AUC) and half-lives. Secondary pharmacokinetic parameters, clearance and volume of distribution can then be defined to describe elimination and distribution of the compound. For the practising clinician, most of these parameters can largely be ignored beyond the simple relationship between dose, AUC and clearance.

The first one, when a drug is administered by a bolus or a short intravenous route, is the following :

$$\text{Dose} = \text{AUC} \times \text{Clearance.}$$

For drugs given at a constant rate via iv infusion, once the steady state plasma concentration has been achieved, the relationships between clearance, plasma concentration and dose rate is as follows :

$$\text{Plasma Css} = \text{Dose rate} / \text{clearance.}$$

Platinum and Other Metal Coordination Compounds in Cancer Chemotherapy 2
Edited by H.M. Pinedo and J.H. Schornagel, Plenum Press, New York, 1996

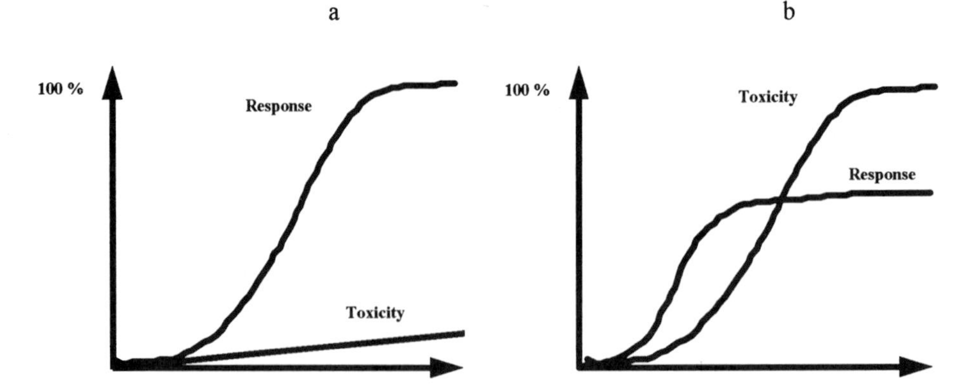

Figure 1 : Relationships between drug dose and drug effects a : drug with no toxicity at doses yielding maximum therapeutic response; b : drug with increasing toxicity when concentrations increase above those needed for maximum therapeutic response.

With knowledge of the plasma clearance of a drug, the dose required for a given AUC or steady state plasma level may readily be calculated.

Pharmacodynamics describes the effects of a drug on the body. The effects of any drug are two fold : activity and toxicity.

The potential pharmacodynamic consequences of that drug in the patient to be treated can be described by a two dimensional graph wherein intensity of drug effect is displayed along the ordinate while the drug dose or time course of drug itself in the plasma is displayed along the abscissa.

For drugs that do not produce toxicity at dosages or serum concentrations close to those required for therapeutic effects, there is little interest for dose optimisation or individualisation. Under these circumstances, patients are treated with dosages high enough to insure achievement of therapeutic concentrations (Fig 1 a).

In contrast, certain drugs, such as antineoplastic chemotherapeutic agents, which frequently produce toxicity at dosages close to those required for a therapeutic effect provide great interest for dose optimisation in individual patients and require establishment of pharmacokinetics-pharmacodynamics relationships (Fig 1 b).

When considering candidate drugs for pharmacokinetically guided dosing, three major criteria must be identified (Egorin 1992):

First, a relationship between drug concentration in serum and response must be established : the response must be optimal in most of the patients when serum drug concentrations are maintained within a therapeutic range.

Second, large interpatient variability in distribution and elimination of the drug was observed as a result of genetic and pathophysiologic conditions. There may also be large differences in sensitivity of patients to a given drug concentration.

Finally, there may be wide differences among patients regarding the dose required to achieve an optimal clinical response.

The obvious goal of considering pharmacokinetics-pharmacodynamics relationships for individual patients is to maximise the probability of producing a desired therapeutic effect while minimising the probability of a toxic event. With anticancer drugs, this goal is often modified to seek the maximum probability of producing a desired therapeutic effect while producing acceptable toxicity. The theoretical pharmacodynamic response will be

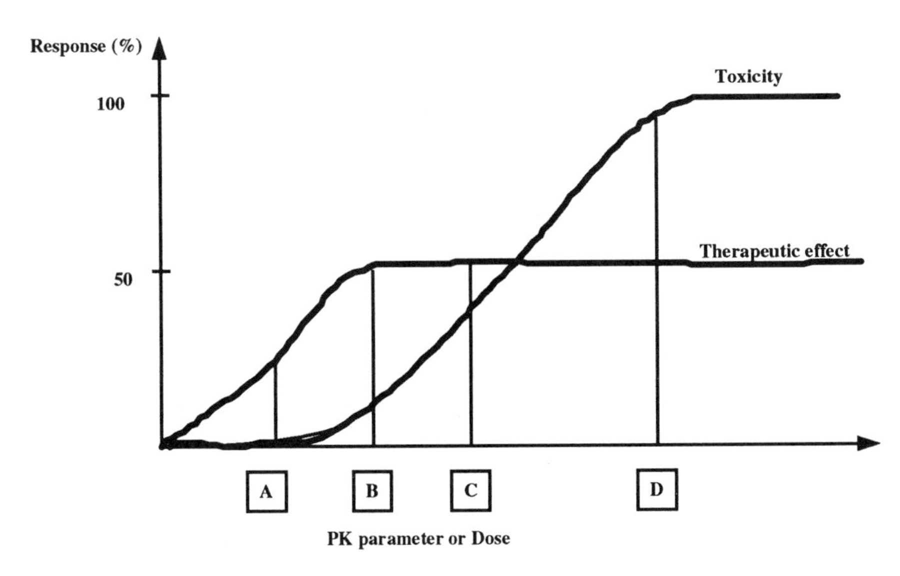

Figure 2 : Theoretical concentration-response curve for an antineoplastic drug.

represented as described in the figure 2 : the slope of the efficacy curve and the maximum effect will be dependent on the sensitivity of the tumor to the drug.

Because pharmacokinetics is known to be quite variable from one patient to another, both toxic and therapeutic responses to drug administration are frequently better correlated with plasma drug concentration or the total amount of drug in the body than with the administered dose (Moore and Erlichman 1987).

Increases in dose or AUC from A to B will increase efficacy with a moderate increase in toxicity, while increases from C to D will give increased toxicity with no increased response. Thus, therapeutic drug monitoring consists of trying to treat all patients between B and C.

On the basis of these criteria, anticancer drugs are clear candidates for pharmacokinetically-guided dosing, but specific problems arise.

In the case of response and delayed toxicity, effects may be seen after multiple courses of therapy. Thus, in cancer chemotherapy, it may not be possible to use conventional pharmacodynamics end-points to determine doses for individual patients.

The other important point for anticancer drugs concerns the relevance of pharmacokinetic-pharmacodynamic relationships. The relevance of pharmacokinetics-toxicity versus pharmacokinetics-activity relationships should be viewed in the context of the use of the drugs (Newell 1994). For palliative treatment, manageable and predictable toxicity is a primary requirement. So, pharmacokinetic-pharmacodynamic relationships are valuable if they can be exploited in order to achieve this. In contrast, if curative therapy is being attempted, more severe toxicity may be accepted, provided that the drug is being used to an optimal therapeutic effect and to achieve this, pharmacokinetic studies may be useful.

Potential pharmacokinetic parameters to be used in pharmacokinetic-pharmacodynamic relationships are generally AUC or Css. Other parameters are Cmax, duration of concentration above a threshold or AUC intensity. Pharmacodynamics will be described by either discontinuous parameters (response, no response) or continuous parameters such as time to progression or survival.

When considering toxicity, two types of parameters may also be considered : quantitative parameters such as leucocyte, granulocyte or platelets counts or semi-quantitative parameters defined by WHO grading (Egorin 1994).

MODELS FOR ESTABLISHING PHARMACOKINETIC-PHARMACODYNAMIC RELATIONSHIPS

Mathematical functions have been used to describe pharmacodynamic effects and the most commonly used function is the modified Hill equation which describes a sigmoïdal equation (Figure 3).

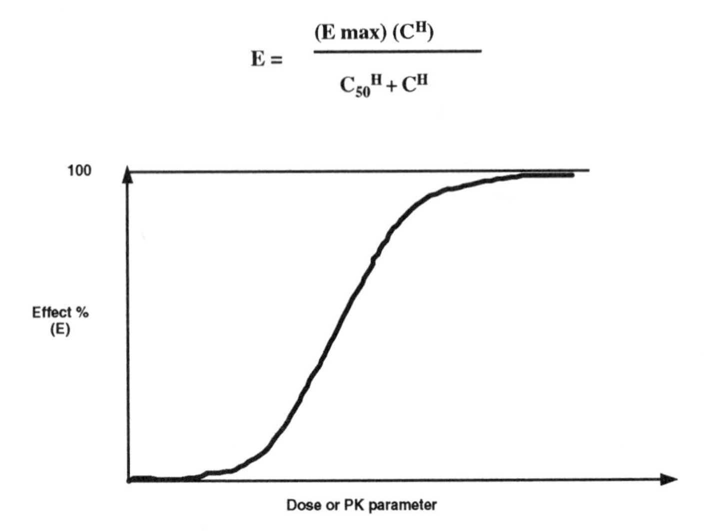

Figure 3: Sigmoïdal dose-effect or concentration-effect relationships in clinical oncology : Hill model.

The mathematical relationship is well described by Hill equation where Emax represents the maximum possible effect, C represents pharmacokinetic parameter or dose and C50 represents the concentration, the AUC or the dose which induces 50% of Emax. H is Hill's constant which defines the degree of sigmoïdicity of the model.
However, pharmacokinetic-pharmacodynamic relationships may also be represented by linear or exponential models with the following equation (Ratain 1990) :

$$\%\ \text{survival fraction} = e^{-kCt}$$

in which % survival fraction or SF (which expresses the ratio between nadir and pretreatment cell counts for example) represents the pharmacodynamic parameter which is proportional to the exponential of the drug concentration or the drug AUC. K is a constant determining the slope of the dose-response curve. This model looks like in vitro cytotoxicity of a drug.

PHARMACOKINETIC - PHARMACODYNAMIC RELATIONSHIPS IN CLINICAL ONCOLOGY

The following table illustrates pharmacokinetic-activity relationships. This list is not exhaustive. Most of the first studies published concern paediatric tumors. But, concerning platinum compounds, particularly encouraging examples of relationships between carboplatin AUC and response have been reported in patients with ovarian or testicular cancer (Horwich, 1991, Jodrell, 1992).
Many examples of pharmacokinetic-toxicity relationships for established agents used in cancer chemotherapy have been reported. They concern principally relationships between hematologic toxicity and pharmacokinetics parameters, but relationships between

Drug	Tumor type	PK parameter	Reference
MTX	ALL	Css	Evans, 1989
Mercaptopurine	ALL	RBC AUC	Lennard, 1987a
Teniposide	Paediatric solid	AUC	Rodman, 1987
Cyclophosphamide	Breast	AUC	Ayash, 1992
Etoposide	Lung	Css	Desoize, 1990
5FU	Head/Neck	AUC	Milano, 1994
Cytarabine	Relapsed leukaemia	Blast Ara CTP	Kantarjian, 1986,
Carboplatin	Teratoma	AUC	Horwich, 1991
Carboplatin	Ovarian	AUC	Jodrell, 1992

Table I : Pharmacokinetic-response relationships for drugs in clinical oncology.

discontinuous parameters and pharmacokinetics parameters have also been reported such as the degree of mucositis and the AUC of 5FU. For platinum compounds; one can mention the study of Reece (1987) who showed a relationship between free platinum peak plasma concentration and the percentage of increase in creatinine. Interesting also, is the relationship between thrombocytopenia and carboplatin AUC which has allowed the adaptation of carboplatin dosing (Egorin, 1984).

SOURCES OF VARIABILITY IN PHARMACOKINETICS

As previously mentioned, pharmacokinetics describes both distribution and elimination of a drug. Sources of variability will influence these two phenomena (Ratain, 1990).

Concerning distribution, some potential problems such as a third space distribution are widely recognised and easily avoided. The main source of variability affecting drug distribution is plasma protein binding. Pharmacological principles hold that the unbound drug (the drug not bound to plasma proteins) is the active drug : only the non protein bound drug is available to diffuse out of the circulating blood and interact with tissues. Alterations in drug plasma protein binding have direct clinical consequences. Such is the case for drugs extensively bound to proteins as etoposide. A study by Stewart (1990) has shown a four-fold difference in free etoposide concentrations depending on albumin levels.

Concerning platinum compounds, differences in the stability of the leaving groups explain the difference in plasma protein binding which is more important for cisplatin (about 90%)

Drug	Toxicity	PK parameter	Reference
Cyclophosphamide	Cardiotoxicity	AUC	Ayash, 1992
Busulfan	Hepatotoxicity	AUC	Grochow, 1990
Cisplatin	Nephrotoxicity	Peak level	Reece, 1987
Cisplatin	Neurotoxicity	Cmax	Gregg, 1992
Vincristine	Neurotoxicity	AUC	Desai, 1982
5FU	Mucositis	AUC	Santini, 1989
Doxorubicin	Leukopenia	Css	Ackland, 1989
Taxol	Leukopenia	AUC	Rowinsky, 1991
Carboplatin	Thrombopenia	AUC	Egorin, 1984,
Etoposide	Leukopenia	AUC, Css	Miller1990, Bennett 1987
Topotecan	Leukopenia	AUC	Van Warmerdam, 1995

Table II : Pharmacokinetic-toxicity relationships for drugs in clinical oncology.

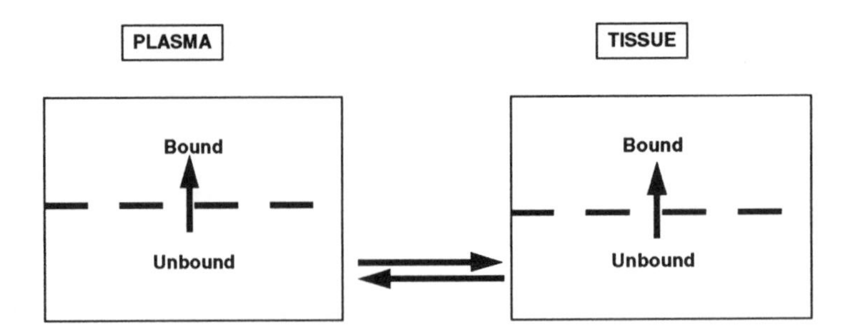

Figure 4 : Mechanisms of drug distribution and plasma and tissue protein binding : example of platinum compounds.

than for carboplatin (70%)(Van der Vijgh 1991). However, the plasma protein binding of platinum compounds is very special since the binding of platinum to albumin is covalent. Thus, the equilibrium which exists for most drugs does not exist for platinum compounds.
Under these conditions, the essential form for evaluating pharmacokinetics of platinum compounds is free or ultrafilterable platinum. Once in circulation, platinum compounds are cleared from the plasma not only by tissue uptake and excretion but also by irreversible binding to plasma proteins (Calvert 1993). Is the irreversible plasma protein binding variable ? This question has until now no answer. It seems that the capacity of binding albumin to platinum is not saturable. So, under these conditions, the variability will be low.
Variability in clearance is a very important phenomenon for most anticancer drugs. Clearance of a drug is a sum of different pathways :
Firstly, renal clearance which may be variable from one patient to another. This clearance is very important for drugs which are mainly eliminated by the kidney. This is the case for e.g. methotrexate, melphalan or carboplatin. Changes in glomerular filtration will induce a modification in exposure of the patient to the drug.
Secondly, non renal clearance will comprise both elimination and metabolism. Variability in hepatic clearance might be due to hepatic dysfunction such as cholestasis, hepatitis, cirrhosis or metastatic involvement. Finally, variability in metabolism may be linked to genetic polymorphisms. This is the case of mercaptopurine (Lennard 1987b), or 5FU (Etienne 1994).
For platinum compounds, different studies showing clearance variabilities are summarised in the table III.

Cisplatin clearance (ml/min)			
Author	Number of patients	Range	Reference
Vermorken	3	307 - 345	1987
Paredes	8	200 - 561	1988
Forastiere	5	320 - 490	1988
Thomas	21	206 - 500	1994

Carboplatin clearance (ml/min)			
Author	Number of patients	Range	Reference
Egorin	22	35 - 135	1984
Calvert	37	58 - 170	1989
Chatelut	70	34 - 202	1995

Table III : Variabilities in total body clearance of cisplatin and carboplatin.

Cisplatin clearance is about 350 ml/min and its factor of variability is about 2.5 ranging between 200 and 500 ml/min. For carboplatin, for which the total body clearance is about 100 ml/min, this parameter varies from 35 to 200 ml/min which corresponds to a factor of 7. The elimination pathways of these two drugs differ due to a difference in stability of the leaving groups of the two drugs : the non renal clearance, consisting mainly of protein and tissue binding, is predominant for cisplatin. The renal clearance is about one-fourth of the total body clearance. This renal clearance comprises both glomerular filtration, tubular secretion and reabsorption. Carboplatin is mainly eliminated by the kidney and as demonstrated by Egorin (1984) and Calvert (1989), the renal clearance of carboplatin is directly related to the glomerular filtration rate. The non renal clearance consisting of protein and tissue binding represents one-fourth the total body clearance.
One can suppose that it will be possible to predict carboplatin clearance from glomerular filtration rate and that adaptive dosing of this drug is possible.
For cisplatin, the problem will be more difficult because we do not know precisely the reasons for the variability in non renal clearance.

METHODS OF DRUG DOSING ADJUSTMENT IN CLINICAL ONCOLOGY

Three dosing strategies can be employed (Egorin 1992).
The first one is an **empiric method or non adaptive dosing**, which is the method most frequently used by oncologists. An agent is used at a mg/m^2 or mg/kg dose recommended for phase II trials. This recommended dose is based on results of phase I clinical trials and its recommendation may involve no pharmacokinetic consideration at all. Dose reductions or delays are implemented in the face of unacceptable toxicity, but doses are seldomly escalated in the absence of toxicity. Due to the poor therapeutic indexes of anticancer drugs, this means that patients not showing signs of toxicity are being underdosed.
Adaptive dosing attempts to take into account the pretreatment patient characteristics known to affect the pharmacokinetics or pharmacodynamics of a drug. That is the case for example of dose reduction of anthracyclines in case of hepatic dysfunction.
Recent developments in the field of pharmacogenetics should also be noted in the context of adaptive dosing; It is the case for amonafide, 5FU and mercaptopurine (Lennard 1987b, Etienne 1994, Ratain, 1987).
The most firmly established example for adaptive dosing according to patient characteristics is the use of renal function to guide carboplatin dosing. To this purpose, different dose adjustments have been proposed.
The first formula has been developed by Egorin (1984). Egorin formula includes two physiological parameters which can be obtained before the treatment : measured creatinine clearance and body surface area and a pharmacodynamic end-point : a target platelet nadir.

$$\text{Dose} = 0.091 \, \frac{\text{Cr Cl}}{\text{BSA}} \times \left[\frac{\text{Prtt plts} - \text{plts nadir}}{\text{Prtt plts}}\right] + 86$$

for previously untreated patients.

$$\text{Dose} = 0.091 \, \frac{\text{Cr Cl}}{\text{BSA}} \times \left(\left[\frac{\text{Prtt plts} - \text{plts nadir}}{\text{Prtt plts}}\right] - 17\right) + 86$$

for previously treated patient

in which Cr Cl is the measured creatinine clearance, BSA the body surface area, prtt plts the pretreatment platelets count and plts nadir, the platelets nadir.

This formula was subsequently validated, showing a good correlation between observed and expected percentage of reduction in platelet counts (1985). The main advantage of the

Egorin formula is to take into account a pharmacodynamic end-point : % decrease in platelet counts. It should be noted that in the Egorin formula, the carboplatin dosage is still influenced by body surface area.

The second formula has been developed by Calvert et al (1989). They derived a formula for patients to be exposed to a particular AUC rather than aiming for a specific platelet nadir.

$$\text{Dose (mg)} = \text{Target AUC (mg/ml x min)} \times (\text{GFR} + 25)(\text{ml/min})$$

In this formula, clearance was the sum of renal clearance determined by the glomerular filtration rate of a patient by 51-chromium-EDTA clearance and of the non renal clearance equal to 25. The authors assumed that the non renal clearance is constant. This formula has been validated and widely used (Sorensen 1991). The main advantage of the Calvert formula is its simplicity to apply. However, any potential benefit from the use of these formulas depends on an accurate method for estimating the GFR. Many centers do not have ready access to isotope clearance techniques and the 24 hour urine collection necessary for an accurate creatinine clearance is difficult to achieve in practice (Calvert 1994).

This is the reason why we try to establish a formula based on pharmacokinetic population analysis and to predict the carboplatin clearance from physiological parameters of patients which are very easy to obtain.

The NONMEM computer program, we have used, allows the treatment of a population as the unit of analysis (Beal 1985, 1992). It is a powerful and accurate program for testing relationships between co-variables and pharmacokinetics. Thus, it allows us to predict the carboplatin clearance from standard morphological and biological patient characteristics (Chatelut 1995).

For establishing the formula, adult patients were used receiving carboplatin as part of different established protocols for various tumor types. The population was very heterogeneous with important differences in renal function, age and carboplatin dose.

In all patients, the pharmacokinetics of carboplatin administered as a 1 hour iv infusion was determined and fitted to a two compartment linear mamillary model with first-order elimination from the central compartment.

The following co-variables were analysed : age, sex, weight, height, body surface area, serum creatinine, serum proteins, serum albumin, cisplatin pretreatment.

The best fit for carboplatin clearance estimation was given by the following equation which considers only four co-variables : weight, age, sex and serum creatinine. The other co-variables studied did not enhance the final regression formula.

$$\text{carboplatin clearance}_{(\text{ml/min})} = 0.134.weight + \frac{218.weight.[(1-0.00457).age.[(1-0.0314).sex]}{serumcreatinine(\mu M)}$$

with weight in kg, age in years and sex = 1 if female.

The clearances calculated according to this formula were closely correlated to the observed clearances. The percent error ranged between -28 and 27% with a median of -4%.

In this formula, the first part corresponds to the non renal ultrafilterable carboplatin elimination. This term is only dependent upon the weight. The second part of the equation corresponds to the renal clearance of carboplatin. The co-variables considered did not differ from those retained in the Cockroft equation : weight, age, sex, creatinine level (Cockroft, 1976). However, coefficients assigned to the different parameters differ.

In a second part of our work, we have validated our formula in an equivalent population : 36 patients with a variety of renal function, age and carboplatin dosage.

Figure 5 shows the correlation between predicted and observed carboplatin clearance. The slope of the correlation was very close to 1 and it seems that no bias exists in our formula.

Finally, the performance of the NONMEM formula was compared with the Calvert formula since in all the patients considered, 51 chromium EDTA clearance had been performed as

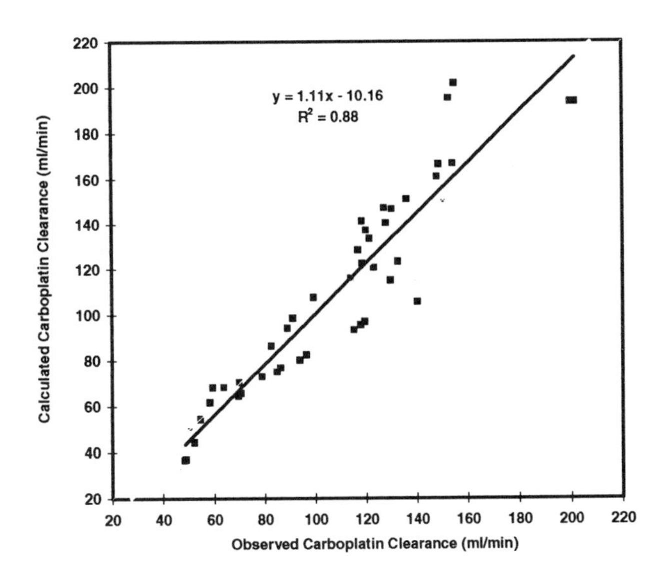

Figure 5 : Relationships between the observed ultrafilterable carboplatin clearance and the clearance predicted by the NONMEM formula.(Chatelut 1995).

recommended by Calvert. The performances of the two formulas are comparable with a median bias of -3% and 2% for the Calvert and NONMEM formulas, respectively.

In conclusion, the formula we propose allows us to predict the administered dose of carboplatin as a function of patient characteristics easily obtained. It did not require the determination of a special parameter such as 51-chromium EDTA clearance. It is well correlated with the Calvert formula and will allow retrospective studies to calculate carboplatin exposure. However, this formula is complicate but only requires a calculator to be done. Finally, this formula cannot be used for categories of patients not explored such as paediatric patients.

In paediatric patients, two formulas have been proposed. The Rodman formula which looks like Calvert formula is normalised for the body surface area (Rodman 1993).

$$\text{Dose (mg/m}^2\text{)} = \text{target AUC (mg/ml x min) x [(0.93 GFR (ml/min/m}^2\text{)} + 15].$$

This equation has been prospectively validated in AUC escalation studies. In the majority of patients, the observed AUC was within +/- 30% of the desired AUC (Marina 1993, 1994).

The second one proposed by the team of Newcastle was derived from the equation established by Calvert for adults (Newell 1993).

$$\text{Dose (mg)} = \text{target AUC (mg/ml x min) x [GFR (ml/min)} + \text{(0.36 x BW) (kg)]}$$

The non renal clearance is determined as a function of body weight. This formula has been validated prospectively with a median bias of 20%.

There is a third approach to adapt dosing of anticancer drugs : **adaptive dosing with feedback control**. It is the most complicated and most complex method. In this approach, population based predictive models are initially used, but contain feasibility for dose alteration based on feedback revision.

Most of these studies have used a continuous infusion therapy and have been based on previously published models. Patients were treated at a standard dose and during treatment, the pharmacokinetics of the considered drug was estimated by limited sampling strategy and compared to that predicted from the population model with which dosing was initiated.

Based on the comparison, more patient-specific pharmacokinetic parameters are calculated and dosing is adjusted accordingly to maintain the drug Css or the exposure desired to produce the desired pharmacodynamic effect.

Despite its mathematical complexity, this approach may be the only way to deliver a desired, precise exposure of an anticancer drug. This methodology has been used for several years to administer methotrexate therapy. More recently, it has been developed more extensively and applied to continuous intravenous infusions of 5FU (Santini 1989), cisplatin (Desoize 1994) and etoposide (Ratain 1991b). But it may also be used to adapt the dose of anticancer drugs from one cycle to another (Thomas 1994).

CELLULAR PHARMACOKINETICS

Now, with these different methodologies, we are able to reduce interpatient pharmacokinetic variability. But, is there any relationship between plasma drug concentration or plasma exposure to a drug and its concentration or exposure to the site of action of this drug? Different factors might modify the quantity of drug reaching the tumor cells : tumor vascularisation, heterogeneity of the tumor and cell kinetics. Finally, an other important factor is linked to the development of mechanisms of drug resistance. Under these circumstances, the cellular pharmacokinetics of anticancer drugs have begun to provide important new insights into cell heterogeneity and determinants of cytotoxicity for several common anticancer drugs.

This is the case for instance for Ara C for which we know the importance of the Ara CTP AUC in leukemic blasts to predict the clinical outcome of ALL (Kantarjian 1986). Indeed, the importance of MTX polyglutamates in blasts of children suffering from ALL has been established as a determinant of response in ALL (Whitehead 1991). Finally, the importance of red blood cell thioguanine nucleotide concentrations for efficacy of the treatment (Lennard 1987b).

Cellular pharmacology of platinum compounds is well known and cell death appears clearly to be related to the formation of interstrand and intrastrand DNA cross-links. Thus, different teams have tried to predict DNA-platinum adducts formation in tumor cells by determining these adducts in peripheral blood cells.

The rationale for such a comparison is summarised in the following table, comparing the different steps of the cellular pharmacology of platinum in peripheral blood cells and tumor cells.

If the uptake of a drug, the formation and removal of DNA-platinum adducts are explored in the two systems, differences arise especially regarding the mechanisms of resistance and the absence of cellular heterogeneity. However, adduct formation in WBC appears to be correlated with the cisplatin dose. Adduct formation appears to be correlated with the

Factor	Blood cells	Tumor cells
Vascularisation	no	yes
Cellular heterogeneity	no	yes
Uptake	yes	yes
Formation of adducts	yes	yes
Removal of adducts	yes	yes
Resistance	no	yes

Table IV : Comparison of factors influencing cellular pharmacology of platinum compounds in blood cells and tumor cells.

pharmacokinetics of ultrafilterable platinum. No relation exists between in vivo and in vitro adducts levels. Finally, adducts formation in WBC might correlate with tumor response (Fichtinger-Schepman 1987, 1990, Reed 1987, 1988, 1990, 1993, Parker 1991, Motzer 1994). However, in all these studies, we have no idea of the kinetics of formation or removal of DNA-platinum adducts formation since the level of DNA-platinum adducts was determined only once during a cycle of chemotherapy. Thus, the idea of Ma et al (1994) to try to determine an AUA must be very interesting to be considered and they found a correlation with response (see presentation of J. Schellens).

CONCLUSIONS AND FUTURE DIRECTIONS

Anticancer drugs are excellent candidates for therapeutic drug monitoring with adaptive control of dosing to optimise efficacy and minimise toxicity. Due to the clinical results obtained with platinum compounds, and the importance of dose-intensity for these drugs, this concept must be applied to these drugs. Adaptive dosing is possible and clinically useful for carboplatin. Adaptive dosing with feedback control might be used for cisplatin.

However, we must now try to develop more extensively cellular pharmacokinetic studies to try to improve pharmacokinetic- response relationships.

Can pharmacokinetic-pharmacodynamic studies improve cancer chemotherapy ? This question requires randomised clinical trials comparing dose adjustment versus empiric dosing of anticancer drugs. On the other hand, pharmacokinetic-pharmacodynamic studies might be included in the early phase of drug development : during phase I, II and III clinical trials.

Finally, determination of pharmacokinetic-pharmacodynamic parameters in tumor biopsies might be of potential benefit. Development of the polymerase chain reaction permits quantitative analysis of DNA or messenger RNA in extremely small biological biopsies. Of course, tumor biopsies present even greater logistic and resource problems than blood sampling. Highly sensitive molecular assays can be developed to allow the analysis of small needle biopsies, but there are obvious concerns about heterogeneity which will need to be addressed.

REFERENCES

Ackland S.P., Ratain M.J., Vogelzang N.J., et al., 1989, Pharmacokinetics and pharmacodynamics of long-term continuous infusion doxorubicin. Clin. Pharmacol. Ther. 45, 340-347.

Ayash L.J., Wright J.E., Tretyakov O. et al. ,1992, Cyclophosphamide pharmacokinetics : correlation with cardiac toxicity and tumour response. J. Clin. Oncol. 10, 995-1000.

Beal S.L., and Sheiner L.B, 1985.: Methodology of population pharmacokinetics, in : Drug fate and metabolism Garrett E.R., Hirtz J. (eds) New York, NY, Marcel Decker, (Vol. 5)., pp 135-183.

Beal S.L., Boeckmann A.J., and Sheiner L.B, 1992., NONMEM users guides, Part VI : PREDPP guide. Technical Report of the Division of Clinical Pharmacology, University of California, San Francisco.

Bennett C.L., Sinkule J.A., Schilsky R.L. et al., 1987, Phase I clinical and pharmacological study of 72-hour continuous infusion etoposide in patients with advanced cancer. Cancer Res., 47, 1952-1957.

Calvert A.H., Newell D.R., Gumbrell L.A., et al. 1989, Carboplatin dosage : prospective evaluation of a simple formula based on renal function. J. Clin. Oncol. 7:1748-1756.

Calvert A.H., Judson I., Van der Vijgh W.J.F., 1993, Platinum complexes in cancer medicine : pharmacokinetics and pharmacodynamics in relation to toxicity and therapeutic activity. Cancer Surveys 17, 189-217.

Calvert A.H., 1994, Dose optimisation of carboplatin in adults. Anticancer Res. 14, 2273-2278.

Chatelut E., Canal P., Brunner V., et al., 1995, Prediction of carboplatin clearance from standard morphological and biological patient characteristics. J. Natl. Cancer Inst. 87, 573-580.

Cockroft D.W., and Gault M.H. ,1976, Prediction of creatinine clearance from serum creatinine. Nephron 16, 31-41.

Desai Z.R., van der Berg H.W., Bridges J.M., Shanks R.G., 1982, Can severe vincristine neurotoxicity be prevented ? Cancer Chemother. Pharmacol. 8, 211-214.

Desoize B., Dumont P., Manot L., et al., 1994, Comparison of two dose prediction models for cisplatin. Anticancer Res. 14, 2285-2290.

Desoize B., Marechal F., Cattan A., 1990, Clinical pharmacokinetics of etoposide during 120 hours continuous infusions in solid tumours. Br. J. Cancer 62, 840-841

Egorin M.J., Therapeutic drug monitoring and dose optimisation in oncology. In Workman P (ed): New Approaches in Cancer Pharmacology : Drug Design and Development. Berlin : Springler-Verlag, 1992, pp 75-91.

Egorin M.J., Reyno L.M., Canetta R.M., et al., 1994, Modeling toxicity and response in carboplatin-based combination chemotherapy. Semin. Oncol. 21, 7-19.

Egorin M.J., Van Echo D.A, Olman E.A., et al., 1985, Prospective validation of a pharmacologically based dosing scheme for cis-diamminedichloroplatinum (II) analogue diamminecyclobutane dicarboxylatoplatinum. Cancer Res. 45, 6502-6506,

Egorin M.J., Van Echo D.A, Tipping SJ, et al ., 1984, Pharmacokinetics and dosage reduction of cis-diammine (1,1-cyclobutane dicarboxylato) platinum in patients with impaired renal function. Cancer Res 44, 5432-5438.

Etienne M.C., Lagrange J.L; Dassonville O., Fleming R., Thyss A., Renée N., Schneider M., Demard F., Milano G. ,1994, Population study of dihydropyrimidine dehydrogenase in cancer patients. J. Clin. Oncol. 12, 2248-2253.

Evans W.E., Crom W.E., Abromowitch M. et al., 1989, Clinical pharmacodynamics of high-dose methotrexate in acute lymphoblastic leukemia. N. Engl. J. Med. 314, 471-477.

Fichtinger-Schepman AMJ, van der Velde-Visser SD., Van Dijk-Knijnenburg HCM, Van Oosterom AT., Baan RA., Berends F., 1990, Kinetics of formation and removal of cisplatin-DNA adducts in blood cells and tumor tissue of cancer patients receiving chemotherapy : comparison with in vitro adduct formation. Cancer Res. 50, 7887-7894.

Fichtinger-Schepman A.M.J., van Oosterom A.T., Lohman P.H.M., Berends F., 1987, Cis-diamminedichloroplatinum (II)-induced DNA adducts in peripheral leukocytes from seven cancer patients : quantitative immunochemical detection of the adduct formation and removal after a single dose of cis-diamminedichloroplatinum (II). Cancer Res. 47, 3000-3004.

Forastiere A.A., Belliveau J.F., Goren M.P., 1988, Pharmacokinetic and toxicity evaluation of five-day continuous infusion versus intermittent cis-diamminedichloroplatinum (II) in head and neck cancer patients. Cancer Res. 48, 3869-3874.

Gregg R.W., Molepo J.M., Monpetit V.J.A., et al., 1992, Cisplatin neurotoxicity : the relationship between dosage, time, and platinum concentration in neurologic tissues and morphologic evidence of toxicity. J. Clin. Oncol. 10, 795-803.

Grochow L.B., Jones R.J., Brundrett R.B., et al., 1990, Pharmacokinetics of busulfan : correlation with veno-occlusive disease in patients undergoing bone marrow transplantation, Cancer Chemother. Pharmacol. 25, 55-61.

Horwich A., Dearnaley D.F., Nicholls J. et al., 1991, Effectiveness of carboplatin, etoposide and bleomycin combination chemotherapy in good prognosis metastatic testicular non seminomatous germ cell tumours. J. Clin. Oncol. 9, 62-69.

Jodrell DI, Egorin MJ, Canetta RM et al., 1992, Relationships between carboplatin exposure and tumor response and toxicity in patients with ovarian cancer. J. Clin. Oncol. 10, 520-528.

Kantarjian HG., Estey EH., Plunkett W. et al., 1986, Phase I-II clinical and pharmacological studies of high-dose cytosine arabinoside in refractory leukemia. Am. J. Med. 81, 387-394.

Lennard I., Lilleyman JS., 1987, Variable mercaptopurine metabolism and treatment outcome in childhood lymphoblastic leukemia. J. Clin. Oncol. 7, 1816-1823.

Lennard L. van Loon JA., Lilleyman JS, Weinshilboum RM., 1987, Thiopurine pharmacogenetics in leukemia : correlation of erythrocyte thiopurine methyltransferase activity and 6-thioguanine nucleotide concentrations. Clin. Pharmacol. Ther. 41, 18-25.

Ma J., Verweij J., Planting AST., de Boer-Dennert M., van der Burg MEL., Stoter G., Schellens JHM., 1994, Pharmacokinetic-dynamic relationship of weekly high dose cisplatin in solid tumor patients. Proc. Am. Assoc; Clin. Oncol. 13, 133.

Marina N.M., Rodman J., Murry D.J., et al. , 1994, Phase I study of escalating targeted doses of carboplatin combined with ifosfamide and etoposide in treatment of newly diagnosed pediatric solid tumors. J. Natl. Cancer Inst. 86, 544-548.

Marina N.M., Rodman J., Shema S.J., et al. , 1993, Phase I study of escalating targeted doses of carboplatin combined with ifosfamide and etoposide in children with relapsed solid tumors. J. Clin. Oncol. 11, 554-560.

Milano G., Etienne M.C., Renée N. et al., 1994, Relationship between fluorouracil systemic exposure and tumor response and patient survival, J. Clin. Oncol. 12: 1291-1297.

Miller AA., Stewart CF., Tolley EA., 1990, Clinical pharmacodynamics of continuous infusion etoposide; Cancer Chemother. Pharmacol. 25, 361-366.

Moore MJ., Erlichman C., 1987, Therapeutic drug monitoring in oncology, Problems and potential in antineoplastic therapy. Clin. Pharmacokin. 13, 205-227.

Motzer RJ., Reed E., Perera F., Tang D., Shamkhani H., Poirier MC., Tsai WY., Parker R., Bosl GJ., 1994, Platinum-DNA adducts assayed in leukocytes of patients with germ cell tumors measured by atomic absorbance spectrometry and enzyme-linked immunosorbent assay. Cancer 73, 2843-2852.

Newell DR., 1994, Can pharmacokinetic and pharmacodynamic studies improve cancer chemotherapy? Ann. Oncol. 5, S9-S15.

Newell DR., Pearson ADJ., Bolmanno K., et al., 1993, Carboplatin pharmacokinetics in children : the development of a pediatric dosing formula. J. Clin. Oncol. 11, 2314-2323.

Paredes J., Hong W.K., Felder T.B. et al., 1988, Prospective randomized trial of high-dose cisplatin and fluorouracil infusion with or without sodium diethyldithiocarbamate in recurrent and/or metastatic squamous cell carcinoma of the head and neck, J. Clin. Oncol. 6: 955-962.

Parker R.J., Gill I., Tarone R., Vionnet JA., Grunberg S., Muggia FM Reed E., 1991, Platinum-DNA damage in leukocyte DNA of patients receiving carboplatin and cisplatin chemotherapy, measured by atomic absorbance spectrometry. Carcinogenesis, 12, 1253-1258.

Ratain M.J, Schlisky RL, Conley B.A, Egorin M.J., 1990, Pharmacodynamics in cancer therapy. J. Clin. Oncol. 8, 1739-1753.

Ratain M.J., Mick R., Berezin F. et al., 1991a, Paradoxical relationship between acetylor phenotype and amonafide toxicity. Clin. Pharmacol. Ther. 50, 573-579.

Ratain M.J., Mick R., Schilsky RI., et al., 1991 b., Pharmacologically based dosing of etoposide : a means of safely increasing dose intensity. J. Clin. Oncol. 9, 1490-1486.

Reece PA., Stafford I., Russel J., Khan M., Gill P.G., 1987, Creatinine clearance as a predictor of ultrafilterable platinum disposition in cancer patients treated with cisplatin : relationship between peak untrafilterable platinum levels and nephrotoxicity. J. Clin. Oncol. 5, 304-309.

Reed E., Ostchega Y., Steinberg S.M., Yuspa .H., Young R.C., Ozols RF., Poirier M.C., 1990, Evaluation of platinum-DNA adduct levels relative to known prognostic variables in a cohort of ovarian cancer patients. Cancer Res. 50, 2256-2260.

Reed E., Ozols R.F., Tarone R., Yuspa S.H., Poirier M.C., 1987, Platinum-DNA adducts in leukocyte DNA correlate with disease response in ovarian cancer patients receiving platinum-based chemotherapy. Proc. Natl. Acad. Sci. USA, 84, 5024-5028.

Reed E., Ozols R.F., Tarone R., Yuspa S.H., Poirier M.C., 1988, The measurement of cisplatin-DNA adduct in testicular cancer patients. Carcinogenesis 9, 1909-1911.

Reed E., Parker R.J., Gill I., Bicher A., Dabholkar M., Vionnet J.A., Bostick-Burton F., Tarone R., Muggia F.M., 1993, Platinum-DNA adduct in leukocyte DNA of a cohort of 49 patients with 24 different types of malignancies. Cancer Res. 53, 3694-3699.

Rodman J.H., Abromowitch M. Sinkule J.A., Hayes F.A., Rivera G.K., Evans W.E., 1987, Clinical pharmacodynamics of continuous infusion teniposide : systemic exposure as a determinant of response in a Phase I trial. J. Clin. Oncol. 5, 1007- 1014.

Rodman J.H., Maneuval D.C., Magill J.H. et al., 1993, Measurement of Tc-99m DTPA serum clearance for the estimation of glomerular filtration rate in children with cancer. Pharmacotherapy 13, 10-16.

Rowinsky E.K., Gilbert M., McGuire W.P., et al., 1991, Sequences of taxol and cisplatin : a phase I and pharmacologic study. J. Clin. Oncol. 9, 1692-1703.

Santini J., Milano G., Thyss A. et al., 1989, FU therapeutic monitoring with dose adjustment leads to an improved therapeutic index for head and neck cancer. Br. J. Cancer. 59, 287-290.

Sorensen B.T., Strömgren A., Jakobsen P., et al., 1991, Dose-toxicity relationship of carboplatin in combination with cyclophosphamide in ovarian cancer patients. Cancer Chemother. Pharmacol. 28, 337-341.

Stewart C.F., Arbuck S.G., Fleming R.A., Evans W.E., 1991, Relation of systemic exposure to unbound etoposide and hematologic toxicity. Clin. Pharmacol. Ther. 50, 385-393.

Thomas D.J., Clifford S.C., Aherne W., et al., 1994, Pharmacokinetic determinants of antitumour activity in bladder cancer patients. Br. J. Cancer 69, supll XXI, 30.

Van der Vijgh W.J.F., 1991, Clinical pharmacokinetics of carboplatin. Clin. Pharmacokinet. 21, 242-261.

Van Warmerdam L.J.C., Verweij J., Schellens J.H.M., et al., 1995, Pharmacokinetics and pharmacodynamics of topotecan administered daily for 5 days every 3 weeks. Cancer Chemother. Pharmacol. 35, 237-245.

Vermorken J.B., Van der Vijgh W.J.F.., Klein I., et al., 1984, Pharmacokinetics of free and total platinum species after short-term infusion of cisplatin. Cancer Treat. Rep. 68, 505-513.

Whitehead V.M., Rosenblatt D.S., Vichich M.J., et al., 1990, Accumulation of methotrexate and methotrexate polyglutamates in lymphoblasts at diagnosis of childhood acute lymphoblastic leukemia : a pilot prognostic factor analysis. Blood, 76, 44-49.

CLINICAL PHARMACOLOGY OF CARBOPLATIN ADMINISTERED IN ALTERNATING SEQUENCE WITH PACLITAXEL IN PATIENTS WITH NON-SMALL CELL LUNG CANCER: A EUROPEAN CANCER CENTRE (ECC) STUDY

LJC van Warmerdam, MT Huizing, G Giaccone, PJM Bakker, JB Vermorken, PE Postmus, N van Zandwijk, MGJ Koolen, WW ten Bokkel Huinink, RAA Maes, WJF van der Vijgh, CHN Veenhof, and JH Beijnen

Correspondence should be addressed to:
JH Beijnen, Dept of Pharmacy, Netherlands Cancer Institute/Antoni van Leeuwenhoek Hospital and Slotervaart Hospital, Louwesweg 6, 1066 EC, Amsterdam, The Netherlands

ABSTRACT

The clinical pharmacology of carboplatin, co-administered with paclitaxel, was investigated in a dose-finding and sequence-finding study in 56 previously untreated patients with non-small cell lung cancer (NSCLC). Carboplatin was administered over 30 minutes and paclitaxel over 3 hours every 4 weeks. Patients were randomized for the administration sequence, being first carboplatin (C) then followed by paclitaxel (P) or vice versa. Each patient received the alternate sequence during the second and subsequent courses. Total platinum concentrations in plasma and plasma ultrafiltrate (pUF) were measured applying flameless atomic absorption spectrometry. Ninety-five concentration-time curves were obtained in plasma and pUF.

Paclitaxel doses were initially escalated from P:100 mg/m^2+ C:300 mg/m^2 with 25 mg/m^2 increments to P:225 mg/m^2+ C:300 mg/m^2. During the course of the study after six dose levels, it became clear that no sequence-depending effect on the toxicity was present, and it was decided to perform the last escalation steps with four patients per dose level. These dose levels were: level 7 (P:225 mg/m^2+ C:350 mg/m^2), 8 (P:250 mg/m^2+ C:350 mg/m^2), and 9 (P:225 mg/m^2+ C:400 mg/m^2). The mean pUF area under the concentration-time curve (AUC) per 300 mg/m² carboplatin was for the sequence C→P 3.52 (range 1.94- 5.83) mg/mL·min, and for the sequence P→C 3.62 (range 1.91- 5.01) mg/mL·min, which is not significantly different ($p = 0.55$). Also for other pharmacokinetic parameters no sequence depending effect was observed. Non-hematological toxicities consisted mainly of myalgia and bone pain. The most important hematologic toxicity was neutropenia, without, interestingly, any thrombocytopenic periods. The AUCs of carboplatin were probably too low to induce thrombocytopenia, although antagonistic properties of paclitaxel cannot be excluded.

INTRODUCTION

Carboplatin (*cis*-diammine 1,1-cyclobutane dicarboxylate platinum (II), CBDCA, JM8, NCS-241240, Paraplatin®) is a second-generation platinum-containing chemotherapeutic compound with established activity against a variety of solid tumors [1,2,3,4]. Its dose-limiting toxicity is myelosuppression, predominantly thrombocytopenia. Carboplatin is much less nephrotoxic, neurotoxic and emetogenic than its parent compound cisplatin [1,2]. Although cisplatin is among the most active agents available for the therapy of non-small cell lung cancer (NSCLC), carboplatin as initial therapy resulted in the longest 1 year survival in a large randomized study in metastatic NSCLC patients [5]. In an attempt to improve the palliation and survival of patients with NSCLC we combined carboplatin with another active agent in NSCLC, paclitaxel. Paclitaxel (NCS-125973, Taxol®) is a new plant-derived anticancer agent which stabilizes and promotes the assembly of microtubules [6,7]. Paclitaxel has shown activity in a variety of solid tumors [6,8], including NSCLC, where the drug has induced response rates between 21% and 24% [9,10,11]. Its dose-limiting toxicities are myelosuppression, particularly neutropenia, and neurotoxicity. The relatively non-overlapping toxicities of carboplatin and paclitaxel, their different mechanisms of action, and their activity in NSCLC as single agent, made the combination of both drugs attractive for further clinical exploration.

Generally, when two drugs are combined, it is important to make allowance for possible drug-drug interactions and schedule-dependency. For example, the toxicity of the combination of paclitaxel and cisplatin was schedule dependent, namely, cisplatin given before paclitaxel was more cytotoxic than the alternate sequence in both *in vitro* [12,13,14] and in clinical studies [15]. Pharmacokinetic monitoring revealed that this was due to the decreased clearance of paclitaxel, leading to an increased exposure to the drug [15]. Besides the interaction between cisplatin and paclitaxel, drug-drug and sequence interactions have also been reported for cisplatin and other drugs as lithium [16], phenytoin [17,18], bleomycin [19], docetaxel [20], etoposide [21], topotecan [22], vincristine [14]. As carboplatin is also a platinum containing compound with the same mechanism of action as cisplatin [23], such drug-drug interactions cannot be excluded *a priori* in the combination carboplatin-paclitaxel. Sequence-dependent interactions have been reported for the combination carboplatin and 5-fluorouracil, where the sequence 5-fluorouracil followed by carboplatin showed higher antitumor activity *in vitro* and in nude mice than the reverse sequence [24]. For the combination of carboplatin with phenytoin, it has been reported that phenytoin levels decreased after carboplatin treatment [25].

Based upon this knowledge, we initiated a pharmacokinetic/ pharmacodynamic study with carboplatin and paclitaxel in patients with NSCLC, with the following aims:

- to determine the pharmacokinetic behavior of carboplatin administered in combination with paclitaxel;
- to determine whether the pharmacokinetics of carboplatin are dependent on the administration sequence of carboplatin and paclitaxel.

The design of the study also allowed us to investigate the use of a modified Calvert-formula [26], which originally was developed in a single agent regimen with the glomerular filtration rate (GFR) determined by the ^{51}CrEDTA test. To test the modified Calvert-formula using the creatinine clearance as measure for the GFR, we retrospectively calculated the AUC of carboplatin by Dose/ (creatinine clearance + 25) and compared it with measured values. It has been reported that, apart from the response rate [3], the AUC of carboplatin is closely related with the degree of thrombocytopenia [1,3,27,28,29]. Therefore, in the present study, the documented toxicity was correlated with the AUC. Other clinical data, including response rates, will be published elsewhere.

Consequently, additional purposes were:

- to determine the accuracy and imprecision of AUC predictions when using a modified Calvert-formula;
- to evaluate the relationship between the pharmacokinetics of carboplatin (i.e. the AUC) and its pharmacodynamics (thrombocytopenia).

This study has been executed as part of a multi-centre phase I/II clinical trial coordinated by the European Cancer Centre.

PATIENTS AND METHODS

Patient Selection

All patients had locally advanced or metastatic NSCLC (stage IIIB and IV), and did not receive prior chemotherapy-treatment. Eligibility criteria included a WHO-performance status (PS) $\leq$ 2, adequate bone marrow function (absolute neutrophil count (ANC) $\geq 2.5 \times 10^9$/L and platelets $\geq 100 \times 10^9$/L), serum bilirubin $\leq 30\ \mu$M, ALAT and ASAT $\leq$ 2.5 x the normal upper limit, serum creatinine $\leq 140\ \mu$M, and age < 70 years. All patients gave written informed consent. Participating institutes were the Netherlands Cancer Institute/Antoni van Leeuwenhoek Hospital, the Academic Hospital Vrije Universiteit and the Academic Medical Centre, all in Amsterdam.

Treatment Plan

Carboplatin (C) (Paraplatin®) was supplied as lyophilized product containing 150 mg carboplatin and 150 mg mannitol as bulking agent. Immediately before use the content of each vial was reconstituted with 15 mL of Water for Injection, and the total dose was added to 250 mL dextrose 5%. Paclitaxel (P) (Taxol®) was supplied as concentrated sterile solution with 6 mg/mL in a 5 mL vial in polyoxyethylated castor oil (Cremophor EL®) and dehydrated alcohol (1:1 vol/vol). These were diluted before use with 0.9% sodium chloride to achieve a total volume of about 1000 mL. Carboplatin was administered as a 30-minute infusion, and paclitaxel as a 3-hour infusion intravenously (IV). Standard premedication consisted of dexamethason (20 mg per os, 12 and 6 hours before the paclitaxel infusion), clemastine (2 mg iv), and cimetidine (300 mg iv), and antiemetics (which was institute-dependent). Increasing doses of paclitaxel were given to 6 cohorts of patients every four weeks in combination with a fixed dose of carboplatin (300 mg/m²). Intra-patient dose escalation was not performed. The initial dose of paclitaxel was 100 mg/m², and 25 mg/m² dose escalation steps were performed up to C:300 mg/m²+ P:225 mg/m². Six patients at each dose level were randomized for the administration sequence, being paclitaxel followed by carboplatin (P→C) or carboplatin followed by paclitaxel (C→P) (Table 1). The alternate sequence was given the second and each following course. Interim analysis of the data showed that no sequence-depending effect on the toxicity was present, and it was, therefore, decided to perform the last escalation steps with four patients per dose level, employ-

Table 1 *Design of the study and the dose levels*

	Cycle 1	Cycle 2	Cycle 3
Randomization:	Paclitaxel → carboplatin	Carboplatin → paclitaxel	as cycle 2
	Carboplatin → paclitaxel	Paclitaxel → carboplatin	as cycle 2
Dose level:	Paclitaxel (mg/m²)	Carboplatin (mg/m²)	
1 (n= 6)	100	300	
2 (n= 7)	125	300	
3 (n= 6)	150	300	
4 (n= 7)	175	300	
5 (n= 9)	200	300	
6 (n= 7)	225	300	
7 (n= 5)	225	350	
8 (n= 4)	250	350	
9 (n= 5)	225	400	

Note: n is the number of patients participating in the pharmacokinetic study

ing the administration sequence P→C the first course and C→P the second course, if pharmacokinetics were studied. Furthermore, the hematological toxicity was milder than expected and, therefore, it was decided to escalate the carboplatin dose as well to level 7 (P:225 mg/m^2+ C:350 mg/m^2), 8 (C:350 mg/m^2+ P:250 mg/m^2), and 9 (C:400 mg/m^2+ P:225 mg/m^2) (Table 1).

Pharmacokinetic Studies

Complete concentration-time curves were obtained from 56 patients, who were sampled during their first and second courses. Samples were collected at 12 time points: immediately before the infusion, at the end of infusion, and at 0.25, 0.5, 1, 2, 4, 8, 12, 24 and 48 hours after the end of the infusion. Plasma was obtained by immediate centrifugation (5 minutes; 1500·g) of the samples. Part of the plasma was transferred directly to an MPS-1 device equipped with a YMT-30 filter (Amicon Division, W.R. Grace & Co., Danvers, MA, USA) and centrifugated for 10 minutes at 1500·g. Plasma and plasma ultrafiltrate (pUF) were stored at -20°C until analysis. Platinum levels were quantitated using a validated method based on Zeeman atomic absorption spectrometry, and were re-calculated as carboplatin concentrations. The plasma concentration versus time curves of carboplatin were analyzed using the pharmacokinetic software package MW/Pharm (MEDI\WARE BV, Groningen, The Netherlands) [30]. This non-linear least-squares, iterative regression program determines slopes and intercepts of the logarithmically plotted curves of multi-exponential functions. Initial estimates of the parameters were determined by an automated curve stripping procedure. The Akaike information criterium was used to select the most optimal model. The carboplatin AUCs were determined on the basis of the fitted curve as the exact integral of the concentration versus time plots from 0 to 48 hours for both plasma and pUF. Other pharmacokinetic parameters, such as the half-lives ($t½\alpha$, $t½\beta$, and $t½\gamma$), the total body clearance (CL), the mean residence time (MRT), and the volume of distribution (Vd), were also calculated by this computer program applying standard equations [30]. The maximum concentrations (C_{max}) were the observed experimental values.
The creatinine clearance was calculated from the Cockcroft and Gault equation [31];

$$\text{creatinine clearance (mL/min)} = \frac{\{[140\text{-age(years)}] \times \text{weight (kg)}\}\ [\times 0.85 \text{ if female}]}{[0.813 \times \text{serum creatinine concentration}(\mu M)]}$$

Carboplatin AUC-values (mg/mL·min) were also calculated by rearranging the Calvert-formula into:

$$\text{AUC} = \text{Dose/ (creatinine clearance + 25)}$$

The performance of this prediction of the AUCs was evaluated using the correlation coefficient (r), the relative mean prediction error (MPE%) (a measure of bias) and the relative root mean square prediction error (RMSE%) (a measure of precision) [32]. These are defined as [33]:

$$MPE\% = [N^{-1} \cdot \sum_{i=1}^{N} (pe_i)] \cdot 100\%$$

$$RMSE\% = [N^{-1} \cdot \sum_{i=1}^{N} (pe_i)^2]^{1/2} \cdot 100\%$$

where N is the number of AUC-pairs (i.e. observed with predicted values), and pe is the relative prediction error $[\ln(AUC_{predicted\ value})-\ln(AUC_{true\ value})]$. The smaller the RMSE%, the better the prediction.

Pharmacodynamics

The pharmacodynamics were explored using a plot of percentage decrease (%decr) in ANC and platelets versus the pUF AUC. The percentage decrease (%decr) is defined as:

$$\%decr = \frac{\text{Pretreatment value - value of the nadir}}{\text{Pretreatment value}} \cdot 100\%$$

Full blood counts with differentials were obtained twice weekly, and were used to calculate the %decr during the first two courses. The Student's t-test and linear regression analysis were performed to establish possible relationships between the toxicities (%decr in ANC, %decr in platelets, and the WHO-grade of the non-hematological toxicities) and patient specific data, such as gender, age, PS, weight, height, creatinine clearance, and administration sequence, but also the taxol dose (mg and mg/m^2), and the pharmacokinetic parameters. The computer programs NCSS® (Number Cruncher Statistical System, Kaysville, Utah, USA, 1992) and Quattro Pro® (Borland International, Scotts Valley, CA, USA, 1992) were used for all calculations.

RESULTS

During the course of the study it became apparent that no sequence-depending effect on the toxicity was present, and it was, therefore, decided to perform the last escalation steps with four patients per dose level, employing the administration sequence P→C during the first course, and C→P the following courses. The characteristics of the patients participating in the pharmacokinetic studies were: median PS of 1 (0-2), mean age of 56 (38-74) years, mean creatinine clearance of 85 (range 45 to 199) mL/min, 15 patients were female and 41 patients were male.

Toxicity

In all 56 patients the main hematological toxicity during the first two courses was neutropenia with 27% of the evaluable courses resulting in grade III neutropenia, and 15% resulting in grade IV neutropenia. Mild anemia (grade II) occurred in only 10% of the patients. The mean nadir in platelets was 251 x 10^9/L. Thrombocytopenia (<100 x 10^9/L) occurred in only 1 patient, who died of toxicity one week after the first cycle (P: 250 mg/m²+ C: 350 mg/m²), following an episode of severe neutropenia, thrombocytopenia, associated hemorrhagic diathesis and sepsis. This patient had gross liver involvement and liver function disturbances (ALAT and ASAT were 2.4x upper normal limit, alkaline phosphatase was 588 U/L) before treatment. Nausea and vomiting were mild and occurred infrequently with standard prophylactic antiemetic medication. Myalgia and bone pain appeared more frequently and their severity increased with the paclitaxel dose (especially >200 mg/m²), as did peripheral neurotoxicity. All patients developed alopecia (grade II or III).

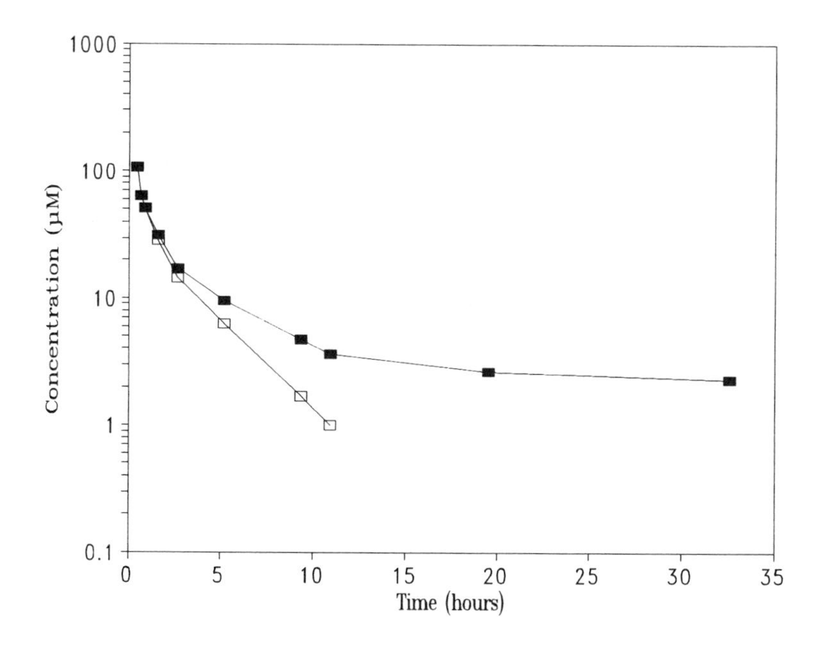

Figure 1. Typical plasma (■) and plasma ultrafiltrate (□) concentration-time curves for carboplatin (300 mg/m²+ paclitaxel 100 mg/m²).

Pharmacokinetics

A total of 56 patients participated in our pharmacokinetic studies during the first course (n= 53) and second course (n= 42). Figure 1 depicts typical concentration-time curves for carboplatin (300 mg/m^2+ P:100 mg/m^2); the shapes of the curves at the alternate sequence and at other dosages were similar. The plasma pharmacokinetics of carboplatin could be described best with a standard open three-compartment model. The pharmacokinetic parameters are tabulated in Table 2 (sequence C→P) and Table 3 (sequence P→C). Overall, for the sequence C→P the mean plasma AUC normalized for 300 mg/m^2 was 6.02 (range 2.79 to 8.92) mg/mL·min, the mean $t\frac{1}{2}\alpha$ was 21 (range 6.0 to 102) minutes, the mean $t\frac{1}{2}\beta$ was 1.8 (range 0.8 to 5.5) hours and the mean $t\frac{1}{2}\gamma$ was 6.5 (range 0.3 to 37) days. The mean Vd was 168 (range 55 to 533) L/m^2, and the mean MRT was 7.8 (range 0.2 to 40) days. For the alternate sequence P→C, the mean values and ranges were plasma AUC per 300 mg/m^2: 6.35 (range 3.08 to 10.7) mg/mL·min, $t\frac{1}{2}\alpha$: 20 (range 5.9 to 102) minutes, $t\frac{1}{2}\beta$: 2.1 (range 0.7 to 19.9) hours, $t\frac{1}{2}\gamma$: 7.1 (range 0.2 to 68) days, Vd: 169 (36 to 390) L/m^2, and the MRT: 8.8 (range 0.2 to 95) days. The maximum concentrations (C_{max}) were achieved at the end of infusion, and were the same as the C_{max} concentrations in pUF (Tables 2 and 3). The pUF pharmacokinetics of carboplatin could be best described with a standard open two-compartment model, for the sequence C→P resulting in a mean pUF AUC per 300 mg/m^2 of 3.52 (range 1.94 to 5.83) mg/mL·min, a C_{max} per 300 mg/m^2 of 87.9 (range 55.7 to 146.5) μM, a mean $t\frac{1}{2}\alpha$ of 35 (range 6.0 to 124) minutes, and a $t\frac{1}{2}\beta$ of 5.0 (range 0.7 to 79) hours. The mean Vd was 23.9 (range 2.6 to 87.9) L/m^2, and the mean MRT was 4.4 (range 1.0 to 70) hours. For the alternate sequence P→C, the mean values and ranges were for the pUF AUC per 300 mg/m^2: 3.62 (range 1.91 to 5.01) mg/mL·min, C_{max} per 300 mg/m^2: 87.2 (range 54.2 to 136.6) μM, $t\frac{1}{2}\alpha$: 30 (range 3.2 to 96) minutes, $t\frac{1}{2}\beta$: 2.7 (range 1.0 to 9.6) hours, Vd: 18.9 (9.3 to 58.9) L/m^2, and the MRT: 2.6 (range 1.3 to 4.3) hours. Thus, the mean values and ranges of the pharmacokinetic parameters were similar for both sequences (Student's t-test; $p > 0.5$), and the 95% confidence intervals were overlapping (F-test; $p < 0.0001$). This appears also from Figure 2, which displays the relation between the AUCs of the sequence P→C versus the AUCs of the sequence C→P ($r = 0.72$; $AUC_{P \to C} = 0.98 \times AUC_{C \to P}$). The 24-hour urinary excretion of carboplatin (measured as total platinum) was measured in 17 patients. The mean percentage excreted was 75% (range 42% to 115%).

To test a modified Calvert-formula using the creatinine clearance as a measure for the GFR, we retrospectively calculated the pUF AUC by Dose/ (creatinine clearance + 25). It appeared that the AUCs could not be reliably estimated using this formula since the mean estimated AUC was 4.82 (range 2.46 to 10.9) mg/mL·min (Figure 3), which was 33% higher (MPE%) than the measured AUC (3.71 mg/mL·-

Table 2 *Pharmacokinetics of carboplatin administered before paclitaxel*

Dose level	C/P dose	Plasma mean values AUC	$t\frac{1}{2}\alpha$	$t\frac{1}{2}\beta$	$t\frac{1}{2}\gamma$	CL	Vd	Plasma ultrafiltrate mean values AUC	C_{max}	$t\frac{1}{2}\alpha$	$t\frac{1}{2}\beta$	CL	Vd	MRT
1	**300/100**	**5.75**	**13.4**	**1.4**	**1.8**	**4.40**	**107.4**	**3.78**	**94.0**	**42.0**	**4.0**	**8.56**	**27.3**	**2.7**
	min:	2.79	6.6	1.0	0.3	2.19	54.9	2.77	71.7	12.0	1.7	6.44	11.6	1.8
	max:	6.88	32.4	2.2	3.5	9.64	155.3	4.92	122.7	60.6	9.7	11.6	61.9	4.9
2	**300/125**	**6.42**	**34.5**	**2.3**	**9.5**	**2.67**	**195.5**	**4.00**	**83.6**	**46.0**	**15.1**	**6.65**	**28.4**	**14.5**
	min:	4.95	6.6	1.1	0.5	0.31	57.1	3.22	64.3	7.0	2.1	0.92	14.8	2.2
	max:	8.92	102	5.5	29.8	5.92	300.0	5.83	100.7	81.5	78.6	10.5	72.1	69.9
3	**300/150**	**5.54**	**14.4**	**1.5**	**6.3**	**2.52**	**220.8**	**3.12**	**94.2**	**16.7**	**1.8**	**10.76**	**15.9**	**2.0**
	min:	3.96	7.8	0.9	1.3	0.60	103.3	1.94	75.5	7.9	1.4	8.40	12.0	1.5
	max:	6.82	23.8	2.2	12.8	4.14	290.3	5.83	111.4	27.2	2.1	17.6	22.6	2.3
4	**300/175**	**6.41**	**23.3**	**2.1**	**6.3**	**2.92**	**159.1**	**3.49**	**90.2**	**33.5**	**3.8**	**10.12**	**27.5**	**2.8**
	min:	5.16	7.0	1.1	0.9	0.48	75.4	1.94	69.0	15.9	1.8	7.60	10.9	1.3
	max:	8.40	65.5	4.8	22.0	5.00	236.3	5.83	121.5	68.4	12.8	13.3	87.9	6.1
5	**300/200**	**5.18**	**19.8**	**1.8**	**6.2**	**3.57**	**187.0**	**3.36**	**97.0**	**26.9**	**2.7**	**9.48**	**20.7**	**2.6**
	min:	4.43	7.8	1.2	1.1	0.66	114.3	1.94	70.7	7.4	1.2	7.60	9.5	1.4
	max:	6.20	38.1	2.2	24.0	5.21	332.7	4.71	146.5	49.1	5.2	11.3	41.6	4.3
6	**300/225**	**6.67**	**31.9**	**2.2**	**13.5**	**2.45**	**161.1**	**3.80**	**85.2**	**45.3**	**4.1**	**9.08**	**26.8**	**3.2**
	min:	5.35	8.4	1.2	1.2	0.03	83.9	1.94	59.4	6.0	1.4	8.32	9.9	1.8
	max:	8.20	91.7	5.1	37.5	4.99	234.6	4.71	101.6	124	9.4	10.3	57.5	4.9
7	**350/225**	**6.66**	**10.5**	**1.4**	**4.0**	**2.68**	**184.5**	**3.55**	**91.6**	**26.3**	**2.7**	**11.25**	**19.9**	**2.3**
	min:	5.26	6.0	0.8	0.8	0.21	87.2	2.53	69.5	6.0	0.7	8.19	2.6	1.0
	max:	7.87	15.8	1.9	7.2	5.46	285.0	4.71	124.4	65.9	7.6	15.1	55.7	4.4
8	**350/250**	**7.22**	**18.0**	**1.5**	**1.0**	**4.43**	**73.7**	**3.99**	**96.0**	**41.4**	**3.00**	**7.91**	**16.0**	**0.15**
9	**400/225**	**9.60**	**26.2**	**1.8**	**3.6**	**2.62**	**183.6**	**5.19**	**102.9**	**52.8**	**4.3**	**8.49**	**43.1**	**3.4**
	min:	8.59	12.7	1.7	0.9	1.58	71.0	2.53	74.3	34.2	2.3	6.97	14.7	2.5
	max:	10.8	44.1	1.8	7.7	3.36	332.7	5.85	128.4	88.0	8.1	9.80	87.9	4.9

Abbreviations: P: paclitaxel; C: carboplatin; dose (mg/m^2); AUC: area under the concentration-time curve (mg/mL·min); $t\frac{1}{2}(\alpha, \beta, \gamma)$: first, second and third phase half-lives (in minutes, hours, days, respectively); CL: total body clearance ($L/h/m^2$); Vd: volume of distribution (L/m^2); C_{max}: maximum concentration; MRT: mean residence time (hours); min: minimum; max: maximum

Table 3 *Pharmacokinetics of carboplatin following paclitaxel administration*

Dose level	P/C dose	Plasma mean values AUC	t½α	t½β	t½γ	CL	Vd	Plasma ultrafiltrate mean values AUC	C_{max}	t½α	t½β	CL	Vd	MRT
1	**100/300**	**6.90**	**13.9**	**1.9**	**20.7**	**2.00**	**191.9**	**3.72**	**87.2**	**45.7**	**3.0**	**8.89**	**21.0**	**2.5**
	min:	5.05	5.9	1.2	1.0	0.15	84.7	2.51	57.4	20.2	1.8	6.44	12.6	1.8
	max:	10.1	34.2	2.7	68.1	4.00	254.1	4.52	116.4	96.2	5.3	12.19	33.4	3.4
2	**125/300**	**6.71**	**23.5**	**1.6**	**9.9**	**2.93**	**161.7**	**3.66**	**84.4**	**28.8**	**2.6**	**9.29**	**18.4**	**2.8**
	min:	4.19	6.0	1.0	1.2	0.13	96.3	2.51	55.4	15.1	1.7	5.85	12.1	2.1
	max:	10.7	60.6	2.9	37.8	5.75	269.5	4.93	136.6	75.6	4.3	12.90	24.3	4.1
3	**150/300**	**5.49**	**9.1**	**1.3**	**2.9**	**4.66**	**159.2**	**3.07**	**84.2**	**15.4**	**1.9**	**11.33**	**17.4**	**2.2**
	min:	3.08	6.0	0.7	0.4	1.20	49.7	1.91	57.6	7.2	1.5	7.75	10.5	1.4
	max:	8.41	16.6	1.8	6.2	10.6	325.2	4.93	103.2	22.7	2.9	17.92	25.2	3.9
4	**175/300**	**7.04**	**15.9**	**4.0**	**6.0**	**2.73**	**162.4**	**3.67**	**83.5**	**44.8**	**3.8**	**9.74**	**25.6**	**3.0**
	min:	4.90	6.0	1.4	0.7	0.48	36.2	1.91	65.0	7.3	1.6	7.70	11.6	1.8
	max:	10.5	42.6	17	24.9	3.90	319.5	4.93	102.6	93.0	9.6	13.90	58.9	4.3
5	**200/300**	**5.45**	**16.7**	**1.5**	**3.6**	**3.40**	**181.5**	**3.32**	**88.8**	**21.8**	**2.1**	**9.86**	**15.7**	**2.3**
	min:	3.38	6.0	1.0	1.0	1.97	72.0	1.91	77.6	3.2	1.0	7.33	9.3	1.3
	max:	7.91	32.0	2.4	11.3	5.65	389.9	4.28	102.0	44.3	3.8	11.40	22.9	4.1
6	**225/300**	**6.68**	**32.0**	**1.6**	**3.6**	**3.32**	**149.8**	**4.13**	**96.4**	**18.0**	**2.3**	**8.01**	**14.5**	**2.7**
	min:	5.26	7.0	1.2	1.1	1.70	75.3	2.74	71.8	7.0	2.1	6.35	11.3	2.3
	max:	8.69	71.4	2.1	8.3	5.51	264.6	5.01	117.6	34.8	2.7	10.32	17.0	3.6
7	**225/350**	**6.19**	**15.9**	**1.6**	**10.7**	**3.49**	**191.6**	**3.82**	**108.4**	**24.8**	**2.0**	**9.92**	**15.1**	**2.1**
	min:	4.69	8.4	1.0	0.5	0.56	65.7	2.74	93.3	14.4	1.5	9.00	12.2	1.8
	max:	8.05	25.8	2.4	27.0	6.65	305.5	5.01	116.9	44.4	2.8	10.65	18.6	2.7
8	**250/350**	**8.02**	**12.9**	**1.4**	**8.1**	**1.35**	**206.0**	**4.60**	**95.8**	**22.5**	**2.5**	**8.6**	**17.0**	**2.9**
	min:	6.82	9.6	1.0	6.1	1.22	151.2	3.04	63.2	7.9	2.2	7.3	14.3	2.8
	max:	9.21	16.2	1.8	10.1	1.49	260.9	5.01	127.2	38.4	3.2	9.6	20.2	3.0
9	**225/400**	**10.0**	**14.8**	**8.1**	**2.4**	**2.76**	**125.0**	**6.00**	**119.2**	**50.5**	**4.2**	**7.0**	**24.0**	**3.7**
	min:	8.22	6.6	1.7	0.0	1.85	49.6	3.04	106.4	30.0	3.4	6.9	19.0	3.3
	max:	12.2	23.8	20	5.3	3.29	217.0	6.45	143.8	62.7	5.4	7.0	28.4	4.3

Abbreviations: C: carboplatin; P: paclitaxel; dose (mg/m^2); AUC: area under the concentration-time curve (mg/mL·min); t½(α, β, γ): first, second and third phase half-lives (minutes, hours, days, respectively); CL: total body clearance ($L/h/m^2$); Vd: volume of distribution (L/m^2); C_{max}: maximum concentration; MRT: mean residence time (hours); min: minimum; max: maximum

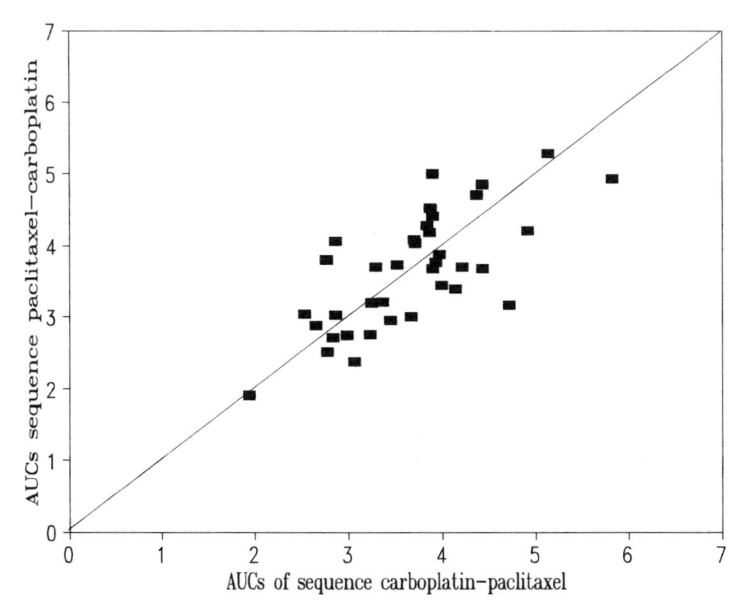

Figure 2. Relation between the AUCs of the sequence paclitaxel-carboplatin versus the AUCs of the sequence carboplatin-paclitaxel (r= 0.72). The solid line represents the line Y= X.

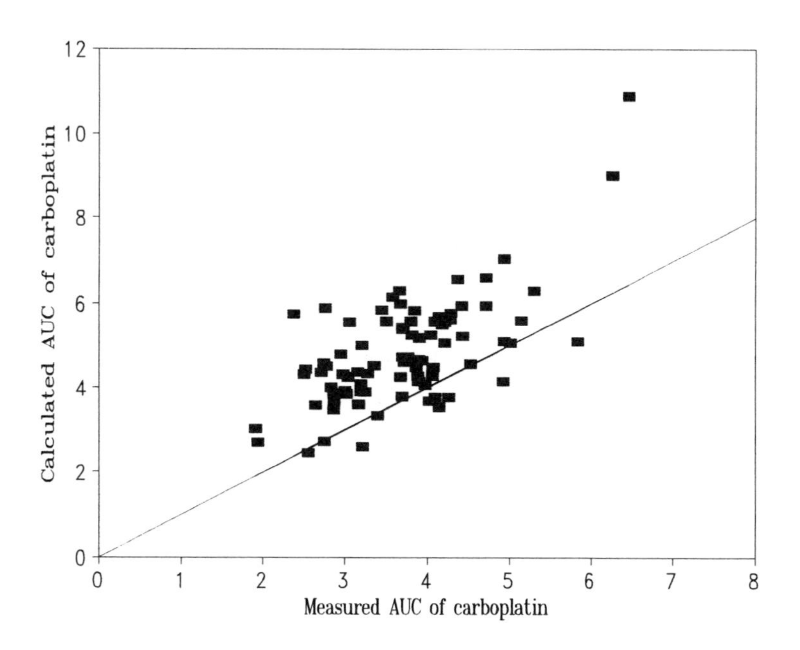

Figure 3. Relation between the calculated AUC and the measured AUC of carboplatin (mg/mL·min). The solid line represents the line Y= X.

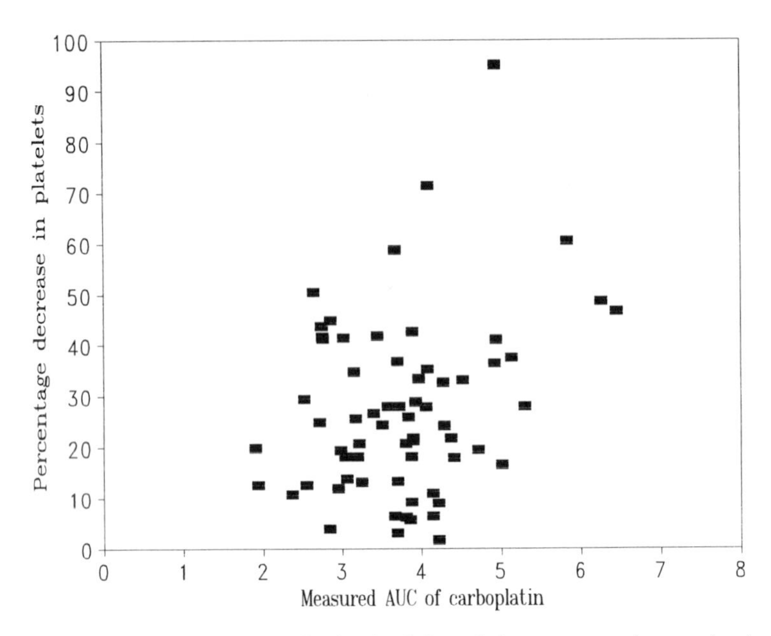

Figure 4. Relation between the AUC of carboplatin and the percentage decrease in platelets.

min, range 1.9 to 6.5 mg/mL·min) ($p < 0.00001$). Furthermore, the prediction of the AUC was associated with a RMSE% of 33.0% and a correlation coefficient of 0.66.

Pharmacokinetic-Pharmacodynamic Relationships

There was a modest, but significant correlation between the pUF AUC of carboplatin and the creatinine clearance ($r = 0.44$, $p < 0.001$) (Table 4). Both parameters were also correlated to the %decr in platelets, although these correlations were weak. A plot of the pUF AUC versus the %decr in platelets yielded a correlation of only $r = 0.29$ ($p = 0.012$) (Figure 4). Another modestly predictive parameter for the %decr in platelets was the number of the course (first or second) ($p = 0.055$) (Table 4). Other parameters, including the pharmacokinetic parameters, were not related to the %decr in platelet count. The same holds true for the %decr in ANC, and the grade of the non-hematological toxicities (alopecia, nausea, neurotoxicity, myalgia, arthralgia), where none of the investigated parameters, except for the paclitaxel dose, were related to these toxicities (all p-values > 0.2).

DISCUSSION

The observation that the myelotoxicity for the combination cisplatin-paclitaxel was sequence-dependent [12,13,14,15] raised the question whether this would also be true for the combination of carboplatin-paclitaxel. The present study shows that there is neither a sequence-dependent toxicity nor a sequence-dependent pharmacokinetic interaction.

Table 4

Relationships between patient characteristics and the pUF AUC and %decr in platelets

Patient Characteristic	pUF AUC		%decr in platelets	
Linear regression	r	p-value	r	p-value
Age	0.15	0.14	0.05	0.68
Weight	0.01	0.88	0.12	0.10
Height	0.31	0.80	0.12	0.11
Creatinine clearance	0.44	<0.001	0.22	0.08
Paclitaxel dose (mg)	0.10	0.31	0.01	0.93
Paclitaxel dose (mg/m²)	0.02	0.80	0.08	0.48
Student's t-tests				
Number of courses received		0.48		0.055
Gender		0.49		0.38
Performance Status		0.51		0.20
Administration Sequence		0.55		0.52

Abbreviations: pUF AUC: area under the plasma ultrafiltrate concentration-time curve of carboplatin; %decr: percentage decrease; r: correlation coefficient

The postulated mechanism for the sequence-dependent interactions between cisplatin and paclitaxel is that cisplatin inhibits cytochrome P-450 dependent paclitaxel-metabolizing enzymes [7]. This mechanism has also been proposed to account for the interaction between cisplatin and etoposide [21]. The lack of sequence-dependent interactions in the present study might be explained by the fact that only cisplatin, and not carboplatin, modulates cytochrome P-450 enzymes [25,34].

Interestingly, thrombocytopenia did not occur, while generally at dosages of 300-400 mg/m² grade I or more of this toxicity is encountered [1,2,23,26,35], especially in combination regimens [3,29]. This was also observed in a phase II study with carboplatin and paclitaxel administered to patients with NSCLC at three weeks intervals, where at a carboplatin dosage of 434 mg/m² only 9% of the courses induced thrombocytopenia [36]. Also other studies with the paclitaxel-carboplatin combination reported only minimal thrombocytopenia at conventional doses [37,38], and even at higher dosages [39]. The underlying mechanism for this phenomenon is, however, not clear. For the present study a pharmacokinetic explanation seems likely, since most patients (65%) had an AUC< 4 mg/mL·min, which is generally not associated with thrombocytopenia [26,3,29,40]. On the other hand, AUCs up to 6.5 mg/mL·min were also measured, while thrombocytopenia did not occur. Although speculative, a possible explanation might be the release of hematopoietic growth factors induced by paclitaxel in the bone marrow prohibiting severe myelosuppression: it has been demonstrated that paclitaxel induces the release of granulocyte-macrophage colony-stimulating factor

(GM-CSF) *in vitro* [41]. Another, also speculative explanation might be antagonistic intracellular interactions [14,42]. For example, stabilization of microtubules by paclitaxel in megakaryocytes.

Although the toxicity of the combination of carboplatin and paclitaxel was milder than expected, it must be noted, however, that at the highest paclitaxel dose reached (250 mg/m^2), lethal toxicity was encountered in a patient with gross liver involvement and liver function disturbances, for which reason the study was discontinued. Evidently, a dose reduction should be considered for patients with a decreased hepatic function. For patients with an adequate hepatic function, the proposed paclitaxel dose for phase II evaluation is 225 mg/m^2 in combination with 400 mg/m^2 carboplatin.

One of the aims of the present study was to investigate the relationship between the pUF AUC and the toxicity (%decr in platelets). This relationship has been observed for carboplatin administered as single agent [23,26,40] and in combination regimens [3,29,40]. The lack of this relationship in the present study might partially be explained by the relatively low AUC-values, but also by the putative pharmacodynamic interactions of carboplatin with paclitaxel. It can be expected that at higher carboplatin dosages and thus at higher AUC-values the relationship between the pUF AUC and toxicity may become apparent again. Indications for this can be found in another study using the carboplatin-paclitaxel combination [43]. Here, thrombocytopenia occurred at an AUC of 8 and 10 mg/mL·min [39]. However, as is clear from the present study, it is not possible to calculate accurately the dose needed to achieve a certain AUC by using a modified Calvert-formula (Figure 3). In contrast, if patients are dosed using the creatinine clearance, they will be underexposed. The actual AUC is, in fact, 30% lower than the target AUC ($p < 0.00001$). Besides this method is biased, it is also imprecise (RMSE% = 33%). This is in agreement with the phase I study of Ozols et al.[44], in which paclitaxel was combined with carboplatin targeted to achieve an AUC of 5.0, 7.5, or 10.0 mg/mL·min. Here, the creatinine clearance has been used in the Calvert-formula to calculate the dose needed to achieve the target AUC. As in the present study, preliminary results ($n = 5$) indicated that the measured AUC was 30% lower than aimed [44]. Whether this finding has implications for carboplatin-paclitaxel studies in which the GFR is calculated by the ^{51}CrEDTA method [36,37,38] is not clear, but it seems recommendable to validate the formula in those cases.

In conclusion, there is no sequence depending effect on the pharmacokinetics of carboplatin, nor on the toxicity. Furthermore, the values of the carboplatin AUCs at the investigated dosages are well in agreement with literature data. However, these AUCs were probably too low to induce thrombocytopenia, although antagonistic properties of paclitaxel cannot be excluded.

REFERENCES

1. Wagstaff AJ, Ward A, Benfield P, et al: Carboplatin, a preliminary review of its pharmacodynamic and pharmacokinetic properties and therapeutic efficacy in the treatment of cancer. *Drugs* 37: 162 (1989)
2. Vermorken JB, ten Bokkel Huinink WW, Eisenhower EA, et al: Carboplatin versus cisplatin. *Ann Oncol* 4: S41 (1993)
3. Jodrell DI, Egorin MJ, Canetta RM, et al: Relationships between carboplatin exposure and tumor response and toxicity in patients with ovarian cancer. *J Clin Oncol* 10: 520 (1992)
4. Gatzemeier U, Heckmayr M, Hossfeld DK, et al. Phase II study of carboplatin in untreated inoperable non-small cell lung cancer. *Cancer Chemother Pharmacol* 26: 369 (1990)
5. Bonomi PD, Finkelstein DM, Ruckdeschel JC, et al. Combination chemotherapy versus single agents followed by combination chemotherapy in stage IV non-small cell lung cancer: a study of the eastern cooperative oncology group. *J Clin Oncol* 7: 1602 (1989)
6. Spencer CM, Faulds D. Paclitaxel: a review of its pharmacodynamic and pharmacokinetic properties and therapeutic potential in the treatments of cancer. *Drugs* 48: 794 (1994)
7. Rowinsky EK. Clinical pharmacology of taxol. *J Natl Cancer Inst* 15: 25 (1993)
8. Huizing MT, Keung ACF, Rosing H, et al. Pharmacokinetics of paclitaxel and metabolites in a randomized comparative study in platinum-pretreated ovarian cancer patients. *J Clin Oncol* 11: 2127 (1993)
9. Murphy WK, Fossella FV, Winn RJ, et al. Phase II study of taxol in patients with untreated advanced non-small cell lung cancer. *J Natl Cancer Inst* 85; 384 (1993)
10. Chang AY, Kim K, Glick J, et al. Phase II study of taxol, merbarone, and piroxantrone in stage IV non-small cell lung cancer: the eastern cooperative oncology group results. *J Natl Cancer Inst* 85: 388 (1993)
11. Ettinger DS. Taxol in the treatment of lung cancer. *Mongraphs Natl Cancer Inst* 15: 177 (1993)
12. Parker RJ, Dabholkar MD, Lee KB, et al. Taxol effect on cisplatin sensitivity and cisplatin cellular accumulation in human ovarian cancer cells. *J Natl Cancer Inst* 15: 83 (1993)
13. Liebmann JE, Fisher J, Teague D, Cook JA. Sequence dependence of paclitaxel (taxol) combined with cisplatin or alkylators in human cancer cells. *Oncol Res* 6: 25 (1994)
14. Rowinsky EK, Citardi MJ, Noe DA, et al. Sequence-dependent cytotoxic effects due to combinations of cisplatin and the antimicrotubule agents taxol and vincristine. *J Cancer Res Clin Oncol* 119: 727 (1993)
15. Rowinsky EK, Gilbert MR, McGuire WP, et al. Sequence of taxol and cisplatin: a phase I and pharmacologic study. *J Clin Oncol* 9: 1692 (1991)
16. Beijnen JH, Bais EM, Ten Bokkel Huinink WW. Lithium pharmacokinetics during cisplatin-based chemotherapy: a case report. *Cancer Chemother Pharmacol* 33: 523 (1994)
17. Grossman SA, Sheidler VR, Gilbert MR. Decreased phenytoin levels in patients receiving chemotherapy. *Am J Med* 87: 505 (1989)
18. Neef C, De Voogd-van der Straaten I. An interaction between cytostatic and anticonvulsant drugs. *Clin Pharmacol Ther* 43: 372 (1988)
19. Balis FM. Pharmacokinetic drug interactions of commonly used anticancer drugs. *Clin Pharmacokinet* 11: 223 (1986)
20. Schellens JHM, Ma J, Brumo R, et al. Pharmacokinetics of cisplatin and taxotere and WBC DNA-adduct formation of cisplatin in the sequence taxotere/ cisplatin and cisplatin/ taxotere in a phase I/II study in solid tumor patients. *Proc Am Soc Clin Oncol* 13: 132 (1994)
21. Relling MV, McLeod HL, Bowman LC, Santana VM. Etoposide pharmacokinetics and pharmacodynamics after acute and chronic exposure to cisplatin. *Clin Pharmacol Ther* 56: 503 (1994)
22. Rowinsky E, Grochow L, Kaufmann S, et al. Sequence-dependent effect of topotecan and cisplatin in a phase I and pharmacologic study. *Proc Am Soc Clin Oncol* 13: 142 (1994)
23. Van der Vijgh WJF: Clinical pharmacokinetics of carboplatin. *Clin Pharmacokinet* 21: 242 (1991)
24. Saikawa Y, Kubota T, Kuo TH, et al. Combined effect of 5-fluorouracil and carboplatin against human gastric cancer cell lines in vitro and in vivo. *Anticancer Res* 14: 461 (1994)
25. Dofferhof ASM, Berendsen HH, Van de Naalt J, et al. Decreased phenytoin level after carboplatin treatment. *Am J Med* 89: 247 (1990)
26. Calvert AH, Newell DR, Gumbrell LA, et al: Carboplatin dosage: Prospective evaluation of a simple formula based on renal function. *J Clin Oncol* 7: 1748 (1989)

27. Egorin MJ, Van Echo DA, Tipping SJ, et al. Pharmacokinetics and dosage reduction of cis-diammine (1,1-cyclobutanedicarboxylato) platinum in patients with impaired renal function. *Cancer Res* 44: 5432 (1984)
28. Marina NM, Rodman J, Shema S, et al. Phase I study of escalating targeted doses of carboplatin combined with ifosfamide and etoposide in children with relapsed solid tumors. *J Clin Oncol* 11: 554 (1993)
29. Reyno LM, Egorin MJ, Canetta RM, et al: Impact of cyclophosphamide on relationships between carboplatin exposure and response or toxicity when used in the treatment of advanced ovarian cancer. *J Clin Oncol* 11: 1156 (1993)
30. Proost JH, Meijer DKF. MW/PHARM, An integrated software package for drug dosage regimen calculation and therapeutic drug monitoring. *Comput Biol Med* 22: 155 (1992)
31. Cockcroft DW, Gault MH. Prediction of creatinine clearance from serum creatinine. Nephron 16: 31 (1976)
32. Sheiner LB, Beal SL. Some suggestions for measuring predictive performance. *J Pharmacokinet Biopharm* 9: 503 (1981)
33. Van Warmerdam LJC, Ten Bokkel Huinink WW, Maes RAA, et al. Limited-sampling models for anticancer agents. *J Cancer Res Clin Oncol* 120: 427 (1994)
34. LeBlanc GA, Sundseth SS, Weber GF, et al. Platinum anticancer drugs modulate P-450 mRNA levels and differentially alter hepatic drug and steroid hormone metabolism in male and female rats. *Cancer Res* 54: 540 (1992)
35. Vadhan-Raj S, Kudelka AP, Garrison L, Gano J, Edwards CL, Freedman RS, Kavanagh JJ. Effects of interleukin-1α on carboplatin-induced thrombocytopenia in patients with recurrent ovarian cancer. *J Clin Oncol* 12: 707 (1994)
36. Langer CJ, Leighton J, Comis R, McAleer C, et al. Taxol and carboplatin in combination in stage IV and IIIB non-small cell lung cancer: a phase II trial. *Proc Am Soc Clin Oncol* 13: 338 (1994)
37. Israel VK, Zaretsky S, Natale RB. Phase I/II trial of combination carboplatin and taxol in advanced non-small cell lung cancer (NSCLC). *Proc Am Soc Clin Oncol* 13: 351 (1994)
38. Paul DM, Johnson DH, Hande KR, et al. Carboplatin and taxol: a well tolerated regimen for advanced non-small cell lung cancer. *Proc Am Soc Clin Oncol* 13: 352 (1994)
39. Shea T, Graham M, Steagall S, et al. Multiple cycles of high dose taxol plus carboplatin with G-CSF (filastrim) and peripheral blood progenitor support. *Proc Am Soc Clin Oncol* 13: 150 (1994)
40. Van Warmerdam LJC, Ten Bokkel Huinink WW, Beijnen JH. Overwegingen voor het gebruik van de Calvert-formule bij het doseren van carboplatine. *Pharm Weekbl* 128: 749 (1993)
41. Pluznik DH, Lee NS, Sawada T. Taxol induces the hematopoetic growth factor granulocyte-macrophage colony-stimulating factor in murine B-cells by stabilization of granulocyte-macrophage colony-stimulating factor nuclear RNA. *Cancer Res* 54: 4150 (1994)
42. Viallet J, Boucher L, Gallant G. In vitro interactions between paclitaxel and other agents in human non small cell lung cancer lines: antagonism with etoposide and doxorubicin. *Proc Am Soc Clin Oncol* 13: 365 (1994)
43. Ten Bokkel Huinink WW, Veenhof CHN, et al. Carboplatin and paclitaxel in patients with advanced ovarian cancer, a dose finding study. *Ann Oncol* 5: 99 (1994)
44. Ozols RF, Kilpatrick D, O'Dwyer P, et al. Phase I and pharmacokinetic study of taxol (T) and carboplatin (C) in previously untreated patients (pts) with advanced epithelial ovarian cancer (oc): a pilot study of the gynaecologic oncology group. *Proc Am Soc Clin Oncol* 12: 259 (1993)

THE CLINICAL DEVELOPMENT OF THE ORAL PLATINUM ANTICANCER AGENT JM216

Ian Judson, Mark McKeage, Janet Hanwell, Claire Berry, Prakash Mistry, Florence Raynaud, Grace Poon, Barry Murrer*, Kenneth Harrap

Cancer Research Campaign Centre for Cancer Therapeutics
The Institute of Cancer Research
15, Cotswold Road, Sutton, Surrey
SM2 5NG, UK
*Johnson Matthey Technology, Reading, UK

INTRODUCTION

The platinum antitumour agents cisplatin and carboplatin have brought significant benefits to the treatment of a wide variety of malignancies[1]. However, drug resistance remains a problem and toxicity can be significant. Where the aim of treatment is symptom palliation, which is more often than not the case, it is important to consider the side effects and convenience of treatment. A collaborative programme of research into new platinum anticancer agents conducted by The Institute of Cancer Research and Johnson Matthey Technology has focused on the twin objectives of developing agents which lack cross-resistance with cis- and carboplatin and agents which can be given by the oral route. A key lead was the identification of the ammine/amine platinum (IV) dicarboxylates[2] which not only exhibit good oral bioavailability but also show activity in certain cisplatin resistant human ovarian cancer cell lines *in vitro*.[3] The specific aims of the oral development programme were to identify an agent with good oral bioavailability, low emetogenicity, no significant specific organ toxicity, such as nephrotoxicity, ototoxicity, or neurotoxicity, and dose limiting myelosuppression. These selection criteria were used to identify bis-acetato-ammine-dichloro-cyclohexylamine-platinum (IV), or JM216 for clinical development[4,5] (Fig. 1). Of these, one of the most stringent was emesis, which was tested in a ferret model in comparison with a number of other agents, especially cisplatin. Preclinical studies of neurotoxicity were conducted in parallel with the clinical trials and while confirming the lack of neurotoxicity compared with cisplatin and ormaplatin they did not play a part in the selection process.[6] The phase I studies with JM216 demonstrated that the preclinical objectives had been amply fulfilled[7] and phase II studies are currently underway to evaluate its antitumour efficacy.

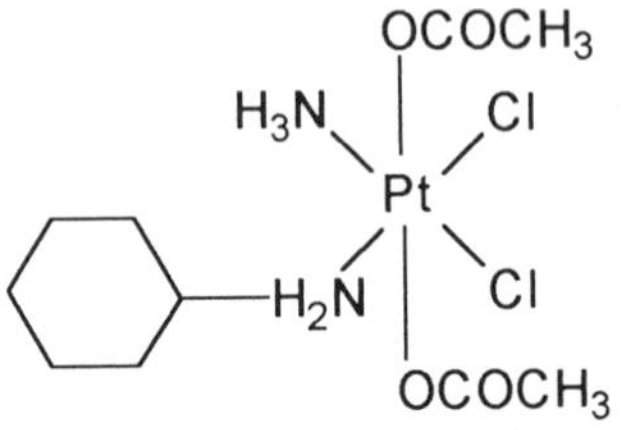

Fig.1 Structure of JM216

SINGLE DOSE PHASE I TRIAL

Methods

In this study JM216 was administered to a typical population of patients with refractory cancer for whom no conventional chemotherapy was available. Inclusion was restricted to patients with adequate WHO performance status ≤2, life expectancy ≥3 months, adequate renal function as defined by ^{51}Cr-EDTA clearance ≥60 ml/min, and no gastrointestinal abnormality likely to compromise absorption. Patients were required to have adequate bone marrow and liver function.

JM216 was supplied by the Johnson Matthey Technology Centre, Reading, UK and formulated by Bristol-Myers Squibb, Buffalo, USA as 10, 50 and 200 mg capsules. Patients were treated with a single oral dose after an overnight fast. The starting dose was 60 mg/m^2 (1/10 the murine maximum tolerated dose, MTD) and escalations were performed according to a modified Fibonacci scheme with ≥3 patients treated at each dose level.

Antiemetics were not given prophylactically in the first instance but patients experiencing emesis were given appropriate treatment to control the symptoms and also to prevent nausea and vomiting with further doses.

In addition to weekly assessment of bone marrow function, patients' renal function was monitored by repeated ^{51}Cr-EDTA clearance before each dose of JM216, potential neurotoxicity was measured using vibration sensation threshold at wrist and ankle[8] and audiometry was performed prior to treatment to act as a baseline for patients experiencing alterations in hearing, tinnitus etc.

Pharmokinetics were studied during the first administration and in some patients treated at higher doses samples were taken on both first and second courses to investigate a possible effect of capsule size on absorption. Blood was taken into heparinised tubes, centrifuged immediately and separated, one portion of plasma being stored in liquid nitrogen for measurement of total platinum and the other placed above Amicon Centrifree filters and centrifuged again at 2000g for 20 min at 4°C to yield an ultrafiltrate for measurement of free platinum. Urine was collected in 8 hr periods for analysis. Platinum was measured using flameless atomic absorption spectrophotometry.

Results

In total 31 patients were treated over the dose range 60 to 740 mg/m^2. Emesis proved to be manageable but occurred in 92% of courses given at doses up to 300 mg/m^2 without prophylactic antiemetics, with median severity CTC grade 2. With prophylactic antiemetics the severity and incidence were reduced to median grade 1 in up to 60% of courses. Although no direct comparison was made, ondansetron plus dexamethasone seemed rather more effective in reducing the incidence of emesis (38% of courses) than metoclopramide + dexamethasone (60% of courses). Nausea was also significantly reduced. Diarrhoea occurred in 58% of courses given without prophylactic antiemetics but this was reduced to 24% with antiemetic therapy, especially ondansetron, and was of median grade 0.

There was no evidence of renal toxicity with no instances of falling EDTA clearance attributable to therapy. In 2 patients a fall in clearance could be attributed to ureteric obstruction. There were no symptoms suggestive of peripheral neurotoxicity and no adverse changes in vibration sensation threshold. Only one patient complained of tinnitus but there was no change in audiometry.

Myelosuppression was variable but generally mild, with only 3 patients experiencing > grade 2 toxicity. Of these one required platelet support and another received antibiotics for neutropenic sepsis.

There was evidence of antitumour activity in 3 patients with ovarian cancer, one of whom experienced a sustained partial remission with a >50% reduction in the size of a subcapsular liver metastasis and significant fall in CA125 from 4950 to 209 IU/ml. Two other patients had falls in CA125 associated with stable disease.

Pharmacokinetic studies showed that absorption of JM216 occurred quite rapidly with peak concentrations being recorded at 1 to 3 hours after ingestion. Interpatient variations in peak concentrations and area under curve (AUC) of both free and total platinum increased with increasing dose and it appeared that absorption was saturable at approximately 300 mg/m^2 (Fig. 2). Urinary platinum levels were low, corresponding to 8.3% of administered dose at the first dose level and only 1.6% at 740 mg/m^2. The majority of platinum recovered was excreted in the first 8 hours. Renal clearance was highly variable (8.8 to 433 ml/min) and did not correlate with glomerular filtration rate or plasma free platinum AUC.

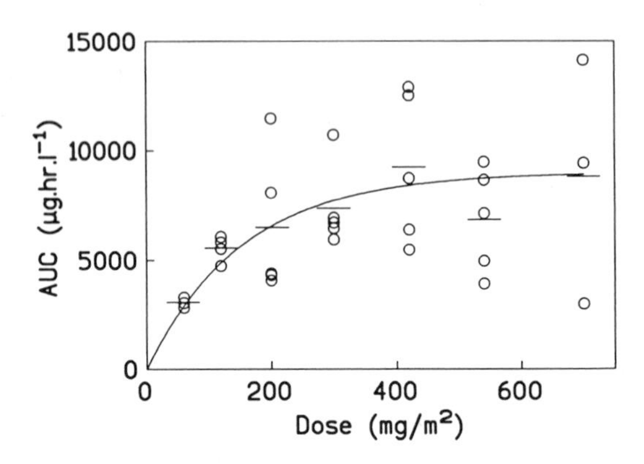

Fig. 2. Pharmacokinetics of single dose JM216

DAILY X 5 PHASE I TRIAL

Having failed to reach the MTD with a single dose of JM216 owing to saturable absorption, it was clear that a repeat dose schedule would be required. Preclinical studies had, in fact, shown that the therapeutic index of JM216 was increased by dose fractionation and hence a daily x 5 schedule was chosen for further evaluation.

Methods

Patients were treated at doses of 30, 60, 100 and 140 mg/m^2/day x 5. Eligibility criteria and monitoring for toxicity were identical to the single dose study. Prophylactic antiemetics were used routinely, usually ondansetron + dexamethasone. Blood was taken for pharmacokinetic studies of free and total platinum following doses 1 and 5 and where possible before intervening doses. Following definition of the MTD in the standard population of previously treated patients a lower dose was used in a large group of patients to evaluated the degree of interpatient variation in toxicity. In addition, a population of patients who had not received prior chemotherapy, many of whom were suffering from mesothelioma, were treated at escalating doses, including an intermediate dose, in an attempt to establish whether a higher dose could be recommended for phase II study in chemotherapy-naive patients.

Results

The fractionated daily x 5 schedule was very well tolerated with good control of emesis. The MTD was 140 mg/m^2 at which dose 2 out of 3 patients experienced grade 4 thrombocytopenia and grade 3 or 4 leukopenia (see Table 1). As with single administration no patients suffered from drug related renal impairment or neurotoxicity, including ototoxicity. Experience was expanded at 100 mg/m^2 in previously treated patients demonstrating that this was a suitable dose for phase II study in this population, being generally associated with myelosuppression of $\leq$2. Re-escalation in untreated patients again established 140 mg/m^2 to be the MTD, since 2 patients both experienced grade 3 neutropenia and 1/2 grade 4 thrombocytopenia. At the intermediate dose of 120 mg/m^2 only 1/3 patients experienced significant myelosuppression (grade 4), which may have been due to a borderline performance status in that individual.

Dose mg/m2	Patients/ Courses	Toxicity by course (CTC grade)									
		Thrombocytopenia					Leucopenia				
		0	1	2	3	4	0	1	2	3	4
30	3/8	7	-	-	1	-	6	1	1	-	-
60	5/18	10	8	-	-	-	14	3	-	1	-
100	15/40	15	15	8	-	2	21	7	7	5	-
120	3/9	3	3	2	-	1	2	6	1	-	-
140	5/11	3	1	2	1	4	3	1	2	4	1

Table 1. Overall Haematological Toxicity

Pharmacokinetic studies showed a linear relationship between dose and platinum concentration on day 1 and 5, over the dose range 30 to 140 mg/m^2/day, with some evidence of accumulation over the 5 days of administration. There were large interpatient variations in both area under curve (AUC) and maximum plasma concentration (C_{max}) for a given dose, but a good correlation was observed between myelosuppression and total Pt C_{max} on day 5 ($r = 0.8$). A full pharmacokinetic profile was obtained following administration of JM216 on day 5 which showed the mean terminal half-life of ultrafiltrable platinum, which corresponds to active metabolites, to be 7.45 ± 4h.

A patient with ovarian cancer who experienced a partial response in the single dose study again experienced significant benefit when rechallenged at relapse using the daily x 5 schedule. In addition, minor responses associated with symptomatic improvement were seen in patients with non-small cell lung cancer and mesothelioma.

DISCUSSION

The experience to date with JM216 has shown that it is indeed possible to give a platinum antitumour agent via the oral route. It might be thought that the gastrointestinal and renal toxicities of cisplatin made this an unlikely prospect since the justification for an oral treatment must be the ease with which it can be given in the out-patient setting. However, nausea and vomiting with JM216 proved easily manageable and no organ specific toxicity has been observed to date.[7] Another serious concern is the increased interpatient variability in pharmacokinetics associated with oral administration. However, the variations in C_{max}or AUC observed were not excessive compared with other cytotoxics. Variable kinetics and small therapeutic index are acknowledged problems with cytotoxics and there may be a case for individualised pharmacokinetically guided, or adaptive, dosing regimens.[9] It proved difficult to show a very good correlation between pharmacokinetic parameters and myelosuppression but the situation is complicated by the fact that JM216 forms 6 platinum complexes (mostly active) in man which have been characterised using techniques such as LC-MS[10] and can be quantified by HPLC/FAAS. However, there was much less variation between dose and myelosuppression, which appeared quite predictable. There was evidence of antitumour activity in both studies and this is now being tested in phase II trials in ovarian, non-small cell and small cell lung cancers. Apart from the amelioration in toxicity associated with dose fractionation of cisplatin, there is little evidence for schedule dependent variations in toxicity or antitumour activity of platinum complexes. The daily x 5 schedule chosen for phase II evaluation appears to be convenient and well tolerated and the results will be awaited with interest.

We should like to acknowledge the support of the Cancer Research Campaign, UK and Johnson Matthey Technology.

REFERENCES

1 Kelland LR New platinum antitumor complexes. *Crit Rev Oncol Hematol* 1993; **15**: 191-219.

2 Giandomenico CM, Murrer BA, Abrams MJ, Vollano JF, et al. Synthesis and reactions of a new class of orally active Pt(IV) antitumour complexes. In: *Platinum and other metal coordination compounds in cancer chemotherapy* (Howell SB, ed), New York:Plenum Press, 1993; 93-100.

3 Harrap KR, Murrer BA, Giandomenico C, Morgan SE, et al. Ammine/amine platinum IV dicarboxylates: a novel class of complexes which circumvent cisplatin resistance.. In: *Platinum and other metal coordination compounds in cancer chemotherapy* (Howell SB, ed), New York:Plenum Press, 1991; 381-399.

4 McKeage MJ, Morgan SE, Boxall FE, Murrer BA, et al Lack of nephrotoxicity of oral ammine/amine platinum (IV) dicarboxylate complexes in rodents. *Br J Cancer* 1993; **67**: 996-1000.

5 McKeage MJ, Morgan SE, Boxall FE, Murrer BA, et al Preclinical toxicology and tissue platinum distribution of novel oral antitumour platinum complexes: ammine/amine platinum(IV) dicarboxylates. *Cancer Chemother Pharmacol* 1994; **33**: 497-503.

6 McKeage MJ, Boxall FE, Jones M, Harrap KR Lack of neurotoxicity of oral bisacetatoamminedichlorocyclohexylamineplatinum(IV) in comparison to cisplatin and tetraplatin in the rat. *Cancer Res* 1994; **54**: 629-661.

7 McKeage MJ, Mistry P, Ward J, Boxall FE, et al A phase I and pharmacological study of an oral platinum complex (JM216): dose-dependent pharmacokinetics with single dose administration.. *Cancer Chemother Pharmacol* 1995; **in press**:

8 Elderson A, Gerritsen van der Hoop R, Haanstra W, Neijt JP, et al Vibration perception and thermoperception as quantitative measurements in the monitoring of cisplatin induced neurotoxicity. *Journal of the Neurological Sciences* 1989; **93**: 167-174.

9 Newell DR Can pharmacokinetic and pharmacodynamic studies improve cancer chemotherapy? *Ann Oncol* 1994; **5(suppl,4)**: 9-15.

10 Poon GK, Raynaud FI, Mistry P, Odell DE, et al Metabolic studies of an orally active platinum anticancer drug (JM216) by liquid chromatography/electrospray ionisation mass spectroscopy (LC-ESI-MS). *J Chromatogr* 1995; **in press**:

MEMBRANE TRANSPORT OF PLATINUM COMPOUNDS

Gerrit Los, Dennis Gately, Michael L. Costello, Franz Thiebaut, Peter Naredi, Stephen B. Howell

UCSD Cancer Center
University of California, San Diego
9500 Gilman Dr.
La Jolla, CA 92093

INTRODUCTION

Small molecules cross membranes by either passive diffusion or are transported actively by transmembrane proteins. If given enough time, however, essentially any molecule will diffuse across a lipid bilayer down its concentration gradient. The rate at which this happens depends on size of the molecule and its hydrophobicity (solubility in oil). Based on these observations most platinum compounds which are small uncharged molecules, are expected to cross membranes relatively easily by passive diffusion. This is, however, only partly true. On the one hand platinum compounds, and specificly cisplatin (cDDP), cross membranes by passive diffusion since the accumulation is not saturable nor is it inhibited by structural analogs. On the other hand, cisplatin transport can be modulated both by a variety of pharmacologic agents that do not cause general permeabilization of the membrane, and by the activation of intracellular signal transduction pathways. Whatever is responsible for the transport modulation, probably more than one mechanism is involved in the transport of platinum compounds across cell membranes. The importance of each of these transport mechanisms is still unclear, however transport of platinum compounds is often down modulated during the emergence of platinum drug resistance. Therefore most of the studies dealing with transport of platinum compounds are performed in platinum drug-resistant variants. In this chapter we will describe the current understanding of membrane transport of platinum compounds in both sensitive and resistant cells and the importance of specific transport molecules in the sequestration of platinum compounds and platinum complexes.

Platinum and Other Metal Coordination Compounds in Cancer Chemotherapy 2
Edited by H.M. Pinedo and J.H. Schornagel, Plenum Press, New York, 1996

PASSIVE AND FACILITATED DIFFUSION

As mentioned, the mechanism by which cisplatin enters the cell is still not fully understood. It has long be assumed that cisplatin enters the cell by passive diffusion[1], however recent work by different groups indicates that mechanisms other than passive diffusion might be involved in the accumulation of cDDP[2]. The latter is supported by the findings that the accumulation of cDDP: (a) can be inhibited by concomitant incubation with certain amino acids[3]; (b) is energy dependent, Na^+ dependent, and ouabain inhibitable[4]; (c) is not greatly affected by changes in pH[5], and (d) is inhibited by aldehydes[6]. In addition, the accumulation of cisplatin is affected by a number of intracellular signaling pathways including protein kinase C (PKC)[7], epidermal growth factor (EGF)[8], and the Ca^{++}/calmodulin pathway[9], suggesting their involvement in the regulation of cisplatin uptake.

A working model that accommodates most of the existing observations concerning cisplatin accumulation into cisplatin sensitive and resistant cells has recently been published by Gately and Howell[10]. This very elegant model, which might be more complicated than described, envisions that approximately one half of the initial drug uptake rate is due to passive diffusion and that the other half is occurring by facilitated diffusion through a gated channel[10]. A schematic diagram of this gated channel is demonstrated in Figure 1. The model would propose that the flux through the channel is regulated by phosphorylation cascades initiated by activation of protein kinase A, protein kinase C, or by the calmodulin dependant kinases[10], which is similar to the regulation of a variety of gated channels. permeability[11].

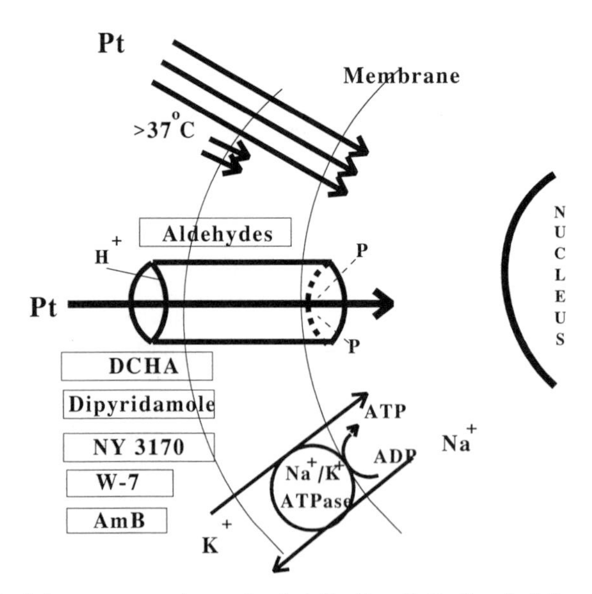

Figure 1. Potential cellular transport pathways for cisplatin. Hypothetically, cisplatin can enter the cell either by way of passive diffusion or through a gated channel. The flux through the channel can potentially be regulated by docosahexaenoic acid (DCHA), dipyridamole, 1-propargyl-5-chloropyrimidin-2-one (NY 3170), calmodulin inhibitors (W-7), amphotericin B (AmB) and phosphorylation by either PKA or PKC. The flux through the channel can potentially be decreased by various aldehydes, while it also seems to depend on functional Na^+/K^+ ATPase, membrane potential, extracellular pH and extracellular osmolality.

This model is further strengthened by: (1) the ability of aldehydes that are unable to permeate the membrane to block 50% of the uptake[12,13], suggesting that critical amino groups of the channel are exposed on the external surface of the cell; (2) the fact that ouabain inhibits 50% of the initial uptake[4] suggesting that ion gradients maintained by this ATPase are crucial to the function of this channel; and, (3) by the fact that the intracellular platinum content increases at higher temperatures due to increased membrane. An important observation is that in no case has it been reported that accumulation can be inhibited by more than 50%[10]. In view of the latter, and presuming that the gated channel is lost during the selection of drug resistant variants, the accumulation of platinum compounds into resistant cells could be inhibited by as much as 50% which is in agreement with the current literature. Furthermore, it has been demonstrated that analogs of cisplatin that are more lipophilic do not exhibit decreased accumulation in resistant cells and show less cross resistance[14,15,16,17]. These findings are in agreement with this theoretical model because such analogs are less dependant on a gated channel for entry into the cell. In addition, the model allows for modulation of the flux through the channel and permits the design of specific experiments addressing the role of membrane transport in the accumulation of cisplatin into the cell.

TRANSPORTERS

Heavy Metal Transport in Yeast

Several transport-protein complexes that mediate the detoxification of heavy metal salts have been identified in bacteria and yeast, and appear to be structurally and functionally well-conserved throughout evolution[18,19,20]. In yeast, three heavy metal resistance genes have been cloned from the budding yeast *S. cerevesiae*, including *cot1*[20], *zrc1*[21] and *cup1*[22]. These proteins appear to be involved in the intracellular detoxification of heavy metals by sequestration of the metal. The deduced amino acid sequence of the *zrc1* gene product predicts a rather hydrophobic protein with six possible membrane-spanning regions, possibly involved in actively pumping zinc out of the cell. In the fission yeast *S. pombe*, an ATP-dependant pump coded by the *hmt1* gene has been shown to confer resistance to cadmium[18]. The HMT1 protein appears to be located in the vacuolar membrane, and the mechanism of resistance appears to involve the sequestration of the metal into vacuoles.

Based on the observation that mechanisms of drug resistance developed by unicellular organisms are often conserved in mammalian cells, we have sought to develop a cisplatin resistant strain of *S. pombe* in anticipation of similarities between the detoxification and sequestration of heavy metals and cisplatin. Using a drug exposure schedule similar to that used for the selection of resistant mammalian cells, it proved possible to select cisplatin resistant strains of *S. pombe*. The characteristics of the cisplatin resistant *S. pombe* strain corresponded well with those found in mammalian cells. First, the cisplatin resistant yeast variant (wtr2) was 4-fold resistant to cisplatin compared to 2 to 5 fold resistance in mammalian cells using similar selections schemes. Second, the resistant phenotype was very stable in *S. pombe* as it is in mammalian cells at such a low level of resistance[2]. Third, the resistant phenotype was expressed in a dominant fashion in the diploid *S. pombe* resulting from mating parental and resistant haploid strains[a] which seemed to be in agreement with data found in the murine L1210[23] and human 2008 cell lines (personal communication S. B. Howell, 1995). A remarkable difference between the mammalian and *S. pombe* resistant phenotype was the fact that rather than having a reduced accumulation of the tritiated cisplatin analog, dichloro(ethylenediamine)platinum(II) ([^{3}H]-DEP) as do many mammalian cells, the resistant yeast cells actually accumulated more drug than the parent *S. pombe* (Fig. 2). The higher [^{3}H]-DEP content was due to an active energy-requiring uptake process as evidenced by the fact that it was reduced by glucose starvation. This observation suggests a detoxification mechanism for cisplatin that is similar to the HMT1/phytochelatin system responsible for cadmium resistance in *S. pombe*.

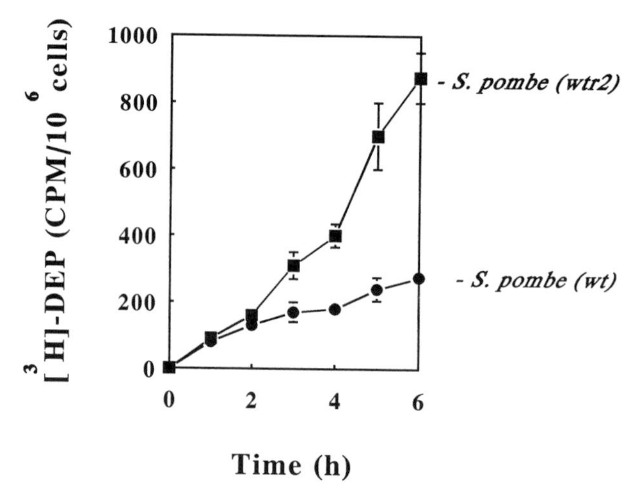

Figure 2. [^{3}H]-DEP accumulation in parental (wt) and cisplatin resistant (wrt2) *S. pombe* strains during continuous incubation.

Based on the amino acid sequence of the *hmt1* gene product, it appears to have 6 transmembrane spanning regions and substantial sequence homology to the human P-glycoprotein gene *MDR1*. The current hypothesis is that cadmium complexes in the cytoplasm with phytochelatins and that this complex is then sequestered via the pump action of the protein encoded for the *hmt1* gene into an intracellular vacuole.

A similar sequestering mechanism might be in place for cDDP. Incubating the cisplatin resistant strain of *S. pombe* with cisplatin at an IC_{50} level for 72 hours resulted

---

[a]Thibaut, F.B., Jimenez, G., Christen, R.D., Enns, R.E., Hom, D.K., Jones, J.A., Howell, S.B. Isolation and characterization of cisplatin-resistant strain of *Schizosaccharomyces Pombe*. (submitted for publication 1995).

in a specific distribution pattern of platinum-containing complexes in the yeast cell. Using Electron Energy Loss Spectroscopy (EELS)[24,25] and Electron Spectroscopic Imaging (ESI)[24,25], a rapidly evolving sophisticated technique that makes use of a transmission electron microscope with an electromagnetic prism built into its electron beam path, platinum was detected between the cell wall and plasma membrane, in portions of the nucleus and in cytoplasmatic vacuoles (Fig. 3). The latter strongly suggests a sequestration of platinum into these vacuoles. This hypothesis is further strengthened by the fact that sulfur, a component of glutathione and phytochelatins, was found at the same locations as platinum (Fig. 3), suggesting a sequestering system similar to that of cadmium in which cisplatin forms sulfur-platinum complexes and capable of being transported into vacuoles.

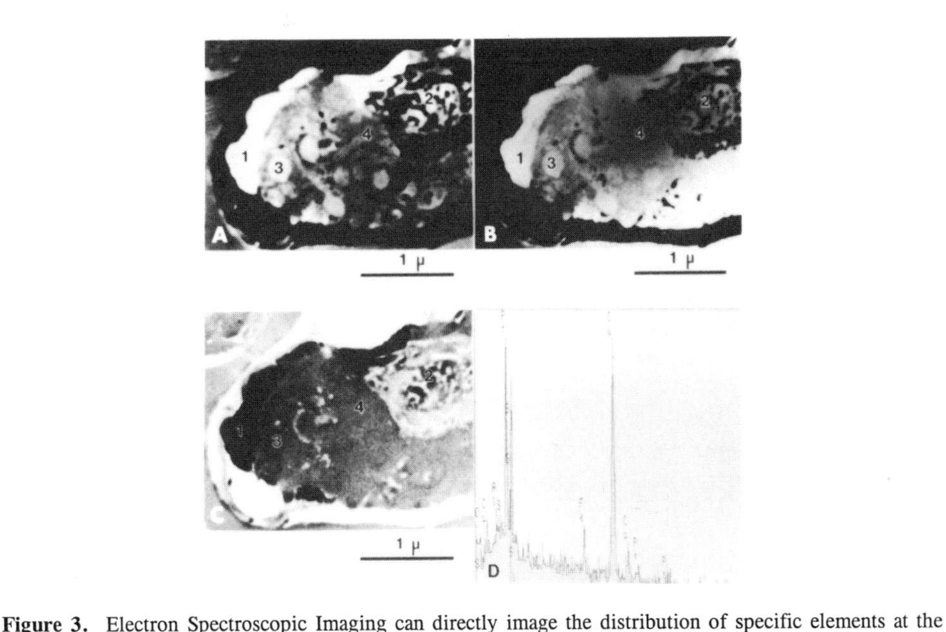

Figure 3. Electron Spectroscopic Imaging can directly image the distribution of specific elements at the ultrastructural level and takes advantage of inelastic scattered electrons. When the monochromatic electron beam strikes the atomic electron orbital of the elements in the section it will reduce the energy of the beam electrons by the amount necessary to remove the specimen electrons from their orbital. The beam then passes into an electromagnetic prism where the electrons are separated according to their energy. For example, useful energy loss values for platinum are 52 and 722 electron volts which correspond to the N and O orbital. From those specific electrons transmitted by the electromagnetic prism an image can be created.
ESI of a thin section of yeast showing examples of detection (bright white areas where the electron beam has been transmitted without deflection) at different energy losses, $\triangle E$. **A:** bright areas are seen between the cell wall and plasma membrane (1,), in portions of the nucleus (2) and in the cytoplasmic vacuoles (3) and mitochondria (4) at $\triangle E = 52$ ev. **B:** at $\triangle E = 229$ ev (sulfur L shell, a similar distribution is seen as with Pt. **C:** at zero energy loss (control), all regions positive in A and B are opaque. **D:** Energy Dispersive Spectrograph. platinum peaks from the M and L electron shells are seen. platinum peaks do not interfere with other peaks present in the spectrum (Cu, Fe, Si, and Al).

To determine whether *hmt1* played a role in the sequestration of platinum into vacuoles the expression was measured by Northern blot analysis. No alteration in expression of *hmt1* was detected in the resistant variants, suggesting that if a transporter regulates the sequestration of platinum into vacuoles, this transport results from something other than an increase in *hmt1* mRNA. Whether we deal with a specific platinum compound transporter like the *cup1* gene for copper or a more general mechanism of sequestration in yeast is unknown and under further investigation.

Heavy Metal Transport in Mammalian Cells

Systems transporting toxic metals are well conserved through evolution. Therefore, it is possible that a previously characterized prokaryotic metal ion transporter is also involved in transport of platinum compounds in platinum drug-resistant mammalian cells. If so, cisplatin resistant cells might be expected to demonstrate cross-resistance to other metal ions that share the same transport system. Hypothesizing that there might be such a transporter, we screened for cross resistance between cisplatin and a variety of metal salts using the human ovarian carcinoma cell line 2008.

The cross-resistance studies yielded three different cisplatin resistance phenotypes: one designated 2008/C13*, selected by chronic exposure to cisplatin at concentrations that were increased stepwise to 5 μM[26], an other designated 2008/MT, selected by exposure to increasing concentrations $CdCl_2$ and $ZnCl_2$ over a period of 2 month[27], and a third one designated 2008/H, selected by chronic exposure to antimony potassium tartrate at concentrations that were increased stepwise to 140 μM[28]. Table 1 shows the resistant characteristics of these variants. It is clear that among the metal salts tested, the cDDP-resistant cells had the highest degree of cross resistance to antimony potassium tartrate. The 2008/MT cells demonstrated low level resistance to cisplatin (2.7-fold) and low level resistance to $CdCl_2$ (1.7-fold), but were not at all resistant to antimony (Table 1). In contrast, the 2008/H cells demonstrated high level of resistance to cisplatin (16-fold) and moderate resistance to antimony (6.6-fold) . In addition, the 2008/H had a 48% reduction in cisplatin accumulation while the 2008/MT did not show a deficit in cisplatin accumulation. The 2008/C13* cells, the ones actually selected with cDDP, demonstrated high levels of cisplatin resistance (15-fold), low level of resistance to $CdCl_2$, moderate resistance to antimony (4.4-fold) and had an impairment in the accumulation of cisplatin almost similar to the 2008/H cells. In other words, $CdCl_2/ZnCl_2$ selection appeared to

Table 1: Cross-resistance in resistant variants of the human 2008 cell line.

Compound	IC_{50} (μM) 2008 Cells	Fold Resistance 2008/C13*	2008/MT	2008/H
Cisplatin	0.1 ± 0.01	15.0 ± 1.0	2.7 ± 0.1	16.0 ± 1.0
Antimony[a]	3.2 ± 0.2	4.4 ± 1.4	1.2 ± 0.2	6.6 ± 1.0
$CdCl_2$	6.9 ± 0.6	2.4 ± 0.4	1.7 ± 0.7	1.8 ± 0.2
$ZnCl_2$	107.0 ± 18	0.9 ± 0.2	1.7 ± 0.6	1.1 ± 0.4
$NiCl_2$	102.0 ± 8	0.8 ± 0.2	0.7 ± 0.1	1.1 ± 0.4
$CoCl_2$	33.0 ± 19	1.1 ± 0.5	1.2 ± 0.1	16.0 ± 1.0

[a]Antimony potassium tartrate; 2008/C13*: cisplatin resistant; 2008/MT: $CdCl_2$ and $ZnCl_2$ resistant; 2008/H: Antimony resistant.

activate a mechanism of resistance to cisplatin that did not involve a reduction in drug accumulation and provided only low level protection.In contrast, selection with either cisplatin or antimony appeared to activate a mechanism of resistance to cisplatin that involved a decrease in drug accumulation and provide higher level of protection to both cisplatin and antimony, suggesting a common mechanism of detoxification between cisplatin and antimony.

One mechanism of resistance to trivalent antimonials in bacteria is mediated by genes that make up the *ars* operon[29]. In *E. coli* and *S. aureus* this operon codes for a specific efflux-transport system that can export antimony but is best known for its ability to confer resistance to arsenic oxyanions[29]. This phenotype also includes cross-resistance to selenite and tellurite. The *ars* system is evolutionarily related to the *nif* genes but not to other classes of transport ATPases, and specifically not to P-glycoprotein and CFRT class transporters[30].

To examine the cross resistance between cisplatin and antimonite further, 2 additional pairs of sensitive and cisplatin resistant human ovarian carcinomas cell lines and 2 pairs of human head and neck carcinoma cell lines were tested for cross-resistance to antimonite in the form of antimony potassium tartrate (Table 2). The 17-fold cisplatin resistant subline 2008/A was 4.9-fold cross resistance to antimonite, and the cisplatin selected A2780/CP was 11-fold resistant to cisplatin and 7.2-fold cross resistance to antimonite. The same observation was made for the 2 head and neck carcinoma cell lines. The cisplatin selected variant UMSCC10b/cDDP was 2 fold resistant to cisplatin and 2.1 fold cross resistance to antimonite. The cisplatin selected UMSCC5/cDDP cells were 2.3 fold resistant to both cisplatin and antimonite. Thus all 4 cDDP-resistant variants demonstrated cross-resistance to antimonite (Table 2) and demonstrated a good correlation between the magnitude of the cisplatin resistance and the magnitude of cross resistance to antimonite (r=0.75).

Table 2: Cross resistance in human cisplatin and antimony resistant cell lines.

Resistant variants	Fold resistance Cisplatin	Antimony	Stibogluconate	Arsenite
2008/A	17	4.9	1.1	3.0
2008/H	16	6.6	1.0	4.3
A2780/CP	11	7.2	nd	nd
UMSCC10b/cDDP	2	2.1	nd	nd
UMSCC5/cDDP	2.3	2.3	nd	nd

All data differ significantly with the parent cell line (p,0.05) except the data for Stibogluconate which do not differ significantly from the control. nd= not determined.

In the protozoa *Leishmania*, the *lmpgp* gene has been reported to confer resistance to antimony potassium tartrate containing antimony in its trivalent form, but not to stibogluconate which contains antimony in the pentavalent state[31]. In contrast, resistance of Leishmania to pentavalent antimonials has been associated with the expression of P-glycoproteins-like protein coded for the *ltpgpA* gene[32], and such resistance can be reverted by verapamil[33]. Coming across this information in *Leishmania*, the question was asked whether stibogluconate would also lack cross resistance in the cisplatin resistant 2008 cells. Using antimony potassium tartrate containing antimony in its trivalent form, stibogluconate which contains antimony in its pentavalent form, and arsenic which contains arsenic in its trivalent form, cross resistance was detected to antimonite and arsenite but not to stibogluconate (Table 2).

If cDDP, antimonite, and arsenite share a common mechanism of resistance, then one might expect that cells selected for resistance to antimonite would also be resistant to cisplatin and arsenite, but not to stibogluconate. As demonstrated in Table 2, this was the case, arguing strongly for a common underlying mechanism of resistance. As indicated above and further emphasized by the data in Table 3, such a common shared mechanism could be impaired uptake. If this is true, then one might expect that both the cisplatin and antimonite-resistant sublines would also demonstrate an impairment in the accumulation of arsenite. Table 3 demonstrates that the accumulation of both the radiolabeled cisplatin analog [^{3}H]-DEP, and the radioactive $^{73}AsO_2^-$ was decreased in both the cisplatin (2008/A) and antimony (2008/H) resistant variant of the 2008 cell line. In addition, no difference in efflux was detected between the 2008, 2008/A and 2008/H for either [^{3}H]-DEP or $^{73}AsO_2^-$, indicating that the mechanism underlying the cisplatin/antimonite/arsenite-resistant phenotype impacted exclusively on accumulation and had no discernable effect on efflux.

Table 3: Impairment of [^{3}H]-DEP or $^{73}AsO_2^-$ accumulation into sensitive and resistant 2008 cells

Drug	Intracellular drug content (pmol/mg protein)		
	2008	2008/A (% of 2008)	2008/H (% of 2008)
[^{3}H]-DEP	88 ± 20	59 ± 4 (67%)	46 ± 8 (52%)
$^{73}AsO_2^-$	332 ± 35	160 ± 14 (52%)	146 ± 10 (56%)

2008/A: cisplatin resistant; 2008/H: Antimony resistant.

Recently Howell *et al* demonstrated that the cDDP/antimonite/arsenite-resistant phenotype (Pt/Sb/As resistant phenotype) also included cross resistance to selenite and tellurite (Howell 1995, personal communication). These findings indicate a striking similarity to the phenotype conferred on bacteria by the *ars* operon, which also includes resistance to selenite and tellurite.

In summary the similarity between the arsenite/antimonite/selenite/tellunite resistance phenotype in bacteria and the Pt/Sb/As resistant phenotype in mammalian cells, strongly suggests the involvement of a mammalian platinum transporter similar to that of the *ars* operon in bacteria. Like cDDP, arsenite accumulation does not show saturation kinetics, and the pump activity conferred by the *ars* operon is phenotypically observed as a decrease in drug accumulation either as an outwardly-direct extrusion or possibly an inwardly-directed vesicle sequestering pump[34].

Vesicle Mediated Transport of Platinum Compounds

Recent studies of Ishikawa *et al.* have indicated that an ATP-dependent export system, referred to as the GS-X pump, might play a key role in the accumulation of G S H drug conjugates in intracellular vesicles and elimination of glutathione-platinum (GS-Pt) complexes from tumor cells[35,36]. Ishikawa found that the *GS-X* pump was functionally overexpressed in cisplatin resistant human HL-60 cells, that transport of GS-Pt complexes, measured with membrane vesicles was ATP dependant and significantly increased in cisplatin resistant human HL-60 cells and that the γ-glutamylcysteine synthetase (γ-GCS) mRNA level was significantly increased in resistant cells[37], which points to increased GSH biosynthesis, a critical determinant in tumor cell resistance to chemotherapeutic agents. Based on these results it was hypothesized that platinum compounds are conjugated with cellular GSH to form GS-Pt conjugates which are subsequently transported into intracellular vesicles via the GS-X pump. The vesicles are fused with the plasma membrane, and GS-Pt conjugates are released from the cell by exocytosis (see also Fig. 4).

The molecular structure of the GS-X pump is not known. Three membrane proteins with apparent molecular masses of 200, 110 and 70 kDa were overexpressed in the cisplatin resistant HL-60 cells, whereas P-glycoprotein was not immunologically detectable in the membrane preparations from resistant and sensitive HL-60 cells[36]. Recently, however, Müller et al. demonstrated that overexpression of the *MRP* (Multidrug Resistance-associated Protein) gene in human cancer cells increased the ATP-dependant glutathione S-conjugate carrier activity in plasma membrane vesicles isolated from these human tumor cells[38], suggesting MRP-mediated efflux of glutathione S-conjugates from cells. A possible link between MRP and the GS-X pump would hypothetically link MRP, a 180- to 195- kDa glycoprotein mainly located in the plasma membrane and associated with multidrug resistance in human cells[39,40,41], to GSH S-conjugate mediated resistance such as resistance to cDDP, to alkylating agents and to arsenite. The latter could partly explain the cross resistance between cisplatin and arsenite as described by Naredi et al[27,28]. Further proof for the link between MRP and the GS-X pump was obtained in HL60 cells by Ishikawa (personal communication, 1995). In the cisplatin resistant HL60 cells in which the *GS-X* pump was functionally overexpressed, the expression of *MRP* was significantly increased, indicating that MRP is either identical to the GS-X pump or able to activate an endogenous GS-X pump activity.

In addition to the MRP/GS-X pump another transport protein has recently been identified, the 110 kDa Lung Resistance-related Protein (LRP)[42]. The overexpression of LRP is often, but not always, parallelled by overexpression of MRP[43,44]. A study screening 61 unselected human tumor cell lines of a NIH panel demonstrated that expression of LRP, more than MDR or MRP, correlated with a broad pattern of cytostatic drug resistance, including the non-MDR drugs like cDDP. The deduced LRP amino acid sequence shows 56.9% identity with the 10 kDa major vault protein σ from *Dictyostellium discoideum.* Vaults are multi-subunit organelles, associated with cytoplasmatic vesicle structures and nuclear pore complexes, suggesting that they constitute a transporter unit for a wide variety of substrates. Immunohistochemical staining of LRP in human tumor cells indicated that LRP was primarily cytoplasmatic, producing a coarsely granular staning pattern, suggesting association with vesicular/lysosomal structures. Furthermore, LRP is an important prognostic factor for poor clinical performance in ovarian cancer, a cancer most commonly treated with platinum drug-based chemotherapy[45].

DISCUSSION

Impaired accumulation of cisplatin has been identified frequently in cells selected for cisplatin resistance. However, the nature of this process is unknown. A better understanding of how cisplatin enters the cell could at least clarify some of the phenomena involved in the process of impaired drug uptake. Cisplatin enters the cell relatively slowly compared to the anticancer drugs that participate in the multidrug resistance phenotype and at least one component of the cisplatin accumulation is likely to be mediated by a transport mechanism or channel[10]. The efflux of cisplatin is even slower and it has been difficult to distinguish any major differences between sensitive and resistant cell lines[46,47]. At this point in time the picture of how platinum compounds cross membranes either to enter the cells or to efflux the cell is rather complex (Fig. 4). Most of the transport work has been performed in platinum drug-resistant cell lines with often multifactional mechanisms of resistance including reduced drug accumulation and increased intracellular detoxification[2,48,49]. Several transport-protein complexes that mediate detoxification of heavy metal salts have been identified in bacteria and yeast and appear to be structurally and functionally well conserved throughout evolution[18,19,20], providing a tool to search for similar patterns in mammalian cells.

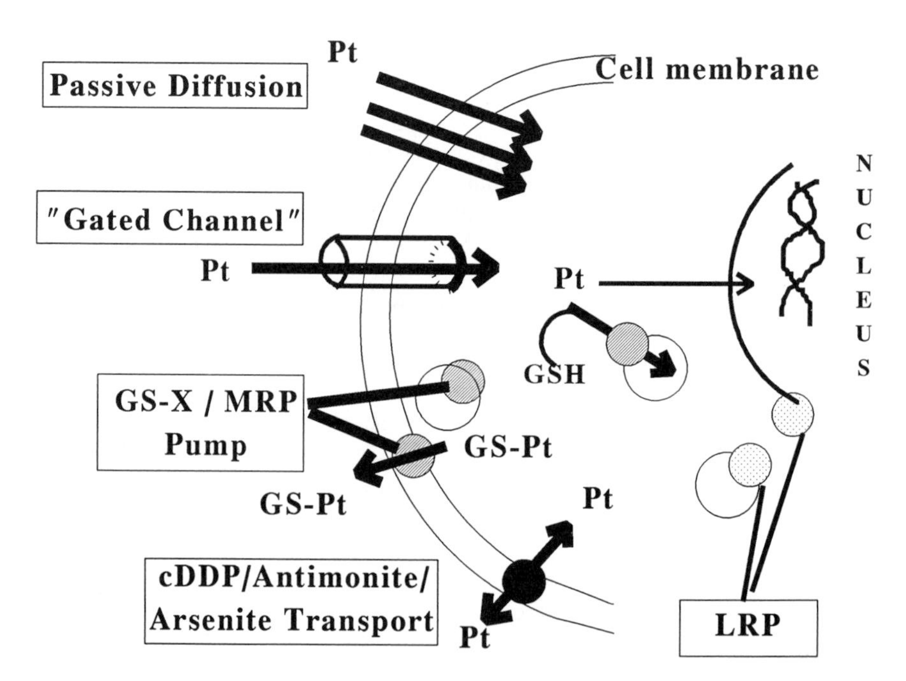

Figure 4. Schematic overview of mechanisms involved in the transportation of platinum compounds across membranes in sensitive and platinum drug-resistant tumor cells.

During studies directed at identifying metal salts that demonstrated a pattern of cross resistance pattern similar to that of cDDP, we found that cisplatin resistant sublines were cross resistant to antimonite and arsenite, while an antimonite selected variant was cross resistant to cisplatin[27,28], providing evidence for a common mechanism of resistance to cDDP, antimonite and arsenite based on a defect in a shared accumulation mechanism system. This pattern of cross resistance is known to be mediated by an ATP-dependant membrane export pump coded for by the *ars* operon in bacteria. The similarities in cross resistance profiles between bacteria and mammalian cells might indicate the presence of a human homolog of the *ars* operon as part of a transport system transporting platinum compounds into platinum drug-resistant cells (Fig. 4). This hypothesis is further strengthened by our finding of an ATP dependant sequestration of [^{3}H]-DEP into membrane vesicles of a cisplatin resistant strain of *S. pombe*, suggesting a mechanism of sequestration similar to that of other heavy metals in bacteria. A similar transport concept is thought to be involved in the sequestration of GSH-platinum complexes in mammalian cells. The GSH-platinum complexes are transported into intracellular vesicles via a GSH *S*-conjugate transporter either similar to the GS-X pump and MRP transporter or associated with the pump activity. The vesicles are fused with the plasma membrane, and GSH-platinum complexes are released from the cell by exocytosis, translocating the transport proteins such as the GS-X pump and MRP proteins to the plasma membrane (Fig. 4).

In summary, platinum compounds enter the cell by either passive diffusion, facilitated diffusion or active transport. The latter might involve a gated channel or specific transport proteins such as gene products involved in the detoxification of platinum compounds. In view of this, the recent discovery of the GS-X/MRP pump, the cDDP/Antimony/Arsenite transport mechanism and the LRP protein will focus attention on functional and, more important, structural aspects of these proteins to determine their role in transport of platinum compounds.

REFERENCES

1. Gale, G., R. Morris, C. R., Atkins, L. M., Smith, A. B. Binding of an antitumor platinum compound to cells as influenced by physical factors and pharmacologically active agents. *Cancer Res.* 33: 813-817 (1973).

2. Andrews, P.A., Howell, S.B. Cellular pharmacology of cisplatin: perspectives on mechanisms of acquired resistance. *Cancer Cells* 2: 35-42 (1990).

3. Byfield, J.E., Calabro-Jones, P.M. Further evidence for carrier mediated uptake of cis-dichlorodiammine platinum. *Proc. Am. Assoc. Cancer Res.* 23: 167 (1982).

4. Andrews, P.A., Velury, S., Mann, S. C., Howell, S.B. Cis-diamminodichloroplatinum-(II) accumulation in sensitive and resistant human ovarian cells. *Cancer Res.* 48: 68-73 (1988).

5. Atema, A., Buurman, K.J.H., Noteboom, E., Smets, L.A. Potentiation of DNA-adduct formation and cytotoxicity of platinum-containing drugs by low pH. *Int. J. Cancer* 54: 1-7, 1993.

6. Dornish, J.M., Peterson, E.O. Modulation of cisdichlorodiammineplatinum by benzaldehyde derivatives. *Cancer Lett.* 46: 63-68 (1989).

7. Basu, A., Lazo, J.S. Sensitization of human cervical carcinoma cells to cis-diamminedi-chloroplatinum(II) by bryostatin. *Cancer Res.* 52: 3119-3124 (1992).

8. Christen, R.D., Hom, D.K., Porter, D.C., Andrews, P.A., MacLeod, C.L., Halfstrom, L., Howell, S.B. Epidermal growth factor regulates the in vitro sensitivity of human ovarian carcinoma cells to cisplatin. *J. Clin. Invest.* 86: 1632-1640 (1990).

9. Kikuchi, Y., Iwano, I., Miyauchi, M., Sasa, H., Nagata, I., Kuki, E. Restorative effects of calmodulin antagonists on reduced cisplatin uptake by cisplatin-resistant human ovarian carcinoma cells. *Gynecol. Oncol.* 39: 199-203 (1990).

10. Gately, D.P. and Howell, S.B. Cellular accumulation of the anticancer agent cisplatin: A review. *Br. J. Cancer.* 67: 1171-1176 (1993).

11. Los, G., Vugt van, M., Vlist van der M., den Engelse, L., Pinedo, H.M. The effect of heat on the interaction of cisplatin and carboplatin with cellular DNA. *Biochem. Pharmacol.* 46: 1229-1237 (1993).

12. Dornish, J. M., Melvik, J. E., Pettersen, E. O. Reduced cellular uptake of cis-diamminedichloroplatinum(II) by benzaldehyde. *Anticancer Res.* 6: 583-588 (1986).

13. Dornish, J. M., Pettersen, E. O., Oftebro, R. Modifying effect of cinnamaldehyde and cinnamaldehydes derivates on cell inactivation and cellular uptake of cis-diamminedichloroplatinum(II) in human NHIK 3025 cells. *Cancer Res.* 49: 3917-3921 (1989).

14. Kraker, A.J. & Moore, C.W. Accumulation of cis-diamminedichloroplatinum(II) and platinum analogs by platinum resistant murine leukemia cells in vitro. *Cancer Res.* 48: 9-13 (1988).

15. Kelland, L.R., Mistry, P., Abel, G., Loh, S.Y., O'Neill, C.F., Murrer, B.A., Harrap, K.R. Mechanism-related circumvention of acquired cis-diamminedichloroplatinum(II) resistance using two pairs of human ovarian carcinoma cell lines by ammine/amine platinum(IV) dicarboxylates. *Cancer Res.* 53: 3857-3864 (1992).

16. Los, G., Mutsaers, P.H.A., Ruevekamp, M., McVie, J.G. The use of oxaliplatin versus cisplatin in intraperitoneal chemotherapy in cancers restricted to the peritoneal cavity. *Cancer Letters* 51: 109-117 (1990).

17. Sharp, S. Y., Mistry, P., Valenti, M.R., Bryant, A. P., Kelland, L.R. Selective potentiation of platinum drug cytotoxicity in cisplatin-sensitive and -resistant ovarian carcinoma cell lines by amphotericin B. *Cancer Chemother. Pharmacol.* 35: 137-143 (1994).

18. Ortiz, D.F., Kreppel, L., Speiser, D.M. Scheel, G., McDonald, G., Ow, D.W. Heavy metal tolerance in the fission yeast requires an ATP-binding cassette-type vacuolar membrane transporter. *EMBO J.* 11: 3491-3499 (1992).

19. Nies, D.H. CzcR and CzcD, gene products affecting regulation of resistance to cobalt, zinc and cadmium (crc system) in *Alcaligenes eutrophus*. *J. Bacteriol.* 174: 8102-8110 (1992).

20. Conklin, D.S., McMaster, J.A., Culberston, M.R., Kung, C. COT1, a gene involved in cobalt accumulation in Saccharomyces. *Mol. Cell. Biol.* 12: 3678-3688 (1992).

21. Kamizono, A., Nishizawa, M., Teranishi, Y., Murata, K., Kimura, A. Identification of a gene conferring resistance to zinc and cadmium ions in the yeast Saccharomyces cerevesiae. *Mol. Cell Genet.* 219: 161-167 (1989).

22. Welch, J., Fogel, S., Buchman, C., Karin, M. The CUP2 gene product regulates the expression of CUP1 gene, coding for yeast metallothionein. *EMBO J.* 8: 255-260 (1989).

23. Richon, V.M., Schulte, N., Eastman, A. Multiple mechanisms of resistance to cis-diamminedichloroplatinum(II) in murine leukemia L1210 cells. *Cancer Res.* 47: 2056-2061 (1987).

24. Stearns, R. C., Katler, M., Godleski, J. J. Contribution of osmium tetroxide to the image quality and detectability of iron in cells studied by electron spectroscopic imaging and electron energy loss spectroscopy. *Micros. Res. Tech.* 28: 155-163 (1994).

25. Xie, X., Yokel, R. A., Markesbery, W. R. Application of electron energy loss spectroscopy and electron spectroscopic imaging to aluminum determination in biological tissue. *Biol. Trace Element Res.* 40: 39-48 (1994).

26. Andrews, P.A., Murphy, M.P., Howell, S.B. Metallothionein-mediated cisplatin resistance in human ovarian carcinoma cells. *Cancer Chemother. Pharmacol.* 19: 149-154 (1987).

27. Naredi, P., Heath, D. D., Enns, R. E., Howell, S.B. Cross-resistance between cisplatin and antimony in human ovarian carcinoma cell line. *Cancer Res.* 54: 6464-6468 (1994).

28. Naredi, P., Heath, D. D., Enns, R. E., Howell, S.B. Cross-resistance between cisplatin, antimonite and arsenite in human tumor cells. *J. Clin. Inv.* 95: 1193-1198 (1995).

29. Kaur, P. & Rosen, B.P. Metallo-regulated expression of the ars operon. *J. Biol. Chem.* 268: 52-58 (1993).

30. Koonin, E.V. A superfamily of ATPase with diverse functions containing either classical or deviant ATP binding motif. *J. Mol. Biol.* 229: 1165-1174 (1993).

31. Callahan, H.L. and S.M. Beverly. Heavy metal resistance: a new role for P-glycoproteins in Leishmania. *J. Biol. Chem.* 266: 18427-18430 (1992).

32. Papadppoulou, B., Dey, S., Roy, G., Grondi, K., Dou, D., Rosen, B.P. and Ouellette, M. Oxyanion resistance and P-glucoprotein gene amplification in Leishmania. *Gen Motors Cancer Res. Found. Meeting*, Toronto, Abstract (1993).

33. Neal, R. A., van Buren, J., McCoy, N.G. and Iwobi, M. Reversal of drug resistance in Trypansoma cruzi and Leishmania donovani by verapamil. *Transactions of the royal Society of Tropical Medicine and Hygiene* 83: 197-198 (1989).

34. Rosen, B.P., Dey, S., Dou, D., Ji, G., Kaur, P., Ksenzenko, M, Yu., Silver, S., Wu, J. Evolution of an ion translocating ATPase. *Annual. N Y Acad. Sci.* 671: 257-272 (1992).

35. Ishikawa, T. & Ali-Osman, F. Glutathione-associated cis-diamminedichloroplatinum(II) metabolism and ATP-dependant efflux from leukemia cells. *J. Biol. Chem.* 268: 20116 - 20125 (1993).

36. Ishikawa, T., Wright, C.D., Ishizuka, H. *GS-X* pump is functionally overexpressed in cis-diamminedichloroplatinum(II)-resistant human leukemia HL-60 cells and down regulated by cell differentiation. *J. Biol. Chem.* 46: 29085-29093 (1994).

37. Ishikawa, T. & Wright, C.D. *GS-X* pump and γ-glutamylcysteine synthetase are co-overexpressed in cisplatin-resistant human leukemia HL-60 cells. *Proc. Am. Assoc. Cancer Res.* 36, abstract 1863 (1995).

38. Müller, M., Meijer, C., Zaman, G. J. R., Borst, P., Scheper, R. J., Mulder, N. H., de Vries, E. G. E., Jansen, P. L.M. Overexpression of the gene encoding the multidrug resistance-associated protein results in increased ATP-dependent glutathione S-conjugate transport. *Proc. Natl. Acad. Sci. USA.* 91: 13033-13037 (1994).

39. Cole, S.P., Bhardwaj, G., Gerlach, J.H., Mackie, J.E., Grant, C.E., Almquist, K.C., Stewart, A.J., Kurz, E.U., Duncun, A.M.V. and Deeley, R.G. Overexpression of a transporter gene in multidrug resistant human lung cancer cell line. *Science* 258, 1650 - 1654 (1992).

40. Zaman, G.J.R., Flens, M.J., van Leusden, M.R., de Haas, M., Mulder, H.S., Lankema, J., Pinedo, H.M., Scheper, R. J., Baas, F., Broxterman, H.J. and Borst, P. The human multidrug resistance-associated protein MRP is a plasma drug efflux pump. *Proc. Natl. Acad. Sci. USA.* 91: 8822-8826 (1994).

41. Flens, M.J., Izquierdo, M.A., Scheffer, G.L., Fritz, J.M., Meijer, C. J. L. M., Scheper, R. J. and Zaman, G. J. R. Immunochemical detection of multidrug resistance-associated protein MRP in human multidrug resistance tumor cells by monoclonal antibodies. *Cancer Res.* 54: 4557-4563 (1994).

42. Scheper, R. J., Broxterman, H. J., Scheffer, G. L., Kaaijk, P., Dalton, W. S., van Heijningen, T. H. M., van Kalken, C. K., Slovak, M. L., de Vries, E. G. E., van der Valk, P., Meijer, C. J. L. M., Pinedo, H. M. Overexpression of M_r 110,000 Vesicular Protein in non P-glycoprotein-mediated multidrug resistance. *Cancer Res.* 53: 1475-1479 (1993).

43. Scheffer, G. L., Wijngaard, P. L. J., Flens, M. J., Izquierdo, M, A., Slovak, M. L., Pinedo, H. M., Meijer, C. J. L. M., Clevers, H. C., Scheper, R. J. The drug resistance related protein LRP is a major vault protein. *Proc. Am. Assoc. Cancer Res.* 36, Abstract 1921 (1995).

44. Izquierdo, M. A., Schoemaker, R. H., Flens, M. J., Scheffer, G. L., Wu, L., Prater, T. L., Scheper, R. J. Overlapping phenotype of multidrug resistance among disease-oriented panels of human cancer cell lines. *Proc. Am. Assoc. Cancer Res.* 36, Abstract 1923 (1995).

45. Scheper, R. J., Scheffer, G. L., Flens, M., Izquierdo, M. A., van der Valk, P., Broxterman, h. J., Pinedo, H. M., Meijer, C. J. L. M., Clevers, H. C. Molecular and clinical characterization of the LRP protein associated with non-P-glycoprotein multidrug resistance. *Proc. Am. Assoc. Cancer Res.* 35, Abstract 2050 (1994).

46. Shinooya, S., Lu, Y., Scanlon, K. J. Properties of amino acid transport systems in K652 cells sensitive resistant to cis-diamminedichloroplatinum(II). *Cancer Res.* 46: 3445-3448 (1986).

47. Waud, W.R. Differential uptake of cis-diamminedichloroplatinum(II) by sensitive and resistant murine L1210 leukemia cells. *Cancer Res.* 47: 6549-6555 (1987).

48. Los, G., Muggia F.M. Platinum resistance; Experimental and clinical status. *Hematology/Oncology Clinics of North America,* 8: 411-429 (1994).

49. Oldenburg, J., Begg, A. C., van Vugt, M. J. H., Ruevekamp, M., Schornagel, J. H., Pinedo, H. M., Los, G. Characterization of resistance mechanisms to cis-diammine-dichloroplatinum(II) in three sublines of the CC531 colon carcinoma cell line *in vitro*. *Cancer Res.* 54: 487-493 (1994).

DETECTION OF ADDUCTS FORMED UPON TREATMENT OF DNA WITH CISPLATIN AND CARBOPLATIN

Anne Marie J. Fichtinger-Schepman, Marij J.P. Welters, Helma C.M. van Dijk-Knijnenburg, Marianne L.T. van der Sterre[1], Michael J. Tilby[2], Frits Berends, and Robert A. Baan.

TNO Nutrition and Food Research Institute
P.O. Box 5815
2280 HV Rijswijk
[1]Free University Hospital
Amsterdam, The Netherlands
[2]Medical Molecular Biology Group
University of Newcastle Upon Tyne, UK.

INTRODUCTION

Since the discovery of the antitumor activity of *cis*-diamminedichloroplatinum(II) (cisplatin), it is generally accepted that the cytotoxic action of platinum drugs is the consequence of their interaction with cellular DNA. Therefore, knowledge about the formation and persistence of DNA damage may help to understand the working mechanism of these compounds and may eventually lead to the development of better Pt-containing antitumor drugs.

To determine platination levels in fluids or tissues, various techniques are available such as atomic absorption spectroscopy (AAS)[1], proton-induced X-ray emission (PIXE)[2], neutron activation analysis[3] , differential pulse polarography[4], electrochemical detection[5], adsorptive voltammetry[6], inductively coupled plasma spectroscopy[7] or by counting the DNA-bound radioactivity after treatment with radiolabelled Pt-compounds. In case of the very low platination levels usually found in *in vivo* experiments, the use of anti-Pt-DNA antibodies has offered a good alternative[8-13]. Antibodies can also be used in a cytochemical assay to detect Pt-DNA damage at the level of the single cell[10]. In such an experiment, we have determined cisplatin-DNA adduct levels in human head and neck tumor (HNSCC) cells with the monoclonal antibody ICR4, raised against platinated DNA[13]. In addition we used a second antibody (MNF) which allowed us to distinguish the tumor cells from other cell types such as white blood cells and fibroblasts (Fig. 1).

However, also with this immunochemical approach it can not be established which

Platinum and Other Metal Coordination Compounds in Cancer Chemotherapy 2
Edited by H.M. Pinedo and J.H. Schornagel, Plenum Press, New York, 1996

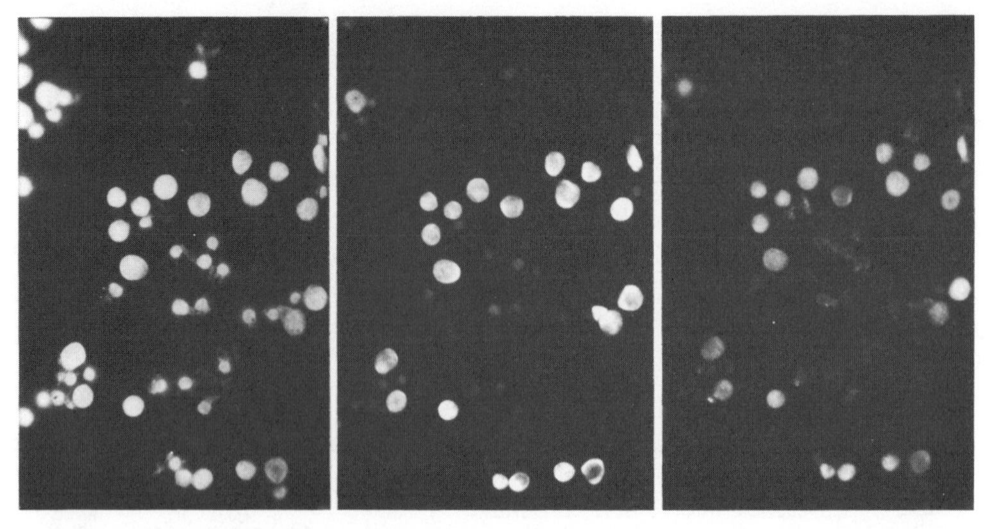

Figure 1. Immunocytochemical detection of cisplatin-DNA adducts in a mixture of cisplatin-treated HNSCC and untreated white blood cells. The pictures show: the nuclear stain in all cells with the DNA-binding dye DAPI (left); the staining of the tumor cells with antibody MNF, which recognizes squamous cell specific keratins, visualized with an FITC-labelled second antibody (middle); and the immunostaining of cisplatin-DNA damage with anti-cisplatin-DNA antibody ICR4, visualized with a TRITC-labelled second, goat anti-rat, antibody (right). Recently, an increased signal could be obtained by use of a third TRITC-labelled antibody (donkey anti-goat).

of the various adducts that can be formed by cisplatin, are present in the DNA sample investigated. In addition to monoadducts, cisplatin forms bifunctional adducts among which the intrastrand crosslinks Pt-GG, Pt-AG and G-Pt-G are the major, and the interstrand crosslinks (G-Pt-G) and DNA-protein crosslinks are the minor ones[14] (see Fig. 2).

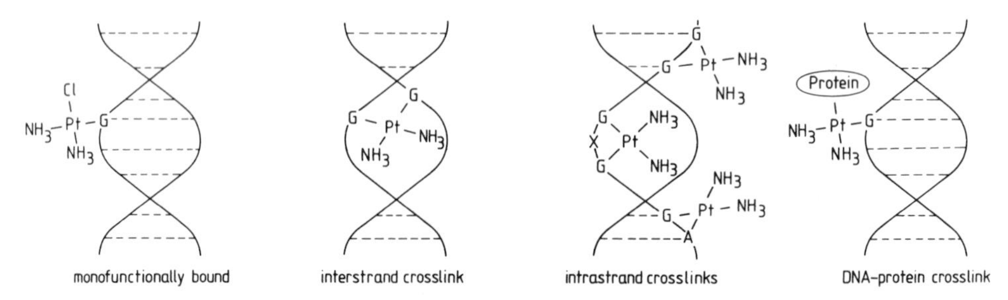

Figure 2. Cisplatin-DNA adducts.

In many studies on Pt-DNA adduct formation only the interstrand crosslinks have been determined, with either the alkaline elution technique[15], an ethidium bromide binding assay[16] or with gel electrophoresis[17]. In our laboratory a method has been developed to detect each of the different types of Pt-DNA adducts. In this assay, platinated DNA is digested to unmodified mononucleotides and Pt-containing products; after separation by column chromatography, the latter are quantified with AAS or with specific antibodies in a competitive ELISA[8,9,14]. A similar system in which a radiolabel is used for the detection of the various DNA-adducts formed by ^{3}H-*cis*-dichloro (ethylenediamine)platinum(II) has been developed by Eastman[18]. Recently, a less laborious and less time-consuming ^{32}P-postlabelling assay has become available for the sensitive detection of the major adducts Pt-GG and Pt-AG[19]. Forsti and Hemminki[20], too, published a ^{32}P-postlabelling assay which, however, is less sophisticated, whereas Sharma et al.[21] have published an approach to detect Pt-DNA adducts after derivatization to fluorescent products. Other methods have been developed to locate Pt-adducts in DNA stretches by taking advantage of the fact that enzymes such as polymerases are unable to bypass Pt adducts[22-24].

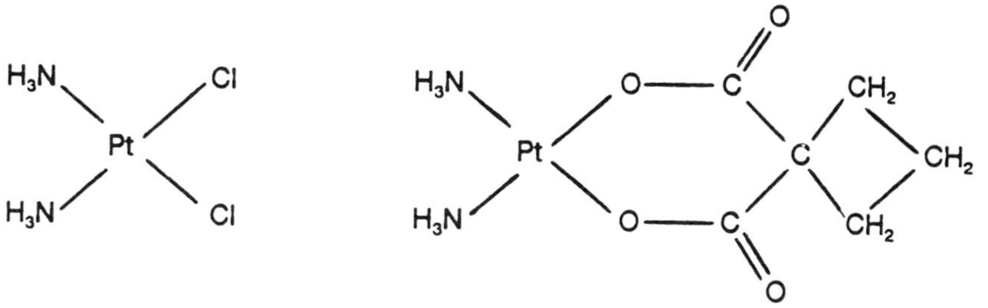

Figure 3. Structure of cisplatin (left) and carboplatin (right).

In general, it can be concluded that adequate tools for the detection of Pt-DNA adducts are available now. In this paper the results are summarized of our studies on DNA-adduct formation by cisplatin and its analog *cis*-(diammine)(1,1-cyclobutanedicarboxylato)-platinum(II) (carboplatin; see Fig.3) *in vitro* and in Chinese hamster ovary (CHO) cells.

INACTIVATION OF MONOFUNCTIONALLY BOUND CISPLATIN-DNA ADDUCTS

After interaction of cisplatin with DNA the majority of adducts found consists of the intrastrand crosslinks Pt-GG, Pt-AG and G-Pt-G. These are formed in two steps: first cisplatin forms monoadducts (Pt-G), which then are slowly converted into bifunctional adducts. Therefore, to accurately determine the presence of the various adducts at a specific time point, it is necessary to prevent the conversion of mono- to bifunctional adducts during DNA isolation or during the analysis procedure. In the past we have tried to inactivate the still reactive monofunctional adducts with 0.1 M NH_4HCO_3 during a 16-h incubation at 37°C. However, in contrast to the positive results obtained with this method in the inactivation of cisplatin monofunctionally bound to GMP[25,26], this approach is not suitable for inactivation of Pt-G adducts present in high-molecular weight DNA, as was demonstrated recently[27,28]. However, a 1-h incubation with 10 mM thiourea at 37°C appears to be adequate[28]. These conditions, which are clearly more vigorous than the 10-mM thiourea incubation for 10 min at 23°C proposed by Eastman[18], were established in studies on freshly *in vitro* cisplatin-treated salmon sperm DNA, which was postincubated with thiourea at 37°C. At different time points, the resulting levels of mono- and bifunctional adducts were compared with those found after a 10-min incubation at 23°C. A biphasic decrease of bifunctional adducts was seen (Fig. 4A): a rapid initial decrease followed by a much slower reduction. Comparable results were obtained for the DNA adducts in CHO cells when the cisplatin treatment was immediately followed by a thiourea postincubation (Fig. 4B). The slower process was also seen in cisplatin-treated DNA in which the monoadducts had already been converted into bifunctional adducts ('aged' DNA). It was concluded that the rapid process represents the inactivation of cisplatin monoadducts, thus preventing the formation of bifunctional adducts, and that the slow decrease of bifunctional adducts is caused by conversion of already formed bifunctional adducts to monoadducts. This conclusion is sustained by the finding that the amounts of (inactivated) monoadduct (Pt-G) increase during the thiourea incubation[28] (Fig. 4A).

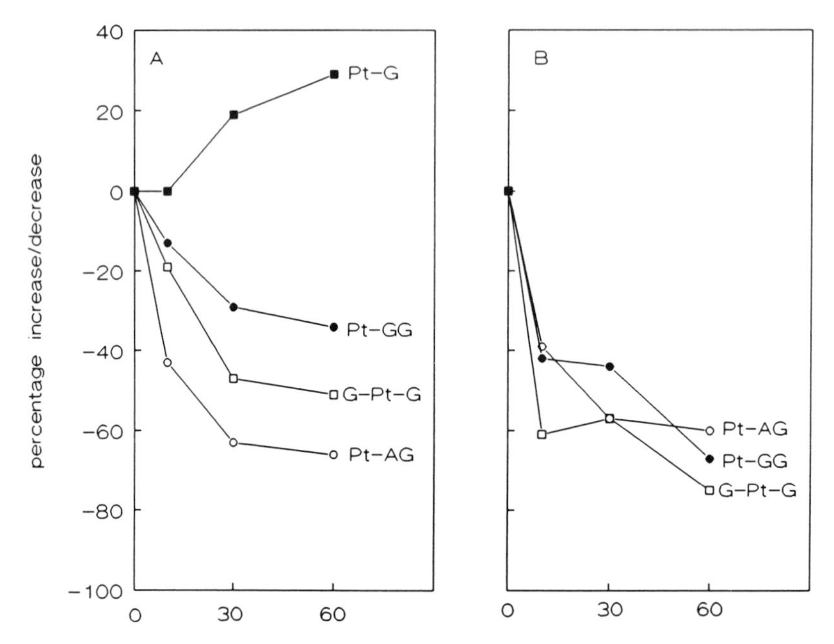

Figure 4. Effects of postincubation of platinated DNA or cells with 10 mM thiourea (TU) at 37°C, immediately after the cisplatin treatment, on the level of the various Pt-adducts[28]. Panel A, *in vitro* platinated DNA (167 μM, 15 min at 37°C); adduct levels were compared with those after a treatment with 10 mM TU for 10 min at 23°C. Panel B, cisplatin (40 μM, 1 h at 37°C)-treated CHO cells; adduct levels were compared with those of cells not postincubated (t=0).

KINETICS OF CISPLATIN-DNA ADDUCT FORMATION

Next the kinetics of the bifunctional adduct formation in *in vitro* cisplatin-treated DNA were investigated[29]. DNA was incubated with cisplatin for 15 min and postincubated in phosphate buffer at 37°C. At intervals samples were taken and further incubated for 1 h after addition of thiourea to a concentration of 10 mM. As shown in Fig. 5, analysis of the samples revealed that under the conditions used it takes 4-6 h before maximum levels of bifunctional adducts and minimum monoadduct levels are reached. The data also indicate a half-life of the monoadduct of about 2 h, which is in agreement with the rate of hydrolysis of monoaquated cisplatin and with published ^{195}Pt-NMR data on the conversion of mono- to bifunctional adducts in ^{195}Pt-cisplatin-treated DNA[30].

In Fig. 6A the cisplatin-DNA adduct formation in CHO cells is shown[29]. The cells were treated for 1 h with 40 μM cisplatin and subsequently postincubated in drug-free medium. After sampling the cells were cultured for 1 h in fresh medium with 10 mM thiourea. The data show that in these cells adduct formation follows similar kinetics as were found *in vitro*, with maximum bifunctional adduct levels present after about 4-6 h of postincubation.

The finding that the formation of the bifunctional adducts, which are supposed to be the cytotoxic lesions, was ongoing during about 4-6 h of postincubation is in excellent

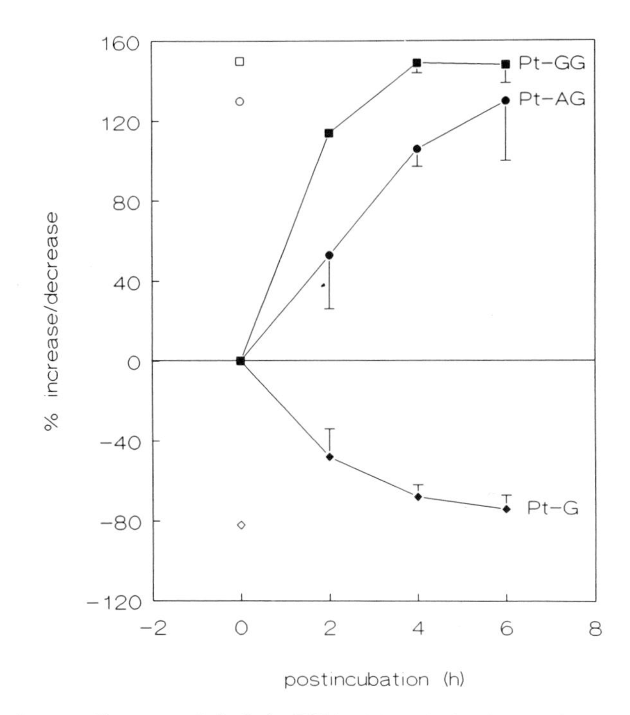

Figure 5. Relative increase/decrease of cisplatin-DNA adducts in *in vitro* platinated salmon sperm DNA during postincubation in phosphate buffer (10 mM, pH 7.2). At the times indicated, samples were incubated for an additional hour with 10 mM thiourea (closed symbols). Additionally, in one experiment a DNA sample was taken immediately after platination and dialysed overnight at 37°C against 0.1 M NH_4HCO_3 (open symbols)[29], showing the failure of this method to prevent mono-adducts from diadduct formation (see appendix).

agreement with the effect of a 1-h 10 mM thiourea postincubation on the survival of these cells[29]. As can be seen from Fig.7A, this effect was strongly dependent on the time between the cisplatin treatment and the start of the incubation with thiourea. When thiourea was applied immediately after the cisplatin treatment, the cytotoxicity of cisplatin was almost completely abolished. When applied after 2 h, the effect was less pronounced and after 4 h only a very small effect was seen, which is consistent with the finding that after 4 h most of the bifunctional adducts had already been formed. Just as in *in vitro* cisplatin-treated salmon sperm DNA, Pt-GG was the major diadduct formed in these CHO cells.

In addition to the intrastrand diadducts, cisplatin also forms a small proportion of interstrand crosslinks. The latter can be assayed separately, with the so-called alkaline elution method. As Fig. 8 shows, their formation proceeds somewhat more slowly, with the maximum level reached after about 7 h postincubation of the 1-h cisplatin-treated CHO cells. At this time point the number of interstrand crosslinks amounted to about 2% of the total DNA adducts[29].

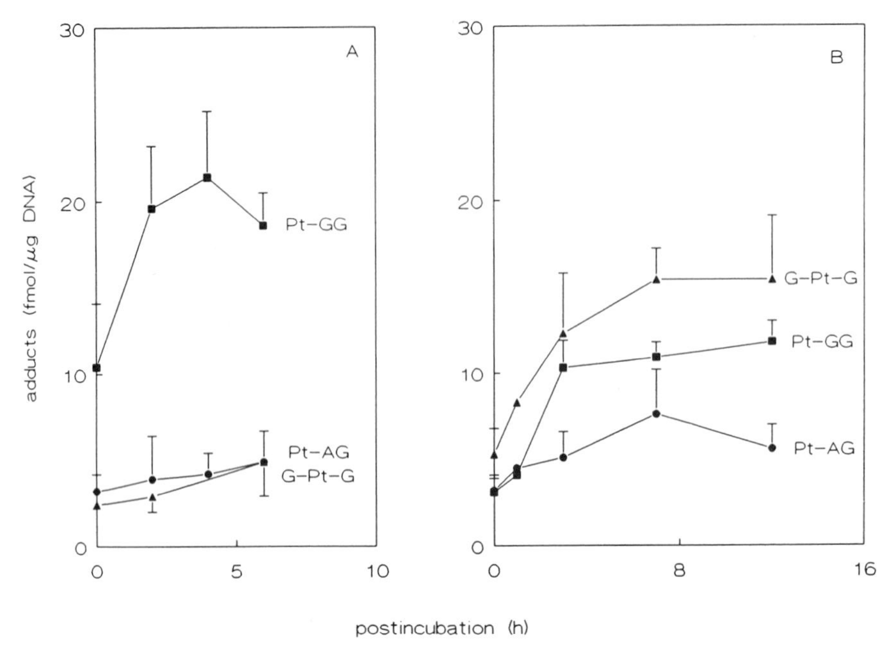

Figure 6. Kinetics of the formation of platinum-DNA adducts in CHO cells during postincubation in drug-free medium after 1-h treatments with equitoxic doses of cisplatin (40 μM; panel A) or carboplatin (0.7 mM; panel B). The cisplatin-monoadducts were inactivated with 10 mM thiourea[29].

KINETICS OF CARBOPLATIN-DNA ADDUCT FORMATION

Because the two Pt-compounds cisplatin and carboplatin differ only in the character of their leaving groups, chloride and cyclobutanedicarboxylate, resp., identical bifunctional adducts will result from their reaction with DNA. To inactivate the mono-

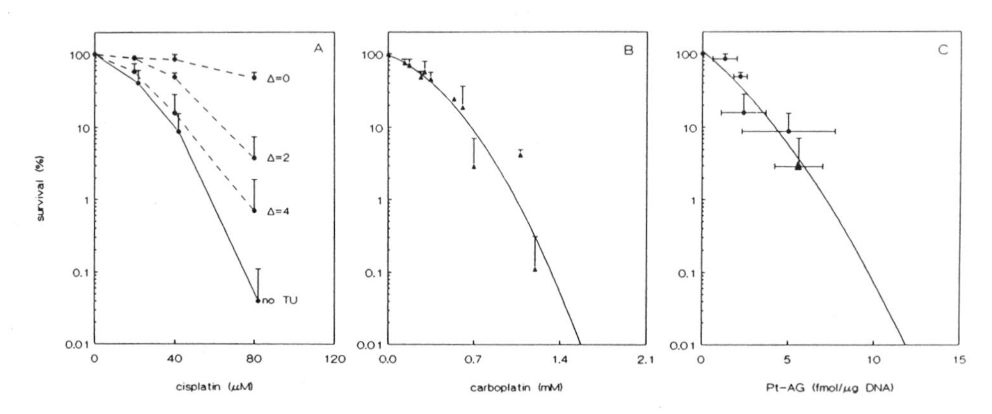

Figure 7. Survival of CHO cells after 1-h treatment with various concentrations of cisplatin (panel A) or carboplatin (panel B) as measured with a clonogenic assay. Part of the cisplatin-treated cells were additionally incubated for 1 h with 10 mM thiourea either immediately (Δ=0), 2 h (Δ=2) or 4 h (Δ=4) after the cisplatin treatment. In panel C survival data of these cells are correlated with the amounts of the Pt-AG adducts present in their DNA at 12 h after the cisplatin or carboplatin treatments[29] (see text).

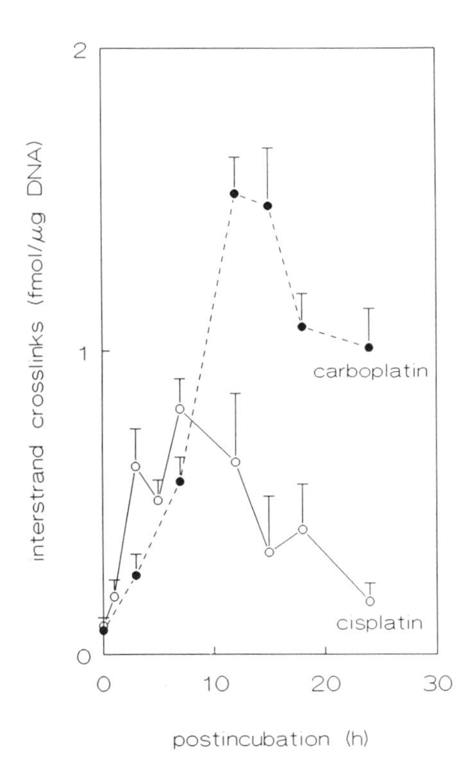

Figure 8. Kinetics of the formation of interstrand crosslinks in CHO cells after 1 h treatments with equitoxic doses cisplatin (40 μM) or carboplatin (0.7 mM). The crosslinks were determined with the alkaline elution method[29].

adducts formed during platination of DNA with carboplatin, however, the 10-mM thiourea treatment for 1 h is inadequate, as we concluded from the fact that survival of carboplatin-treated cells was not improved by such a treatment (not shown). Only much higher thiourea concentrations could prevent cell death, but these conditions also caused significant conversion of bifunctional adduct to monoadduct. Therefore, although possibly not optimal, we inactivated the monoadducts during an overnight dialysis against 0.5 M NH_4HCO_3 at 37°C after removal of the excess of carboplatin already at a very early stage of the DNA isolation procedure[31].

As can be seen from Fig. 9A, after *in vitro* carboplatin treatment of isolated DNA the same spectrum of adducts, with Pt-GG as the major adduct, was found as after cisplatin treatment. It also shows that detection of the adducts with AAS and determination with the specific antibodies in the competitive ELISA yielded similar results. However, to obtain equal platination levels after a 4-h incubation period, about 230-fold more carboplatin than cisplatin was needed[31]. In Fig. 9B the adduct spectrum is given for a DNA sample isolated from carboplatin-treated CHO cells. The strong preference for formation of bifunctional adducts on pGpG sequences as found for cisplatin in treated cells[32-38] (Fig. 6A) was not observed for carboplatin. The adduct spectrum is clearly different from that after *in vitro* DNA platination with carboplatin (Fig. 9A).

To further analyse the formation of bifunctional carboplatin adducts, CHO cells were treated with a dose of carboplatin equitoxic to that used in the cisplatin study. The survival of the CHO cells after 1-h incubations with cisplatin and carboplatin are shown in Fig.7A and B, resp. These data indicate that a dose of 0.7 mM carboplatin led to about the same cell killing (90%) as caused by a 40 μM cisplatin dose, *i.e.* about 17 times more carboplatin than cisplatin was needed.

The kinetics of the formation of the bifunctional adducts during postincubation of CHO cells treated with 0.7 mM carboplatin, as measured with the antibodies in the competitive ELISA[31], are shown in Fig.6B. The formation appeared slower than after cisplatin treatment (Fig. 6A): the maximum adduct levels were reached after about 7 h. However, the major difference between these two adduct spectra was the finding that after carboplatin treatment not the Pt-GG, but the G-Pt-G adducts were the major products formed. Sofar, in all cell types studied after treatment with carboplatin, we found that the Pt-GG adducts were not preferentially formed.

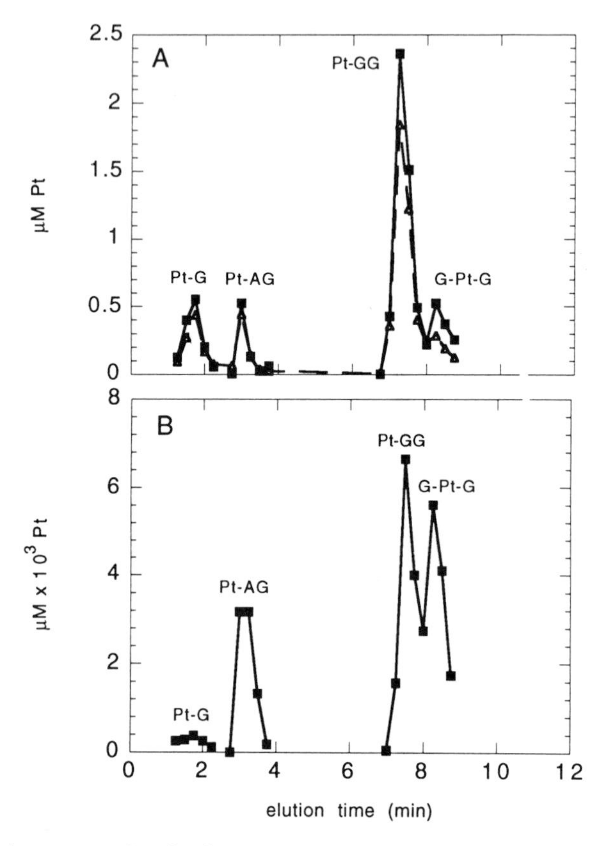

Figure 9. Column chromatography of a digested sample of salmon sperm DNA that had been reacted *in vitro* with 1.35 mM carboplatin for 4 h at 37°C (panel A). The Pt-concentrations in the collected column fractions were determined with AAS (open symbols) as well as with the ELISA (closed symbols). In panel B ELISA data are given for a DNA sample isolated from CHO cells that had been postincubated for 18 h after a 1-h carboplatin treatment[31].

The percentage of interstrand crosslinks formed by carboplatin in the CHO cells was higher than that found after the equicytotoxic cisplatin treatment: at their maximum levels they amounted to about 3-4% of total DNA platination[31]. As found for cisplatin, the formation of the carboplatin-induced interstrand crosslinks was slower than that of the intrastrand crosslinks; it reached its maximum after about 12 h postincubation (Fig. 8).

COMPARISON OF CISPLATIN AND CARBOPLATIN ADDUCTS: WHICH IS THE CYTOTOXIC LESION?

Because exactly the same bifunctional adducts, the supposedly cytotoxic lesions, are formed by cisplatin and carboplatin but in other proportions, it is tempting to speculate about the nature of the key lesion for cytotoxicity. As can be concluded from the data shown in Fig. 8, no direct correlation can be established between the number of interstrand crosslinks formed by these compounds and cytotoxicity, because at equal toxicity the level of this type of adducts was about twice as high after the carboplatin treatment. When the levels of the other bifunctional adducts after treatment with equitoxic doses of the drugs are compared (Fig. 6A and B) it is clear that the same holds for the Pt-GG and G-Pt-G adducts. The only lesion that is formed in comparable amounts is the Pt-AG adduct. To visualize the correlation between cytotoxicity and the Pt-AG adduct levels, survival data of the CHO cells were plotted against the Pt-AG levels present in cellular DNA samples at 12 h after the drug treatment[29] (see Fig. 7C). Two data points were obtained from treatments with the equitoxic doses of cisplatin (40 μM) and carboplatin (0.7 mM) leading to about 10% survival (see Fig.7A and B), the other data are from cells treated with 40 μM cisplatin which subsequently were subjected to a thiourea posttreatment, either immediately or at 2 or 4 h after the cisplatin treatment[29] (Fig. 7A). The relation between cell survival and Pt-AG adduct levels obtained in this way appeared to be in fair agreement with the survival data plotted against the concentrations of cisplatin or carboplatin (Fig. 7A and B), which would not be the case for the other two intrastrand crosslinks.
It should be realized that these findings do not prove that Pt-AG is the cytotoxic lesion. We only determined the adduct levels in total genomic DNA and it may be that the presence of a specific adduct at a specific time at a specific site in the chromosome is causing cell death. Also, the combined action of all bifunctional adducts may be responsible for the cytotoxic effect. However, in that case the contribution of each type of adduct to cell death has to be approximately the same because similar amounts of total DNA adducts were found in the CHO cells at 12 h after treatment with equitoxic doses of cisplatin and carboplatin[29]. *A priori*, this possibility cannot be excluded but it seems unlikely in view of the already known biologically relevant differences between the various adducts[39-45].

APPENDIX

The cisplatin-DNA adduct levels shown in Fig. 6A were obtained after inactivation of the monoadducts with thiourea. In the past we used 0.1 M NH_4HCO_3, which has been shown recently not to be suitable for inactivation of monoadducts in high-molecular weight DNA[27,28]. The formation of the bifunctional DNA adducts in CHO cells determined after the use of NH_4HCO_3 as inactivator of cisplatin-monoadducts is shown in Fig. 10. Compared with the thiourea method, the maximum adduct levels appeared to be reached much faster, *i.e.* almost immediately after the cisplatin treatment; initially, the total amount was also higher than the highest level seen after monoadduct inactivation by thiourea (Fig. 6A). Furthermore, a decrease of the adduct levels was observed with the NH_4HCO_3 method. After about 4 h postincubation, no significant differences between the results obtained with both methods could be observed. The latter can be explained by the fact that after about 4 h the majority of monoadducts had been already converted to bifunctional adducts. Hence no significant differences between the two methods were to be expected at this time. The high levels of bifunctional adducts immediately after the cisplatin treatment as seen with the

NH_4HCO_3 method must be artifacts in the sense that they do not represent the levels present in the cells at that time point. It is likely that monoadducts were still converted to bifunctional adducts during the NH_4HCO_3 incubation. An additional contribution that made these levels so high probably came from the absence of active cellular repair enzymes during DNA isolation. In other words, these high levels would have been found also in the cells after 4-6 h of postincubation provided no repair of adducts had occurred. As can be seen from Fig.5, this is exactly what happened when the formation of the adduct levels in *in vitro* cisplatin-treated DNA was studied with the use of a thiourea posttreatment (closed symbols) and when the maximum levels obtained after 6 h postincubation were compared with those of a sample taken immediately after the cisplatin treatment and dialysed overnight at 37°C against 0.1 M NH_4HCO_3 (open symbols). From these findings we conclude that the decrease in adduct levels during postincubation of cisplatin-treated cells as found with the NH_4HCO_3 method can be considered as a measure of the (excision) repair of the adducts during the first hours after the cisplatin treatment. This explanation is supported by the observation that with the NH_4HCO_3 method no decrease in bifunctional adduct levels during the first 4.5 h of postincubation was found with xeroderma pigmentosum cells, which are known to be deficient in excision repair[32].

In conclusion: the fact that the high bifunctional DNA-adduct levels observed with the NH_4HCO_3 method in cisplatin-treated CHO cells were not seen with the thiourea method, has to be explained by the assumption that with the latter method the net result of adduct formation and repair is determined.

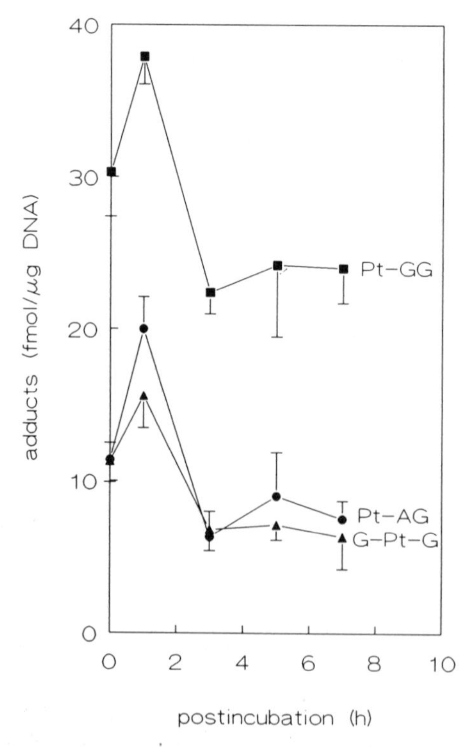

Figure 10. Kinetics of the formation of cisplatin-DNA adducts in CHO cells during postincubation in drug-free medium after 1-h treatment with 40 μM cisplatin. For monoadduct inactivation, DNA was incubated with 0.1 M NH_4HCO_3, for 16 h at 37°C.

ACKNOWLEDGEMENTS

These studies have been financially supported by the Dutch Cancer Society.

REFERENCES

1. A.M.J. Fichtinger-Schepman, P.H.M. Lohman, and J. Reedijk, Detection and quantification of adducts formed upon interaction of diamminedichloroplatinum(II) with DNA, by anion-exchange chromatography after enzymatic degradation, *Nucleic Acids Res.* 10:5345 (1982).
2. T.G.M.H. Dikhoff, J.A. van der Heide, M. Prins, and J.G. McVie, Determination of platinum in human tissues with PIXE, *IEEE Trans Nucl Sci.* 30:1329 (1983).
3. P.S. Tjioe, K.J. Volkers, J.J. Kroon, J.J.M. de Goeij, and S.K. The, Determination of traces of Au and Pt in biological materials as part of a multi-element radiochemical neutron activation analysis system, *Int. J. Environ. Anal. Chem.* 17:13 (1984).
4. V. Brabec, O. Vrana, V. Kleinwächter, and F. Kiss, Modifications of the DNA structure upon platinum binding, *Studia Biophysica* 101:135 (1984).
5. P.J. Parsons, P.F. Morrison, and A.F. LeRoy, Determination of platinum-containing drugs in human plasma by liquid chromatography with reductive electrochemical detection, *J. Chromatogr.* 385:323 (1987).
6. O. Nygren, G.T. Vaughan, T.M. Florence, G.M.P. Morrison, I.M. Warner, and L.S. Dale, Determination of platinum in blood by adsorptive voltammetry, *Anal. Chem.* 62:1637 (1990).
7. J.G. Morrison, D. Bissett, I.F.D. Stephens, K. McKay, R. Brown, M.A. Graham, A.M. Fichtinger-Schepman, and D.J. Kerr, The isolation and identification of *cis*-diamminedichloroplatinum(II)-DNA adducts by anion exchange HPLC and inductively coupled plasma mass spectrometry, *Int. J. Oncol.* 2:33 (1993).
8. A.M.J. Fichtinger-Schepman, R.A. Baan, A. Luiten-Schuite, M. van Dijk, and P.H.M. Lohman, Immunochemical quantitation of adducts induced in DNA by *cis*-diamminedichloroplatinum(II) and analysis of adduct-related DNA-unwinding, *Chem.-Biol. Interactions* 55:275 (1985).
9. A.M.J. Fichtinger-Schepman, A.T. van Oosterom, P.H.M. Lohman, and F. Berends, *cis*-Diamminedichloroplatinum(II)-induced DNA adducts in peripheral leukocytes from seven cancer patients: quantitative immunochemical detection of the adduct induction and removal after a single dose of *cis*-diamminedichloroplatinum(II), *Cancer Res.* 47:3000 (1987).
10. P.M.A.B. Terheggen, B.G.J. Floot, E. Scherer, A.C. Begg, A.M.J. Fichtinger-Schepman, and L. den Engelse, Immunocytochemical detection of interaction products of *cis*-diamminedichloroplatinum(II) and *cis*-diammine(1,1-cyclobutanedicarboxylato)platinum(II) with DNA in rodent tissue sections, *Cancer Res.* 47:6719 (1987).
11. A.M.J. Fichtinger-Schepman, R.A. Baan, and F. Berends, Influence of the degree of DNA modification on the immunochemical determination of cisplatin-DNA adduct levels, *Carcinogenesis* 10:2367 (1989).
12. E. Reed, S. Gupta-Burt, C.L. Litterst, and M.C. Poirier, Characterization of the DNA damage recognized by an antiserum elicited against *cis*-diamminedichloroplatinum(II)-modified DNA, *Carcinogenesis* 11:2117 (1990).
13. M.J. Tilby, C. Johnson, R.J. Knox, J. Cordell, J.J. Roberts, and C.J. Dean, Sensitive detection of DNA modifications induced by cisplatin and carboplatin *in vitro* and *in vivo* using a monoclonal antibody, *Cancer Res.* 51:123 (1991).
14. A.M.J. Fichtinger-Schepman, L.L. van der Veer, J.H.J. den Hartog, P.H.M. Lohman, and J. Reedijk, Adducts of the antitumor drug *cis*-diammine-dichloroplatinum(II) with DNA: formation, identification, and quantitation. *Biochemistry* 24:707 (1985).
15. K.W. Kohn, L.C. Erickson, R.A.G. Ewig, and C.A. Friedman, Fractionation of DNA from mammalian cells by alkaline elution, *Biochemistry* 15:4629 (1976).
16. R. Sriram, and F. Ali-Osman, S1-Nuclease enhancement of the ethidium bromide binding assay of drug-induced DNA interstrand cross-linking in human brain tumor cells, *Anal Biochem.* 187:345 (1990).
17. J.C. Jones, W. Zhen, E. Reed, R.J. Parker, A. Sancar, and V.A. Bohr, Gene-specific formation and repair of cisplatin intrastrand adducts and interstrand cross-links in Chinese hamster ovary cells, *J. Biol. Chem.* 266:7101 (1991).
18. A. Eastman, Reevaluation of interaction of *cis*-dichloro(ethylenediamine)platinum(II) with DNA, *Biochemistry* 25:3912 (1986).

19. F.A. Blommaert, and C.P. Saris, Detection of platinum-DNA adducts by ^{32}P-postlabelling, *Nucleic Acids Res.* in press (1995).
20. A. Försti, and K. Hemminki, A ^{32}P-postlabelling assay for DNA adducts induced by *cis*-diamminedichloroplatinum(II), *Cancer Letters* 83:129 (1994).
21. M. Sharma, R. Jain, and T.V. Isac, A novel technique to assay adducts of DNA induced by anticancer agent *cis*-diamminedichloroplatinum(II), *Bioconjugate Chem.* 2:403 (1991).
22. A.L. Pinto, and S.J. Lippard, Sequence-dependent termination of *in vitro* DNA synthesis by *cis*- and *trans*-diamminedichloroplatinum(II), *Proc. Natl. Acad. Sci. USA* 82:4616 (1985).
23. G. Villani, U. Hübscher, and J.-L. Butour, Sites of termination of *in vitro* DNA synthesis on *cis*-diamminedichloroplatinum(II) treated single-stranded DNA: a comparison between *E.coli* DNA polymerase I and eucaryotic DNA polymerases α, *Nucleic Acids Res.* 16:4407 (1988).
24. M.M. Jennerwein, and A. Eastman, A polymerase chain reaction-based method to detect cisplatin adducts in specific genes, *Nucleic Acids Res.* 19:6209 (1991).
25. A.M.J. Fichtinger-Schepman, J.L. v.d. Veer, P.H.M. Lohman, and J. Reedijk, A simple method for the inactivation of monofunctionally DNA-bound *cis*-diamminedichloroplatinum(II), *J. Inorg. Biochem.* 21:103 (1984).
26. A.M.J. Fichtinger-Schepman, F.J. Dijt, W.H. de Jong, A.T. van Oosterom, and F. Berends, *In vivo cis*-diamminedichloroplatinum(II)-DNA adduct formation and removal as measured with immunochemical techniques, *in*: "Platinum And Other Metal Coordination Compounds In Cancer Chemotherapy", Martinus Nijhoff Publishing, Boston (1988).
27. G.R. Gibbons, J.D. Page, and S.G. Chaney, Treatment of DNA with ammonium bicarbonate or thiourea can lead to underestimation of platinum-DNA monoadducts. *Cancer Chemother. Pharmacol.* 29:112 (1991).
28. A.M.J. Fichtinger-Schepman, H.C.M. van Dijk-Knijnenburg, F.J. Dijt, S.D. van der Velde-Visser, F. Berends, and R.A. Baan, Effects of thiourea and ammonium bicarbonate on the formation and stability of bifunctional cisplatin-DNA adducts: consequences for the accurate quantification of adducts in (cellular) DNA. *J. Inorg. Biochem.* 58:177 (1995).
29. A.M.J. Fichtinger-Schepman, H.C.M. van Dijk-Knijnenburg, S.D. van der Velde-Visser, F. Berends, and R.A. Baan, Cisplatin- and carboplatin-DNA adducts: is Pt-AG the cytotoxic lesion? *Carcinogenesis* in press (1995).
30. D.P. Bancroft, C.A. Lepre, and S.J. Lippard, S.J. ^{195}Pt NMR Kinetic and mechanistic studies of *cis*- and *trans*-diamminedichloroplatinum(II) binding to DNA. *J. Am. Chem. Soc.* 112:6860 (1990).
31. F.A. Blommaert, H.C.M. van Dijk-Knijnenburg, F.J. Dijt, L. den Engelse, R.A. Baan, F. Berends, and A.M.J. Fichtinger-Schepman, Formation of DNA adducts by the anticancer drug carboplatin: different nucleotide sequence preferences in vitro and in cells, *Biochemistry* 34:8474 (1995).
32. F.J. Dijt, A.M.J. Fichtinger-Schepman, F. Berends, and J. Reedijk, Formation and repair of cisplatin-induced adducts to DNA in cultured normal and repair-deficient human fibroblasts. *Cancer Res.* 48:6058 (1988).
33. W.C.M. Dempke, S.A. Shellard, A.M.J. Fichtinger-Schepman, and B.T. Hill, Lack of significant modulation of the formation and removal of platinum-DNA adducts by aphidicolin glycinate in two logarithmically-growing ovarian tumour cell lines *in vitro*, *Carcinogenesis* 12:525 (1991).
34. B.T. Hill, K.J. Scanlon, J. Hansson, A. Harstrick, M. Pera, A.M.J. Fichtinger-Schepman, and S.A. Shellard, Deficient repair of cisplatin-DNA adducts identified in human testicular teratoma cell lines established from tumours from untreated patients, *Eur. J. Cancer* 30A:832 (1994).
35. J. Hansson, A.M.J. Fichtinger-Schepman, M. Edgren, and U. Ringborg, Comparative study of two human melanoma cell lines with different sensitivities to mustine and cisplatin, *Eur. J. Cancer* 27:1039 (1991).
36. G.A.P. Hospers, N.H. Mulder, B. de Jong, L. de Ley, D.R.A. Uges, A.M.J. Fichtinger-Schepman, R.J. Scheper, and E.G.E. de Vries, Characterization of a human small cell lung carcinoma cell line with acquired resistance to *cis*-diamminedichloroplatinum(II) *in vitro*, *Cancer Res.* 48:6803 (1988).
37. C.M.J. de Pooter, P.G. Scalliet, H.J. Elst, J.J. Huybrechts, E.E.O. Gheuens, A.T. van Oosterom, A.M.J. Fichtinger-Schepman, and E.A. de Bruijn, Resistance patterns between *cis*-diamminedichloroplatinum(II) and ionizing radiation, *Cancer Res.* 51:4523 (1991).
38. A.M.J. Fichtinger-Schepman, C.P.J. Vendrik, W.C.M. van Dijk-Knijnenburg, W.H. de Jong, A.C.E. van der Minnen, A.M.E. Claessen, S.D. van der Velde-Visser, G. de Groot, K.L. Wubs, P.A. Steerenberg, J.H. Schornagel, and F. Berends, Platinum concentrations and DNA adduct levels in tumors and organs of cisplatin-treated LOU/M rats inoculated with

cisplatin-sensitive or -resistant immunoglobulin M immunocytoma, *Cancer Res.* 49:2862 (1989).
39. D. Burnouf, M. Daune, and R.P.P. Fuchs, Spectrum of cisplatin-induced mutations in Escherichia coli. *Proc. Natl. Acad. Sci. USA* 84:3758 (1987).
40. J.-S. Hoffmann, N.P. Johnson, and G. Villani, Conversion of monofunctional DNA adducts of *cis*-diamminedichloroplatinum(II) to bifunctional lesions, *J. Biol. Chem.* 264:15130 (1989).
41. K.M. Comess, J.N. Burstyn, J.M. Essigmann, and S.J. Lippard, Replication inhibition and translesion synthesis on templates containing site-specifically placed *cis*-diamminedichloroplatinum(II) DNA adducts, *Biochemistry* 31:3975 (1992).
42. Y. Corda, M.-F. Anin, M. Leng, and D. Job, RNA Polymerases react differently at d(ApG) and d(GpG) adducts in DNA modified by *cis*-diamminedichloroplatinum(II), *Biochemistry* 31:1904 (1992).
43. R. Visse, A.J. van Gool, G.F. Moolenaar, M. de Ruijter, and P. van de Putte, The actual incision determines the efficiency of repair of cisplatin-damaged DNA by the Escherichia coli UvrABC endonuclease, *Biochemistry* 33:1804 (1994).
44. B.A. Donahue, M. Augot, S.F. Bellon, D.K. Treiber, J.H. Toney, S.J. Lippard, and J.M. Essigmann, Characterization of a DNA damage-recognition protein from mammalian cells that binds specifically to intrastrand d(GpG) and d(ApG) DNA adducts of the anticancer drug cisplatin, *Biochemistry* 29:5872 (1990).
45. J.-C. Huang, D.B. Zamble, J.T. Reardon, S.J. Lippard, and A. Sancar, HMG-domain proteins specifically inhibit the repair of the major DNA adduct of the anticancer drug cisplatin by human excision nuclease, *Proc. Natl. Acad. Sci. USA* 91:10394 (1994).

DETECTION OF PLATINUM LESIONS AT THE NUCLEOTIDE LEVEL IN CELLS USING SINGLE STRAND LIGATION PCR

John A. Hartley, Robert L. Souhami, and Keith A. Grimaldi

Department of Oncology
University College London Medical School
91 Riding House Street
London W1P 8BT, UK

INTRODUCTION

The mechanism of action of many anti-cancer drugs, including platinum based drugs, involves direct, covalent adduct formation on nucleotides in DNA. Cisplatin forms mainly intrastrand cross-links between two adjacent guanines in runs of two or more consecutive guanines and between AG pairs.[1,2] Recently, the sequence preferences of lesion formation have been studied in isolated DNA[3] by exploiting the finding that such lesions can block the progression of DNA *taq* polymerases. This and other assays using plasmid DNA have revealed that even relatively simple DNA damaging agents such as cisplatin show a degree of sequence preference in adduct formation.[4,5] Studying drug interactions in artificial systems with oligonucleotides or with plasmid DNA provides useful information but cannot necessarily predict intracellular behaviour where DNA exists in a highly ordered structure, complexed with many proteins, and where other cellular components may affect reactivity.

In intact cells, the formation and repair of DNA single strandbreaks, DNA interstrand crosslinks and DNA-protein crosslinks in the genome as a whole can be measured at pharmacologically relevant doses of drugs using the technique of alkaline elution.[6] For cisplatin, established sensitive immunological methods can measure the overall levels of the major lesions, GpG and ApG intrastrand adducts[7], and atomic absorption spectroscopy can measure overall levels of cisplatin-DNA adduct formation.[8] More recently, modified Southern blotting procedures have allowed the detection of adduct formation and repair at the level of the gene (10-20kb). These studies have shown that for DNA damage such as pyrimidine dimers transcriptionally active genes are repaired more efficiently than inactive genes, that within a gene the coding region is repaired more efficiently than the non-coding, and that the transcribed strand is repaired more rapidly than the non-transcribed (reviewed in [9,10]). In the case of antitumour alkylating agents such as nitrogen mustards both monoadducts[11] and crosslinks[12,13] are repaired preferentially in the coding region of a transcriptionally active gene. Furthermore, the adducts were preferentially located in specific genomic regions compared with the genome as a whole.

The fact that a covalent lesion can block the progress of *taq* polymerase forms the basis of a quantitative PCR assay to examine cellular damage and repair at the sub-gene/gene fragment level (500-2000 base pairs). This methodology has recently been applied to the measurement of damage in cells by the antitumour agent cisplatin in 2000 base pair gene fragments.[14,15] We have recently modified and simplified the technique to allow the accurate determination of damage in cells by several classes of antitumour drugs in small gene fragments (500 base pairs) which thus allows damage and repair to be compared in intron, exon and promoter regions of the same gene.[16] None of these methods, however, is capable of giving information at the level of the individual base in cells.

To carry out such studies it is necessary to have a technique that is sensitive enough to measure DNA damage in a single copy gene within cells. DNA damage at the nucleotide level within cells treated with anti-cancer drugs has been investigated but these limited studies looked at the highly reiterated alpha-DNA and, while providing some useful information, have limited physiological significance.[17,18,19] This methodology is not relevant to the study of single copy genes. A technique known as ligation mediated PCR (LM-PCR) has been recently developed whereby a double stranded oligonucleotide linker is ligated to blunt-ended DNA molecules formed by Sequenase extension of a primer, using as template DNA in which sites of alkylation or UV damage have been converted chemically or enzymatically into a strand break.[20,21] The necessity for production of a stand break excludes its use for the study of many important anti-cancer drugs such as cisplatin.

We have recently developed a method, single strand ligation PCR, which exploits both the observation that covalent DNA lesions can block *taq* polymerase[3], and the property of T4 RNA ligase to ligate[22] single stranded deoxyribo-oligonucleotides to single stranded DNA. This method allows the sequence selectivity of adduct formation by agents such as cisplatin to be determined in single copy genes in cells.

SINGLE STRAND LIGATION PCR (SSLIG-PCR)

The method of sslig-PCR is outlined in figure 1. Full experimental procedures can be found in.[23,24] It involves a first round PCR using a single 5′-biotinylated primer which defines the area of the gene to be investigated. 30 cycles of linear amplification by PCR generates a family of single-stranded molecules of varying length for which the 5′- end is defined by the primer and for which the 3′- ends are defined by the positions of the DNA-drug adducts without the need for a strand break. In order to exponentially amplify these molecules, which are captured and isolated by binding to streptavidin coated magnetic beads, a single stranded, 5′-phosphorylated, oligonucleotide is ligated to their 3′-OH ends using T4 RNA ligase.[22] With both ends of the DNA molecules defined they can then be exponentially amplified with nested primers and detected following a final amplification with an end-labelled primer and visualisation on a sequencing gel. The use of "linear" PCR in the first primer extension should increase its sensitivity compared to LM-PCR

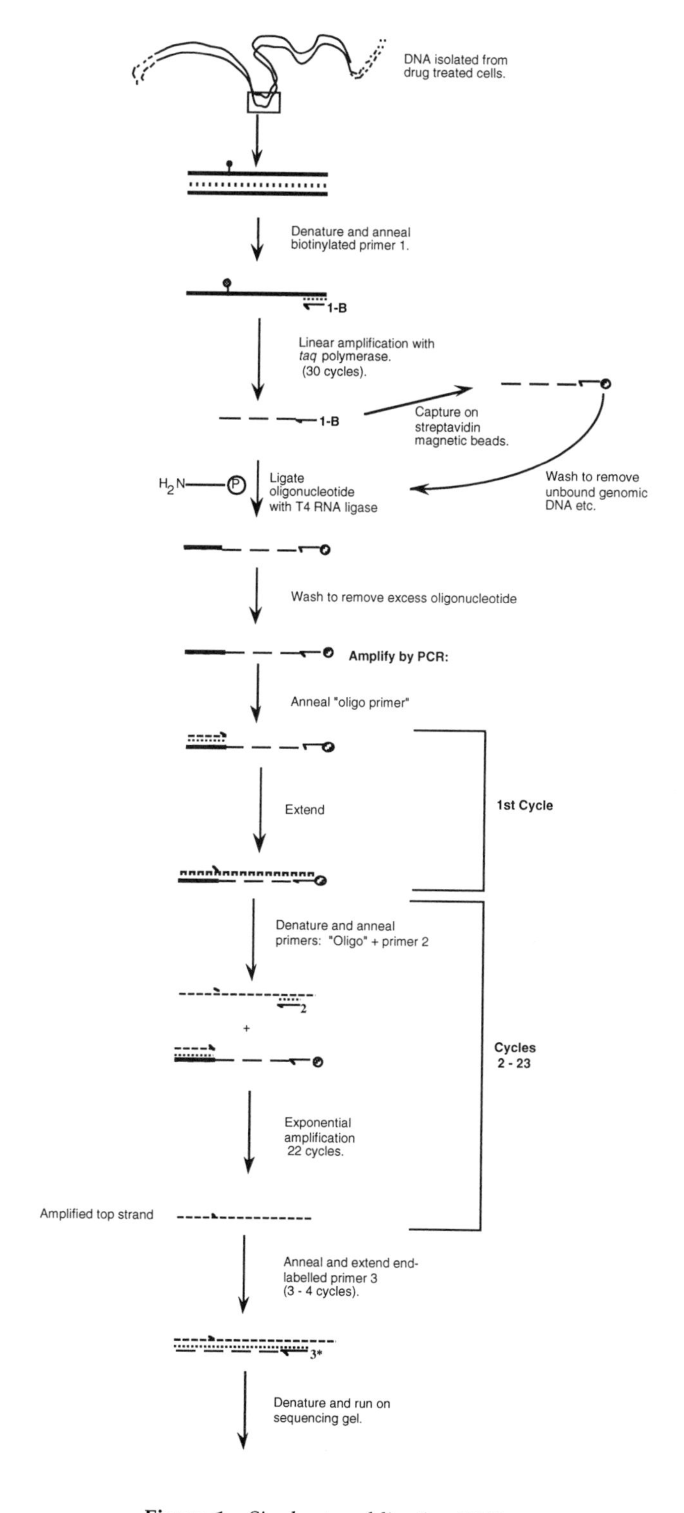

Figure 1. Single strand ligation PCR.

which is limited to one extension reaction of the first primer to produce a double stranded blunt ended molecule.

We have found single-stranded ligation PCR (sslig-PCR) to be a sensitive and reproducible technique and have applied it to study the interaction of various conventional anti-cancer drugs including nitrogen mustards and platinum analogues,[23] and novel DNA sequence selective crosslinking agents,[25] in several human genes including N-*ras*, H-*ras* and p53.

CISPLATIN LESIONS AT THE NUCLEOTIDE LEVEL

An example of the use of sslig-PCR to detect drug binding to the human N-*ras* gene is shown in figure 2. Cisplatin, when incubated with "naked" genomic DNA, binds preferentially to runs of two or more consecutive guanines and to AG pairs and adducts were detected at all occurrences of these sequences. However, an extremely large, but highly reproducible, variation in intensity of binding is observed between different GG and AG sites. Cisplatin also binds weakly to the first guanine in the sequence 5'-GCG (not shown). Figure 3 shows the results obtained from the non-transcribed strand of a portion of the N-*ras* gene from cells treated with cisplatin. Higher doses of drug are needed compared to naked DNA since less drug reaches the DNA compared to drug treated isolated DNA. With the non-transcribed strand as template some background bands appear in the untreated control which seem to represent intrinsic obstructions to *taq* polymerase since they consistently appear at the same positions. They may be due to some secondary structure of the DNA, especially in GC base pair regions, blocking to some extent the progress of the *taq* polymerase in the first round of PCR. However all attempts to overcome the effects of secondary structure, such as the inclusion of co-solvents DMSO, formamide or spermidine, or pre-denaturation with sodium hydroxide failed to remove these bands. Interestingly the background when the transcribed strand is used as template for sslig-PCR is consistently very low and the overall level of damage at a given dose of drug, as indicated by the decrease in "full-length" product, is also lower than on the non-transcribed strand.[23] This is not unique to the N-*ras* gene and has been found for other actively transcribed genes.

The intensity of drug-induced bands show a dependence on drug concentration. As expected, the intensity of the "full-length" band is highest in untreated cells and decreases as the concentration of cisplatin increases. It can also be seen that the intensity of individual lesions increases from 50 μM to 100 μM cisplatin (e.g. the 5′-AGG-3′ sites at position 121-123), but when the concentration is increased to 200 μM the intensity of the bands closest to the primer increase whereas those further away from the primer either show no change (5′-AGG-3′ at position 121-123) or actually decrease in intensity. As the concentration of drug increases the percentage of template molecules bearing more than one lesion will increase and the *taq* polymerase will be blocked at the first lesion, i.e. the one closest to the primer. The reproducibility of the method in cells is demonstrated from the intensities of individual bands in duplicate drug treatments being very similar as can be seen with the two independent 50 μM treatments.

The sequence specificity of cisplatin binding in cells is generally very similar to that in isolated genomic DNA and again adducts are formed at all runs of two or more guanines. The site of strongest binding in both cells and naked DNA is at the sequence 5′-AGG-3′ at position 121-123. However, bands of almost equal intensity are seen in cells at the site 5′-TACT-3′ which are not present at all in treated 'naked' DNA. This finding, which is highly

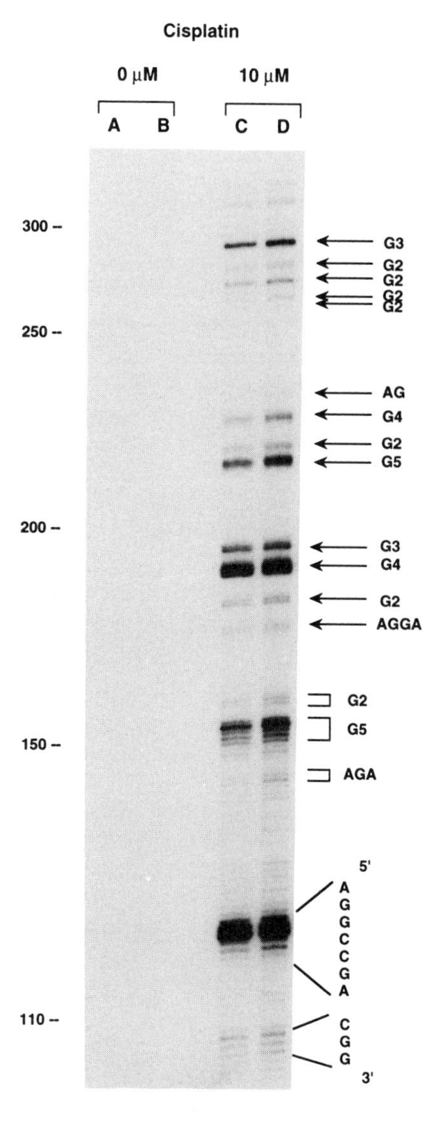

Figure 2. Sslig-PCR on the non-transcribed strand of a region of intron 1 of the human N-*ras* gene using isolated genomic DNA treated with 10 μM cisplatin. The lanes A + B and C + D show the results from duplicate treatments of DNA either with or without cisplatin. The figures on the left of the diagram refer to the distance in bases from the first base of the end-labelled primer. The letters on the right indicate the bases which correspond to the bands on the gel. G2, G4, G5 etc. represent runs of consecutive guanines. Taken from[23].

reproducible, therefore represents a novel, cell-specific, lesion. The sequence in this region, 5′-TACT-3′, occurs only once on the non-transcribed strand in the gene segment under study. On the transcribed strand of the same gene fragment the most intense binding is also observed at an occurrence of the sequence 5′-AGG-3′. Binding is also detected at the single 5′-TACT-3′ sequence on the transcribed strand which although of much less intensity compared to the non-transcribed strand, is of equal intensity compared to other GG pairs on this strand. Again binding at this sequence is not observed in drug treated naked DNA.

The extremely variable and unpredictable nature of cisplatin binding to genomic DNA was also observed with the H-*ras* gene. Figure 4 shows the pattern of binding of cisplatin (in intact cells) to the coding strand of a region comprising the first exon of the gene. The adducts are distributed among GG and AG pairs as expected however there is a GG rich stretch of nucleotides (5'– GGT GGT GGT GGG CGCC <u>GGC</u> <u>GGT</u> –3') containing codons 12 and 13 (underlined) — sites of activating mutations for this oncogene — at which the level of cisplatin binding is surprisingly low. This is presumably due to the secondary structure of the DNA in this region. Whether or not this secondary structure is in any way involved in the susceptibility to mutation of these sites, e.g. by affecting the rate of repair of DNA lesions, may be a worthwhile subject for future study.

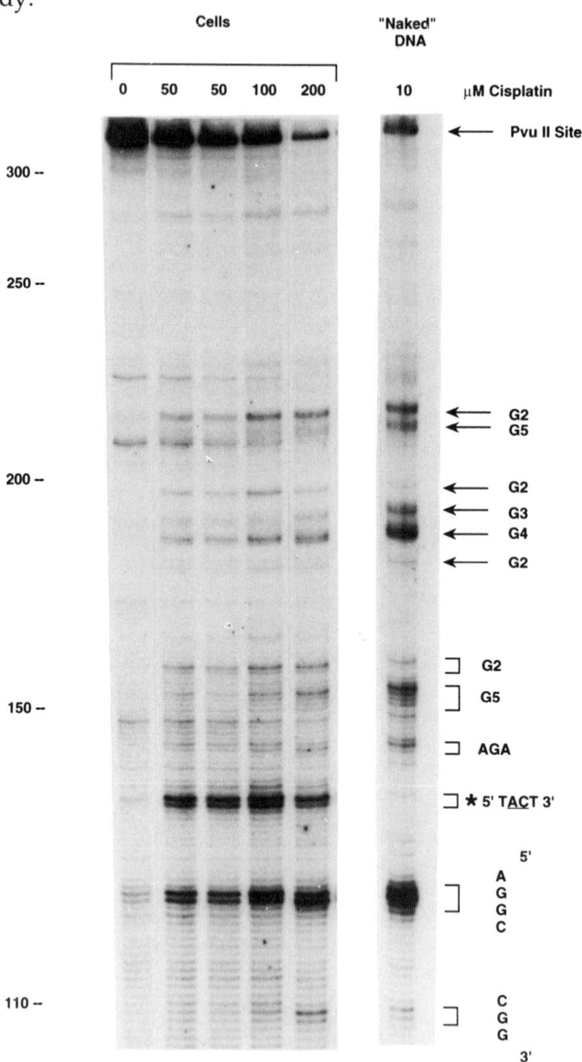

Figure 3. Concentration dependent adduct formation on the non-transcribed strand of N-*ras* by cisplatin in cells and comparison with treatment of "naked" DNA. The bands on the autoradiograph represent cisplatin adducts formed in cells that were treated with the indicated concentrations of cisplatin. DNA extracted from cells was cut with Pvu II before sslig-PCR to create a defined stop site on lesion free templates. The asterisk next to the sequence 5'-TACT indicates lesion site which was cell specific as an adduct was not detected at this position in "naked" DNA. Figure taken from[23].

DISCUSSION

The novel method of sslig-PCR allows the determination of the sequence specificity of covalent adduct formation by DNA damaging agents at the individual nucleotide level in single copy genes. The method is sensitive, so that physiologically relevant doses of drug can be used both with isolated DNA and for the treatment of intact cells. This method should be applicable to any molecule that covalently binds to DNA creating a lesion that blocks *taq* polymerase. It can be used with a wide range of agents since it is not limited to those whose lesion can be converted into a strand break.

The sequence specificities for cisplatin detected by sslig-PCR are in accordance with the intrastrand cross-link formation behaviour of cisplatin previously elucidated using plasmid DNA.[3,4,5,26] It is likely that all cisplatin adducts formed are detected by sslig-PCR. A quantitative PCR assay has demonstrated that *taq* polymerase is completely blocked by one adduct per DNA strand,[14] and Comess *et al*[26] have shown that, in artificial systems, adduct bypass by *taq* polymerase is at most 3% for GG crosslinks and 19% for AG crosslinks. It can be seen that adducts on pairs of guanines are represented by two bands on the autoradiograph whereas only one might be expected. One interpretation is that this is due to the presence of monoadducts at the guanine sites. This is unlikely however, for several reasons. It has been demonstrated in both isolated DNA and in cells that intrastrand crosslinks (both GG and AG) account for more than 90% of lesions on DNA.[1,27] Such intrastrand crosslinks form rapidly[28] and are likely to be the lesions detected by sslig-PCR at the GG sites since similar patterns of binding were observed with isolated DNA following 1 hour or up to 18 hours of incubation (data not shown). Two bands have also been demonstrated in plasmid DNA containing a single crosslink[26] and are probably due to greater distortion of the DNA on the 5′ side of the cross-link than on the 3′ side.[29,30,31] The 3′ adduct of the cross-link would then only partially block *taq* polymerase occasionally allowing bypass and subsequent blockage at the 5′ adduct.

In cells a consistent background of bands appears in the untreated controls. These may be due to spontaneous DNA damage at labile sites.[32] The background is consistently less when the transcribed strand is used as template. A possible reason for this is that transcription coupled repair is removing lesions from the transcribed strand at a greater rate than other repair mechanisms are removing lesions from the non-transcribed strand.[33,34] N-*ras* is a constitutively expressed gene and will be undergoing transcription during the drug treatment. Such a mechanism could also explain the lower background in the untreated template of the transcribed strand if that seen on the non-transcribed strand is indeed due to endogenous damage.

Binding of cisplatin to two occurrences of the sequence 5'-TACT-3' in the N-*ras* gene is observed. This is not a known binding site for cisplatin and is unexpected. It was anticipated that some lesions might be absent in DNA from treated cells due to obstruction by proteins complexed with DNA or by the higher order of structure within the cellular environment. The extra bands are not the result of a point mutation of C to G since the same cells were also the source of the "naked" DNA. The blockages at these 5′-TACT-3′ sequences are not due to either interstrand cross-links or to double strand breaks because no lesions are seen in either the non-transcribed or the transcribed strand at bases opposite the 5′-TACT-3′ sites. They could be the result of cross-linking between DNA and nuclear protein in these regions. Alternatively the local DNA structure in cells is such that it allows cisplatin to bind to cytosine as well as adenine. Although the interpretations remain hypothetical the results clearly demonstrate the importance of a method to determine the sequence binding specificities directly within cells.

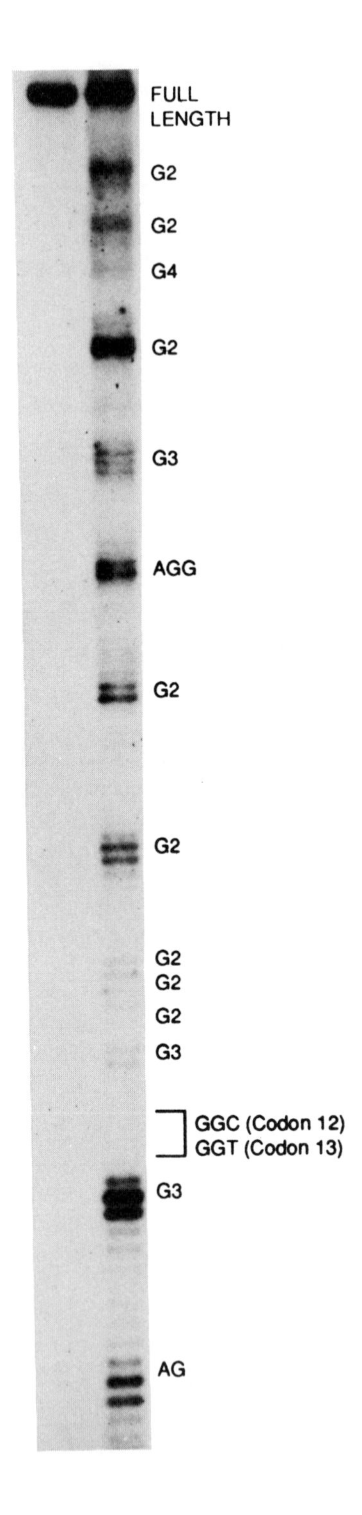

Figure 4. Cisplatin binding to the H-*ras* gene (exon 1) in cells. Human fibroblasts were treated with 100μM cisplatin (lane 2). Sites of GG and AG, and the positions of codons 12 and 13 are indicated.

Repair of drug damaged individual bases in genes has not yet been investigated in eukaryotic cells. DNA repair is known to be heterogeneous, it has been found that active genes are often repaired more efficiently than inactive genes,[35,36,37] and that the transcribed strand is repaired more rapidly than the non-transcribed strand.[33,38,39] These studies have mainly been performed with Southern blotting based methods which investigate damage at the level of the whole gene (10 - 20 kb). It is now clear that the sequence environment affects the rate of repair of individual lesions. For example, in the bacterial *lac* I gene it has been demonstrated that, after UV damage, lesions at certain sites are repaired slowly and that these repair "slow spots" correlate with the positions of mutation "hot spots".[40] The mapping of UV-induced photoproducts has also been achieved by either piperidine or enzymatic cleavage at sites of adducts followed by ligation-mediated PCR to amplify gene-specific fragments.[20,21] A slow rate of repair of UV-induced dimers was found to correspond to mutational sites in the p53 gene.[41] The rates of dimer removal in the PGK1 gene were also mapped and shown to vary at sites of transcription factor binding.[42] Thus the precise position of genetic mutations caused by carcinogens may be determined by both the distribution of the initial damage and by the efficiency of the subsequent repair of individual lesions.

With sets of primers from each strand it will be possible to use sslig-PCR to compare drug-induced damage and its repair at the nucleotide level in both the transcribed and the non-transcribed strands of single copy genes. Zhen *et al*,[43] using Southern blotting, have found differences in the gene specific repair of cisplatin interstrand, but not intrastrand, crosslinks which seemed to correlate with the degree of resistance to the drug. Sslig-PCR will be able to refine such studies to the nucleotide level and will also help to identify either sequences that are particularly susceptible to damage or repair, or molecules capable of creating lesions which are resistant to repair.

Acknowledgements

Work in the authors laboratory is funded by the Cancer Research Campaign.

REFERENCES

1. Fichtinger-Schepman, A. M. J., Van Der Meer, J. L., Den Hartag, J. H. J., Loman, P. H. M. and Reedijk, J. *Biochemistry*, 24:707 (1985).
2. Roberts, J. J. and Friedlos, F. *Biochim. Biophys. Acta*, 655:146 (1981).
3. Ponti, M., Forrow, S. F., Souhami, R. L., D'Incalci, M. and Hartley, J. A. *Nucleic Acids Res.*, 19:2929 (1991).
4. Murray, V., Motyka, H., England, P. R., Wickham, G., Lees, H. H., Denny, W. A. and McFadyen, W. A. *J. Biol. Chem*, 267: 18805 (1992).
5. Cullinane, C., Wickham, G., McFadyen, W. D., Denny, W. A., Palmer, B. D. and Phillips, D. R. *Nucleic Acids Res.*, 21:393 (1993) .
6. Kohn, K. W., Hartley, J. A. and Mattes, W. B. *Nucleic Acid Res*, 15:10531 (1987).
7. Fichtinger-Schepman, A-M J., van Oosterom, A.T., Lohman, P.H.M., and Berends, F.: *Cancer Res.*, 47:3000 (1987).
8. Parker, R.J., Gill, I., Tarone, R., Vionnet, J.A., Grunberg, S., Muggia, F.M., and Reed, E.: *Carcinogenesis*, 12:1253 (1991).
9. Bohr VA, Phillps DH and Hanawalt PC, *Cancer Res* 47:6426 (1987).
10. Bohr VA *Carcinogenesis* 11:1983 (1991) .

11. Wassermann K, Kohn KW and Bohr VA *J Biol Chem* 265:13906 (1990).
12. Futscher BW, Pieper RO, Dalton WS and Erickson LC *Cell Growth and Differentiation* 3:217 (1992).
13. Larminat F, Zhen W and Bohr VA *J Biol Chem* 268:2649 (1993).
14. Jennerwein, M. M. and Eastman, A. *Nucleic Acids Res.*, 19:6209 (1991).
15. Kalinowsky, D. P., Illenye, S. and Van Houten, B. *Nucleic Acids Res.*, 20:3485 (1992).
16. Grimaldi KA, Bingham JP, Souhami RL and Hartley JA *Analytical Biochem.* 222:236 (1994).
17. Hartley, J. A., Bingham, J. P. and Souhami, R. L. *Nucleic Acid Res.*, 20:3175 (1992).
18. Murray, V., Motyka, H., England, P. R., Wickham, G., Lee, H. H., Denny, W. A. and McFadyen, W. D. *Biochemistry*, 31:11812 (1992).
19. Bubley, G.J., Ogata, G.K., Dupuis, N.P., and Teicher, B.A. *Cancer Res.*54:6325 (1994)
20. Pfeifer, G. P., Steigerwald, S. D., Mueller, P. R., Wold, B. and Riggs, A. D. *Science*, 246:810 (1989).
21. Mueller, P. R. and Wold, B. *Science*, 246:780 (1989) .
22. Tessier, D. C., Brousseau, R. and Vernet, T. *Analyt. Biochem*, 158:171 (1986).
23. Grimaldi, K.A., McAdam, S.R., Souhami, R.L., and Hartley, J.A. *Nucleic Acids Res.*, 22:2311 (1994).
24. Grimaldi, K.A., McAdam, S.R. and Hartley, J.A.: In: *Technologies for Detection of DNA Damage and Mutations.*, Pfeifer, G.P. (ed), Plenum Press, New York, in press, 1996.
25. Smellie, M., Grimaldi, K.A., Bingham, J.P., McAdam, S.R., Thompson, A.S., Thurston, D.E., and Hartley, J.A. *Proc.Am. Assoc. Cancer Res.*, 35, 536 (1994)
26. Comess, K. M., Burstyn, J. N., Essigman, J. M. and Lippard, S. L. *Biochemistry*, 31:3975 (1992).
27. Fichtinger-Schepman, A. M. J., Van Oosterom, A. T. and Lohman, P. H. M. *Cancer Res*, 47:3000 (1987).
28. Bernges, F. and Holler, E. *Nucleic Acid Res*, 19:1483 (1991).
29. Marrot, L. and Leng, M. *Biochemistry*, 28:1454 (1989).
30. Sherman, S. E. and Lippard, S. J. *Chem. Rev*, 81:1153 (1987).
31. Shwartz, H., Shavitt, O. and Livneh, Z. *J. Biol. Chem*, 263:18277 (1988).
32. Lindahl, T. *Nature*, 362:709 (1993).
33. Mellon, I., Spivak, G. and Hanawalt, P. C. *Cell*, 51:241 (1987).
34. Selby, C. P. and Sancar, A. *Science*, 260:53 (1993).
35. Bohr, V. A., Smith, C. A., Okumoto, D. S. and Hanawalt, P. C. *Cell* 40:359 (1985).
36. Hanawalt, P. C. *Genome*, 31:605 (1989).
37. Bohr, V. *Carcinogenesis*, 12:1983 (1991).
38. Mellon, I. and Hanawalt, P. C. *Nature*, 342:95 (1989).
39. Lommel, L. and Hanawalt, P. C. *Mol Cell Biol*, 13:970 (1993).
40. Kunala, S. and Brash, D. E. *Proc. Natl Acad. Sci. U.S.A.*, 89:11031 (1992).
41. Tornaletti S and Pfeifer GP *Science* 263:1436 (1994).
42. Gao S, Drovin R and Holmquist *Science* 263:1438 (1994).
43. Zhen, W., Link, C. J., O'Connor, P. M., Reed, E., Parker, R., Howell, S. B. and Bohr, V. A. *Molec. Cell. Biol.*, 12:3689 (1992).

A REVIEW OF THE MODULATION OF CISPLATIN TOXICITIES BY CHEMOPROTECTANTS

Robert T. Dorr, Ph.D.
Associate Professor of Pharmacology
College of Medicine and
The Arizona Cancer Center

The University of Arizona
1515 N. Campbell Avenue
Tucson, AZ 85724

INTRODUCTION

In 1988, Dr. Brian Leyland-Jones questioned "Whither the Modulation of Platinum" in an editorial in the Journal of the National Cancer Institute (Leyland-Jones, 1988). The quest for specific chemoprotectants clearly predates the advent of platinum-based chemotherapy and yet, progress has been slow with only a few chemoprotectants available in the clinic. These include mesna for the oxazophosphorines (Shaw, 1987), amifostine for cisplatin (in certain European countries), and more recently, the cardioprotectant iron chelator, cardioxane (ADR-529 or ICRF-187) [Speyer et al., 1988 and 1992]. For the platinum-based antitumor agents, a large number of thiol-based or sulfur-containing nucleophiles have been tested. These include the endogenous tripeptide glutathione, disulfiram and its metabolite, diethyldithiocarbamate, the inorganic salt, sodium thiosulfate, and the phosphorylated aminothiol, amifostine (Ethyol or WR-2721) (Glover et al., 1987) (Figure 1). Of these, several appear to have promise for clinical utility against platinum-induced nephrotoxicity and possibly the peculiar cumulative-dose limiting neuropathy. For this condition, a unique melanocortin-derived neuropeptide, ORG-2766, appears to have potential utility.

In this chapter, the preclinical and clinical status of these agents will be reviewed. Amifostine, which shows great promise as a cisplatin modulator, will be omitted since a latter presentation focuses on the advanced stage of development with this chemoprotective agent. Most of the current review will be limited to cisplatin since there are a wealth of studies with this agent and since chemoprotection is largely limited to non-myelosuppressive toxicities. In contrast, other chapters will specifically deal with the use of hematopoietic growth factors as *rescue* agents for platinum-based agents and particularly for carboplatin-induced thrombocytopenia. In this regard, the recent discovery of human thrombopoietin (TPO) has added tremendous excitement (de Sauvage et al., 1994) since few existing growth factors have significant clinical effects on platelet proliferation.

Two primary concepts for the development of chemoprotectants mandate: (1) *selective* protection of non-tumorous normal tissues and (2) the addition of little, if any, toxicity. For this reason, most of the potential modulators of cisplatin toxicity have been based on the endogenous defense system for electrophiles, the tripeptide glutathione. Studies in platinum resistance clearly implicate an up-regulation of either glutathione (Freeman and Meredith, 1989) or more specifically, one of the family members of glutathione-S-transferase enzymes (Andrews and Howell, 1990; Andrews et al., 1985). These ubiquitous cytosolic enzyme systems are responsible for conjugating toxic electrophiles to the sulfur moiety in GSH to yield a non-toxic thioether. This conjugate can be further metabolized and excreted in the urine or bile. Most of the platinum modulators are believed to act similarly, as alternative sulfur-based nucleophiles capable of binding activated electrophilic or ligand-exposed platinum species (Dedon and Borch, 1987). Although few such conjugated species have been isolated, it is presumed that this direct mechanism may explain most of the activity of the thiol-based modulators. However, there is some evidence that other mechanisms, such as growth factor stimulation, may also be operant. The following sections will review the different platinum modulating agents with an emphasis on those compounds which have undergone clinical testing. For each agent an attempt was made to include the mechanistic and pharmacokinetic rationale for action as a platinum chemoprotectant.

Glutathione (GSH)

Glutathione (GSH) with its reduced sulfhydryl in the cysteine moiety (Figure 1) is the primary cytosolic thiol which helps maintain cellular defenses against endogenous or xenobiotic electrophiles (Meister, 1985). With the platinum-containing anticancer agents, glutathione and its related conjugative transferase enzymes appear to be responsible for resistance in many cultured tumor cell lines either wholly or partially (Andrews and Howell, 1990). Preclinical studies by Zunino and co-workers have shown that exogenously-administered GSH reduces cisplatin-induced lethality and nephrotoxicity (Zunino, 1983; 1989). A more recent preclinical study has described diminished cisplatin-induced neurotoxicity with large doses of intravenous GSH (Hamers et al., 1993a). At GSH doses ≥ 200 g/mg/kg, cisplatin-induced peripheral neuropathy was prevented, as measured by quantitative nerve conductance studies in the rat. Importantly, there was no reduction in cisplatin's experimental antitumor activity in these trials suggesting that GSH therapy might be useful in cancer patients receiving high-dose cisplatin therapy.

In an initial clinical trial, intravenous GSH was combined with cisplatin and cyclophosphamide in 40 women with advanced ovarian cancer. The GSH dose of 1.5 g/m^2, was infused over a 15 minute period. This was given prior to a 30 minute infusion of cisplatin (40 mg/m^2/d) for four consecutive days (total 160 mg/m^2 per course). Cyclophosphamide, 600 mg/m^2, was given as an IV bolus only on day 4 of the 28 day treatment cycle. The overall response rate (complete plus partial) was 86% with 62% of patients achieving a complete remission documented by second-look laparotomy. The assessment that GSH reduced or prevented cisplatin-induced toxicity in this trial was less clear (Table 1). Nausea and vomiting were severe in all patients but serum magnesium

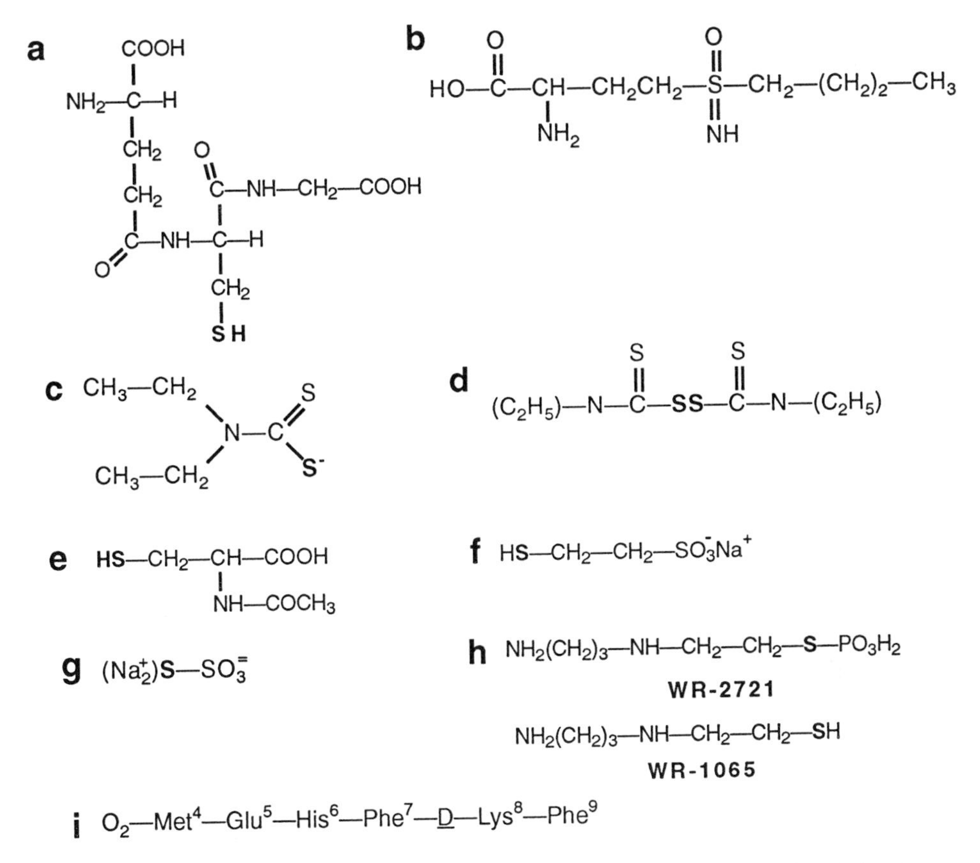

Figure 1: Structures of cisplatin modulatory sulfhydryls and peptide-derivatives: Glutathione (**a**), the GSH-depleting antimetabolite L-buthionine sulfoximine (**b**); the sulfhydryl metabolite of disulfiram, diethyldithiocarbamate or DDTC (**c**); the parent disulfide, disulfiram (**d**); N-acetylcysteine (**e**); mesna (**f**); sodium thiosulfate (**g**); the phosphothiol-amifostine (WR-2721 and its active metabolite actifostine (WR-1065) formed by alkaline phosphatase activity (**h**); and the $ACTH_{4-9}$ peptide analog ORG-2766 which contains an oxidized terminal methionine (O_2-Met) and a racemized, D-phenylalanine at position 8 (**i**). Known active moieties are bolded fur sulfhydryl-based antagonists.

levels were maintained and serum creatinine rose slightly in only a few patients. However, as Table 1 shows, ototoxicity and peripheral neuropathy were commonly seen at higher cumulative cisplatin dose levels even with GSH. Furthermore, 7 of 37 evaluable patients (all responders) could not receive all five planned courses of therapy due to delayed hematopoietic recovery (2 patients), early neurotoxic symptoms (3 patients) and hepatotoxicity (one patient).

The pharmacokinetics of total and free platinum do not appear to be altered by concurrent GSH administration. In 12 patients with non-small cell lung cancer, GSH 2.5 g was administered intravenously 15 minutes prior to the conclusion of a 30 minute cisplatin 80 mg/m^2 infusion. Table 2 compares the mean pharmacokinetic parameters of unbound (active) platinum species in these patients. None of the pharmacokinetic parameters of cisplatin were altered by concurrent GSH therapy including the peak level and the cumulative exposure (Leone et al., 1992).

Table 1. Effect of glutathione on cisplatin toxicity in women with advanced ovarian cancer.

Toxicity	No. Patients	Percent
Intractable vomiting	40	100%
Leukopenia (< 1,000 cells/mm^3)	2	5%
Thrombocytopenia (< 50,000/mm^3)	2	5%
Anemia (< 7.9 g/dL of hemoglobin)	7	7.5%
Liver enzyme elevation	8	20%
Creatinine rise (1.5-2.0 mg/dL)	3	7.5%
Peripheral neuropathy (ECOG Scale)		
Grade 1	17	43%*
Grade 2	7	17.5%*
Grades 3, 4	0	
Ototoxicity		
(Audiogram abnormalities)	11	28%*

*Generally associated with cumulative cisplatin doses ≥ 640 mg/m^2.

From: F. De Re et al., Cancer Chemother Pharmacol 25:1990.

Another recent trial randomized patients with relapsed ovarian cancer to receive weekly cisplatin at 50 mg/m^2 for 9 weeks with (n = 16) or without (n = 17) a concommitant IV infusion of GSH at 2.5 g/m^2 (Colombo et al., 1995). Patients were evenly matched for prior mean cumulative cisplatin doses: 550 mg/m^2 in the cisplatin arm and 538 mg/m^2 in the arm plus GSH. The results of this trial showed that response rates (CR + PR) were roughly comparable in the two arms of the study and there was a trend towards reduced toxicity and maintained cisplatin dose-intensity (Table 3). However, there was little objective evidence

Table 2. Mean (SD) pharmacokinetic parameters of unbound platinum with or without glutathione.

Pharmacokinetic Parameter	First Dose Cisplatin	First Dose Cisplatin Plus GSH
Peak Level (mg/L)	4.76 (0.57)	5.07 (1.58)
Half-Life		
α (hr)	0.43 (0.03)	0.43 (0.04)
β (hr)	39.4 (11.4)	42.7 (33.8)
AUC (mg•hr/L)	7.02 (0.56)	9.04 (2.45)
Clearance (mL/hr/kg)	181 (16.3)	181.6 (33.5)
Volume of Distribution (L/kg)	10.37 (3.74)	7.27 (1.52)

From: R. Leone et al., Cancer Chemother Pharmacol 29:385, 1992.

Table 3. Results of a randomized study of weekly cisplatin plus GSH in relapsed ovarian cancer.

Variable	CDDP (n = 17)	CDDP + GSH (n = 16)
CDDP 100% Dose - Intensity Achieved	27%	56%
Anemia	2	0
Leukopenia (No.)*	1	3
Thrombocytopenia (No.)*	2	0
Nephrotoxicity (No.)*	0	0
Nausea & Vomiting (No.)*	14	10
Nerve Amplitude Change	- 5.58	- 3.4
Response (CR + PR)	9/15	12/16

* ≥ Grade III by WHO Criteria.

N. Colombo et al., 1995.

that ototoxicity was significantly reduced by the GSH regimen and other toxicities were equivalent. No nephrotoxicity was described in either arm. Thus, while GSH did not appear to impair cisplatin antitumor efficacy, evidence for significant protection from toxicity was not evident in this trial.

Several other non-randomized feasibility studies have been performed with cisplatin and GSH. Bohm et al. (1991) showed that an infusion of GSH (5 g), 15 min prior to cisplatin 90 mg/m^2 resulted in no increase in the urinary excretion of N-acetyl-β-glucosaminidase levels (a marker of cisplatin renal tubular toxicity). Responses were observed in 9/11 evaluable ovarian cancer patients and the renal excretion of cisplatin (23% after 24 hrs) was not reduced by the addition of GSH. Similarly, a GSH infusion of 2.5 g, resulted in no evidence of renal toxicity in 11 patients with advanced colorectal cancer who received 5-fluorouracil plus 3 consecutive days of cisplatin 40 mg/m^2/day (Cozzaglio et al., 1990). The clinical response rate was low with 2/10 partial responses in this trial and mild neurotoxicity was observed in 4/11 treated patients. With very high dose cisplatin therapy (160 mg/m^2) plus cyclophosphamide (600 mg/m^2), the addition of 1.5 g/m^2 of GSH was thought to reduce the incidence of severe neuropathy apparent after 3-5 courses (Pirovano et al., 1992). While there was a trend towards more severe neurotoxicity with patients > 50 years of age, there was no disabling neuropathy in any patients, despite the delivery of cumulative cisplatin doses of 800 mg/m^2 (five courses). This clinical finding was reinforced by neurophysiological results showing preservation of sensory amplitude potentials measured serially in median, ulnar, and sural nerves (Pirovano et al., 1992). Similar neuroprotective results were reported with GSH and high dose cisplatin plus bleomycin (Fontanelli et al., 1992) and with cisplatin and cyclophosphamide (De Re et al., 1993).

Finally, a placebo-controlled randomized trial with GSH has been performed in gastric cancer patients receiving weekly cisplatin 40 mg/m^2, fluorouracil 500 mg/m^2, epirubicin 35 mg/m^2 and the 6S-stereoisomer of leucovorin 250 mg/m^2 (Cascinu et al., 1995). Patients were randomized to receive GSH at a dose of 1.5 g/m^2 15 minutes before cisplatin and again on days 2-5 at a dose of 600 mg/m^2 by IM injection. Clinical neurologic evaluations and electrophysiologic studies were performed at baseline, after nine courses

Table 4. Clinical effects of glutathione in a randomized trial with cisplatin in gastric cancer.

Parameter	Incidence (No. Patients)			
	GSH (n = 25)		Placebo (n = 25)	
Response				
Complete (%)	5 (20)		3 (12)	
Partial (%)	14 (56)		10 (40)	
Survival (months)	14		10	
Treatment Delay (weeks)	55		94	
	Week 9 (n = 25)	Week 15 (n = 24)	Week 9 (n = 25)	Week 15 (n = 18)*
Neurotoxicity by WHO Grade	25	20	9	2
I, II	0	4	15	13
III, IV	0	0	1	3
Hemotransfusions	18	14	46	16

*Assessable patients.

Cascinu et al., 1995.

(360 mg/m^2) and after 15 weekly courses (600 mg/m^2 cumulative cisplatin dose). Table 4 summarizes the findings in this pivotal trial. These results show that GSH reduced the incidence of moderate to severe neurotoxicity and the requirement for transfusions without adversely effecting response or survival. Serial electrophysiologic studies of latency and sensory amplitude potentials for median, ulnar and sural nerves also showed preservation of nerve function in patients receiving GSH. This compares to a time (cumulative dose)-dependent loss of nerve function in pateints on the placebo arm of the trial. Other toxicities such as leukopenia, thrombocytopenia, alopecia and mucositis were not significantly different in the two arms.

Curiously, nephrotoxic effects of cisplatin were not described for either arm of this trial, even though exogenously-administered GSH is known to accumulate in the kidney (Litterst et al., 1982). However, other preclinical studies in the rat suggest that GSH also significantly decreases platinum accumulation in dorsal root ganglia (Cavaletti et al., 1992). This may explain the apparent neuroprotective efficacy of GSH seen in the clinical trials reviewed above.

L-Buthionine Sulfoximine

Buthionine sulfoximine (L-BSO) is an inhibitor of the rate-limiting enzyme involved in glutathione synthesis, γ-glutamyl-cysteine synthetase (Griffith and Meister, 1979). Treatment of tumor cells with L-BSO can enhance the antitumor activity of numerous alkylating agents and in some systems, can overcome acquired resistance to cisplatin (Hamilton et al., 1985). While cisplatin nephrotoxicity does not appear to be mediated by renal GSH consumption (Leyland-Jones et al., 1983; Litterest et al., 1982; Maines, 1986), the depletion of renal GSH by diethylmaleate can increase cisplatin-induced lethality and BUN increase (Litterest et al., 1986).

In contrast, Mayer et al. (1987) found that pretreatment of rats with L-BSO diminished the nephrotoxicity of cisplatin as measured by BUN increase and inhibition of renal gamma-glutamyl transpeptidase synthesis. This occurred in the face of a 47% decrease in renal GSH concentrations. Furthermore, there was no effect of the BSO pretreatment on cisplatin antitumor efficacy in C3H mice bearing the MBT-2 bladder cancer.

The mechanism for the BSO-mediated reduction in cisplatin nephrotoxicity is unknown. One possibility is that cisplatin nephrotoxicity requires rapid turnover of GSH by the gamma-glutamyl cycle, as has been speculated by Maines (1986). Alternatively, BSO may alter cisplatin toxicity by a mechanism not involving GSH depletion as has been suggested in studies of different GSH depletors on rabbit proximal tubules (Schnellman and Mandell, 1986). Indeed, studies in rats have shown that L-BSO reduces the level of *trans*-platinum accumulation in the kidney by 50% at a timepoint when renal GSH levels would be unaffected (Mayer and Maines, 1990).

Thus, while the mechanism is unclear, L-BSO may offer an intriquing means of modulating cisplatin nephrotoxicity. Clearly, more preclinical study is needed with this combination, particularly since L-BSO is known to synergize with the toxicities of other alkylating agents, and in particular, cyclophosphamide (Friedman et al., 1990; Soble and Dorr, 1987). In humans, BSO significantly increases melphalan-induced myelosuppression (O'Dwyer et al., 1992).

Diethyldithiocarbamate (DDTC)

Diethyldithiocarbamate or tetraethylthiuram disulfide (DDTC) is a metabolite of the alcohol inhibitor disulfiram (Antabuse). In addition to its active (reduced) sulfhydryl moieties which chelate metals, DDTC can also augment T-cell function (Renoux and Renoux, 1984). The metal chelating properties of DDTC have facilitated its clinical use as a treatment for nickel poisoning (Sunderman, 1971). In the rat, DDTC can remove platinum bound to renal tubules and restore gamma-glutamyl transpeptidase activity in brush border preparations of renal tubule cells (Bodenner et al., 1986). Other preclinical studies in rodents have demonstrated that DDTC significantly ameliorates the nephrotoxicity and myelosuppression of cisplatin without reducing antitumor efficacy (Borch and Pleasants, 1979; Borch et al, 1980; Gale et al., 1982; Bodenner et al, 1986). The cisplatin dose-modification factor for DDTC in nontumor-bearing mice was 3.2 in one study. This signifies that substantial dose escalations of cisplatin could be performed without lethal suppression of normal bone marrow progenitors (Evans et al., 1984).

In contrast, DDTC can produce dose-dependent cytotoxicity in human tumor cell lines. Cells exposed for 1 hour to DDTC concentrations of .001 to 10 μg/mL are significantly inhibited in terms of growth (Cohen and Robins, 1990). Other in vitro studies have also described cytotoxic effects of DDTC in tumor cells (Powell, 1954) and in human T-lymphocytes wherein a biphasic dose-response pattern was demonstrated (Spath et al., 1987; Rigas et al., 1979). These growth inhibitory activities may be pertinent to some of the clinical toxicities with this agent.

In addition to nucleophilic binding of activated platinum species in the bone marrow (Gonias et al., 1984), an indirect mechanism for the myeloprotective effect of DDTC has been postulated. This may involve augmented production of hematopoietic growth factors by DDTC-treated bone marrow stromal cells

(Schmalbach and Borch, 1990). Thus, DNA synthesis and proliferation are required for this DDTC effect following an insult by carboplatin or cisplatin. In this study, murine bone marrow cultures exposed to carboplatin and DDTC were shown to express significantly higher levels of granulocyte/macrophage colony stimulatory activity (Schmalbach and Borch, 1990). The clinical relevance of this finding is that adequate bone marrow reserve may be required for a beneficial effect from DDTC, and therefore, heavily pretreated patients may not respond well to this DDTC effect.

Several clinical trials have been performed with DDTC in patients receiving cisplatin. In a Phase I trial, 19 pateints received 35 courses of cisplatin, 120-160 mg/m^2 given as an IV bolus. This was followed 45 minutes later by DDTC, 4 g/m^2, given as a 1 hour infusion (Berry et al., 1990). This regimen significantly reduced the nephrotoxicity of dose-intensive cisplatin whereas ototoxicity was prominent and became dose-limiting (Table 5). Nausea and vomiting were substantial. In addition, DDTC produced substantial acute toxicities including hypertension, flushing, diaphoresis, agitation and local burning (Berry et al., 1990). In another Phase I trial, two dose levels of DDTC, 75 mg/kg and 150 mg/kg, were administered 45 minutes after cisplatin was infused at doses of 50 mg/m^2 to 120 mg/m^2 (Qazi et al., 1988). No nephrotoxic effects were observed and there was some evidence for amelioration of nausea and vomiting. However, DDTC again produced substantial autonomic toxicities which were much more severe at the higher dose level of 150 mg/kg. These consisted of diaphoresis, chest discomfort, agitation and numbness in the arm used for the DDTC infusions. This precluded DDTC dose escalations above 150 mg/kg. However, this dose produced peak plasma DDTC levels of 1.0 mM which are above those shown to be effective for protection from cisplatin nephrotoxicity in rodents (Qazi et al., 1988). And, these DDTC doses are substantially higher than those used by Paredes et al. in head and neck cancer patients receiving cisplatin and 5-fluorouracil (Paredes et al., 1988). The relatively low 600 mg/m^2 DDTC dose (approximately 15 mg/kg) used in this latter trial may explain the resultant lack of chemoprotective efficacy for DDTC in this population (Paredes et al., 1988).

Diethyldithiocarbamate has also been combined with high-dose carboplatin in women with relapsed ovarian cancer (Ten Boklel Huinink, 1987; Rothenberg et al., 1988). At a dose of 4 gm/m^2, DDTC did not impair antitumor effects but it

Table 5. Cisplatin toxicities with DDTC in a Phase I setting.*

Cisplatin Dose (mg/m^2)	No. Courses	Neurotoxicity ≥ Grade 2	Nausea and Vomiting (Grade 3, 4)	Creatinine Clearance (mL/min)	
				Baseline	Day 21
120	11	0	3	97 ± 7	83 ± 8
140	11	1	5	87 ± 4	77 ± 8
150	9	0	4	102 ± 6	95 ± 10
160	4	1	2	132 ± 24	100 ± 15
Total	35	2	14	99 ± 4	86 ± 4

*Berry et al., 1990.

also did not protect patients from severe carboplatin-induced hematologic toxicity, including 3 treatment-related deaths. A similar lack of protective activity for carboplatin and DDTC was reported in a European trial (Ten Bokkel Huinink, 1987). Again, DDTC produced substantial autonomic toxicities, suggesting that this agent has very little value in combination with carboplatin.

More recently, a large randomized placebo-controlled trial has been reported with DDTC (Gandara et al, 1995). In this trial, patients with ovarian cancer (123), small cell lung cancer, SCLC (53) or non-small cell lung cancer, NSCLC (38) were randomized to receive a cisplatin-based chemotherapy (100 mg/m^2) over 60 minutes. This was administered with hydration (2,700 mL) and mannitol (12.5 g) along with cyclophosphamide at a dose of 750 mg/m^2 on day 1 for ovarian cancer patients, or with etoposide for lung cancer patients at a dose of 100 mg/m^2 IV over 60 minutes on days 1, 2, and 3 of each 4 week treatment cycle. Diethyldithiocarbamate was administered at a dose of 1.6 g/m^2 over 4 hours, beginning 15 minutes prior to cisplatin. The primary endpoints were completion of 6 cycles, toxicity, graded by NCI common criteria, objective response, and the cisplatin dose intensity delivered. Patients were blindly randomized to receive placebo (n = 96) or DDTC (n = 99) prior to cisplatin.

Table 6 compares the overall dose intensity delivered in the 2 study arms. It is apparent that the addition of DDTC significantly reduced the cumulative cisplatin dose due to an increase in toxicity-related early withdrawal from the study. A more detailed analysis of toxicities (Table 7) shows that DDTC did not prevent ototoxicity or peripheral neuropathy and its use was not associated with a lower mean serum creatinine level. Furthermore, DDTC prodced several additional toxicities, notably hyperglycemia, hypertension and dehydration. Clinical response rates were not significantly effected by DDTC and overall 49% responded (CR + PR) in the DDTC arm versus 43% in the placebo arm.

Table 8 summarizes the results of 5 clinical studies with DDTC as a chemoprotectant for cisplatin. There are 3 "positive" studies describing nephroprotection (2) and possible neuroprotection (1). However, there are

Table 6. Randomized placebo-controlled trial of DDTC as a chemoprotectant for cisplatin.

— Methods —			
	Randomization Group		
Characteristic	Placebo (n = 99)	+ DDTC (n = 96)	P-Value
Withdrawal Due to Toxicity	9%	23%	.008
Cisplatin			
• Mean Dose Intensity	23.7 mg/m^2/wk	23.6 mg/m^2/wk	—
• Mean Cumulative Dose	379 mg/m^2	247 mg/m^2	.0001
Completed 6 Cycles	28%	6%	.001

*Gandara et al., 1995.

Table 7. Randomized placebo-controlled trial of DDTC as a chemoprotectant for cisplatin.

— CHEMOTHERAPY TOXICITY RESULTS —

Toxicity	Randomization Group Placebo (n = 99)	 + DDTC (n = 96)	P-Value
Creatinine (mg/dL)	.017	0.71	.008
Ototoxicity • dB Reduction at 3,000 Hz	7.95	13.86	NS
• Clinical* (%)	6	9	NS
Neuropathy* (%)	12	13	NS
Vomiting*	10	16	NS
Leukopenia* (%)	11	14	NS
Thrombocytopenia* (%)	0	4	NS

— TRANSIENT DDTC TOXICITIES —

Toxicity	Randomization Group Placebo (n = 99)	 + DDTC (n = 96)	P-Value
Hyperglycemia (%)	4	19	< .05
Hypertension (%)	10	20	< .05
Dehydration (%)	9	20	< .05
Taste Alteration (%)	1	12	< .05
Flushing	6	9	NS
Injection Site Reaction	3	3	NS

* ≥ Grade III NCI Scales.

* Gandara et al., 1995.

2 clearly negative trials. The study by Paredes et al. (1988) may be criticized by the use of a low DDTC dose, whereas the more recent randomized placebo controlled trial by Gandara et al. used a 4 hour DDTC infusion of 1.6 g/m^2. Company studies by MGI Pharma showed that this dose was both tolerable and could achieve potentially chemoprotectant blood levels of DDTC (Gandara et al., 1995). These target DDTC levels have been estimated to be in the range of 400 µmol/L with an AUC of 15 µmol/L•min^{-1}. However, despite the use of the 1.6 mg/m^2 dose and the 4 hour infusion period, DDTC acute toxicity was still prominent and there was no evidence for a chemoprotective effect. Indeed, the

Table 8. Summary of studies with DDTC as a chemoprotectant for cisplatin.

First Author (Journal, Year)	Cisplatin mg/m^2	DDTC Dose	Conclusion
Qazi (JNCI, '88)	50-120	75, 150 mg/kg	Nephroprotection, Partial myeloprotection
Paredes (JCO, '88)	120	0.6 mg/m^2	No chemoprotection
Berry (JCO, '90)	120-150	4 g/m^2	Nephroprotection, Ototoxicity dose-limiting
Gandara (Plat Text, '91)	100 (days 1, 8)	4 g/m^2	Possible neuroprotection, Otoprotection
Gandara (JCO, '95)	100	1.6 g/m^2	No chemoprotection

Modified after D. Gandara et al., 1995.

use of DDTC appeared to *reduce* cisplatin dose-intensity compared to placebo. Overall, these results suggest that DDTC does not comprise a useful chemoprotectant for either cisplatin, based on 5 clinical trials, or for carboplatin based on two negative trials in animals (Dibble et al., 1987; Francis et al., 1989).

The effect of DDTC on platinum pharmacokinetics has been studied in patients receiving high dose cisplatin therapy (100 mg/m^2 IV on days 1 and 8 or 200 mg/m^2/course [DeGregorio et al., 1989]). There was no apparent alteration in the peak or total platinum ultrafiltrate levels in patients receiving a DDTC dose of 4 gm/m^2. This dose produced peak DDTC levels ranging from 131 to 970 μM, depending on the length of the DDTC infusion. These results suggest that (1) chemoprotectant DDTC levels (> 400 μM) can be achieved, (2) that there is no reduction in cisplatin plasma levels and (3) DDTC does not change the un-bound fraction, which averaged 10% of total platinum levels. However, the over-all nephroprotective benefits of DDTC with platinum-containing agents, is significantly constrained by DDTC's acute toxicity profile. Thus, DDTC does not extend the ability to deliver dose-intensive chemotherapy and there is little evidence that DDTC significantly reduces nonrenal toxicities, particularly the cumulative neuropathies associated with cisplatin. These features reinforce the conclusion that DDTC does not comprise a particularly useful chemoprotectant for platinum-based anticancer agents.

Disulfiram

Disulfiram, the parent molecule of DDTC, has also been evaluated preclinically as a chemoprotectant for cyclophosphamide (Hacker et al., 1982) and cisplatin (Borch and Katz, 1980). The results of an initial Phase I study suggested that disulfiram may have a better side-effect profile than DDTC (Stewart et al., 1987). In this pilot study twelve patients were treated with cisplatin, 100 mg/m^2 and oral disulfiram begun 1 hour before a 2 hour cisplatin infusion. The dose-limiting toxicity of disulfiram was reversible confusion at a dose of 3,000 mg/m^2 and no severe toxic effects were observed at lower doses. Other toxic effects of disulfiram reported in this trial included dyspnea, generalized weakness and facial swelling. However, these disulfiram doses were projected to produce low peak DDTC levels of 1-2 μM, which is too low to provide significant cytoprotection based on the mouse model (Bodenner et al.,1986). However, disulfiram did not appear to increase platinum toxicity nor impair cisplatin antitumor effects. Thus, the overall benefit of disulfiram awaits the results of pivotal randomized clinical trials.

N-Acetylcysteine (NAC)

N-Acetylcysteine (NAC) has been reported to block the normal organ toxicities of doxorubicin-induced cardiomyopathy in the mouse heart (Doroshow et al., 1981) and oxazophosphorine-induced bladder damage in animals and humans. This sulfhydryl-based compound is an established cytoprotectant for the liver in acetaminophen overdosage (Rumack et al., 1981). In one preclinical study in rats, NAC was found to reduce cisplatin-induced renal and gastrointestinal toxicities (Leyland-Jones et al., 1980). However, other preclinical studies showed that the coadministration of cysteine or NAC does not reduce the toxicity or antitumor activity of cisplatin (Speer, Ridgway Hall, 1975; Burchenal et al., 1978; Slater, 1977). Furthermore, another preclinical trial showed that NAC was not beneficial at blocking the myelosuppressive effects of cisplatin in mice (Lerza et al., 1986). These findings, the lack of consistent preclinical evidence of protection, and the inability to extend NAC's cytoprotectant properties to humans, suggests that this agent does not comprise a useful antagonist of platinum-induced toxicity.

Mesna

Mesna is the sodium salt of 2-mercaptoethane sulphonate. It acts as a uroprotectant for the oxazophosphorine class of alkylating agents which includes cyclophosphamide (Scheef et al., 1979) and ifosfamide (Shaw and Graham, 1987). Mechanistically this sulfhydryl agent dimerizes to the inactive disulfide in the plasma. It is then *selectively* activated by enzymatic reduction to the monomer in the kidney tubules (Shaw and Graham, 1987). In this fashion, the drug can act as a selective chemoprotectant for the urinary bladder to counteract acrolein and other electrophilic metabolites of the oxazophosphorines which accumulate in the urine. Whereas mesna can be directly mixed with these alkylating agents, it is reported to be physically incompatible with cisplatin (Asta-Werke Degussa Pharma, 1984).

In non-tumor bearing mice, mesna similarly protects against the lethality of high dose cisplatin in a sequence-dependent fashion (Dorr and Lagel, 1990). Similarly, if mesna is directly mixed with cisplatin in vitro or concurrently administered in vivo, the antitumor efficacy of cisplatin is reduced in mice bearing P-388 leukemia (Dorr and Lagel, 1989). However, by delaying mesna administration for 5 minutes *after* cisplatin is administered, there is no significant impairment in antitumor efficacy (Table 9). This pilot preclinical study suggests that mesna inactivates cisplatin in a time-dependent fashion. Whether this interaction is clinically useful or safely feasible without impairing cisplatin antitumor effects is not known.

Sodium Thiosulfate

Sodium thiosulfate is an inorganic sulfur-containing nucleophile which has an ionized (activated) sulfur at physiologic pH. This agent is used as part of the

Table 9. Effect of mesna on survival of mice treated with cisplatin.

Cisplatin (mg/kg)	Mesna (mg/kg)	Dose-Interval (min)	Survival Median Days	Survival %
Normal Mice				
30	0	—	6	0
30	50	0	7	40
30	50	5	6	10
P-388 Leukemia				
0	50	—	9	—
10	0	—	19*	
10	50	0	15*+	
10	50	5	19*	

Dorr and Lagel., 1989.

*p < .05 different from control (mesna only).

+p < .05 different from cisplatin without mesna.

treatment of cyanide intoxication (Baskin et al., 1992) and to antagonize the toxic actions of mechlorethamine (nitrogen mustard) [Owens and Hatigoglu, 1961; Dorr et al., 1988; Bonadonna et al., 1965]. The drug is eliminated primarily by glomerular filtration and renal tubular secretion with a total body clearance of 190 ± 76 mL/min/m^2 and a renal clearance of 50 ± 11 mL/min/m^2 (Shea et al., 1984). About 29% of the dose is recovered as unchanged thiosulfate in the urine, but overall, 95% of the dose is recovered in the urine over 4 hours and the apparent half life in the plasma is approximately 80 minutes. The relatively small volume of distribution for sodium thiosulfate suggests that extravascular distribution is very limited. This lays the pharmacokinetic groundwork for the use of sodium thiosulfate as a systemic chemoprotectant for cisplatin administered in the intraperitoneal (IP) cavity in women with advanced ovarian cancer.

Mechanistically, thiosulfate is believed to act as an alternative nucleophile for binding to electrophilic platinum coordination sites. Based on results with thiourea which showed an ability to dissociate established platinum-DNA interstrand cross-links in vitro (Filipski et al., 1979), there is a belief that sodium thiosulfate may similarly "rescue" normal tissues such as the kidney, from toxic insult by cisplatin (Burchenal et al., 1978).

In a mouse model, sodium thiosulfate protected from cisplatin-induced nephrotoxicity as measured by BUN levels, kidney weight and medullary hemorrhage (Howell and Taetle, 1980). However, the concurrent application of sodium thiosulfate with cisplatin, both in tumor cell cultures and in vivo, blocked the antitumor activity of platinum . A similar inhibitory effect was observed in 4 mouse tumor models showing that *concurrent intravenous* use of cisplatin and sodium thiosulfate, allows for a larger platinum dose to be safely administered, but with a *decrease* in antitumor activity (Aamdal et al., 1988). Interestingly, rat studies show that concurrent sodium thiosulfate does not substantially reduce cisplatin intracellular concentrations nor subcellular distributions in the kidney cells (Uozumi and Litterst, 1986).

Another early mechanistic trial in bacterial cells and mice, showed that simultaneously-administered thiosulfate did not reduce plasma levels of total platinum or non-protein bound platinum in vivo. In contrast, the in vitro studies showed that thiosulfate inhibited the binding of platinum to serum proteins and caused a commensurate decrease in cellular platinum uptake. Thus, thiosulfate may not change "bulk" platinum pharmacokinetics but at the cellular level, it prevents platinum binding to macromolecules and entry into cells due to the formation of Pt-thiosulfate complexes in the extracellular fluid (Uozumi et al., 1984). This suggests that simple alterations in drug distribution in the renal system, do not fully explain the nephroprotective effect of thiosulfate.

These studies set the stage for the loco-regional use of IP cisplatin and IV thiosulfate in ovarian cancer. In an initial clinical trial, Howell et al. (1982) showed that the maximally-tolerated IP dose of cisplatin could be increased from 90 mg/m^2 to 270 mg/m^2 with concurrent IV thiosulfate. Some clinical responses were obtained in this trial. However, pharmacokinetic studies showed that even with IV thiosulfate, there were still large amounts of unneutralized platinum in the plasma. The resultant total systemic exposure to unneutralized drug following IP cisplatin at 270 mg/m^2 was approximately twice that produced by an IV dose of 100 mg/m^2. This suggests that thiosulfate, at doses up to 2.13 $g/m^2/hr$ for 12 hours, were not effectively neutralizing cisplatin which was redistributing into the plasma from the IP space. Nonetheless, protection of the kidneys with IV thiosulfate was good, and the ratio of the platinum area under the curve (AUC) for peritoneal/plasma compartment was high. This ranged from 13.4 to 15.2 for a 90 mg/m^2 IP dose of cisplatin and 2 dose levels of thiosulfate (Howell et al., 1983) [Table 10]. Thus, there is still a major pharmacokinetic advantage for IP cisplatin in localizing drug exposure to the tumor site. Other advantages of thiosulfate included the elimination of thrombocytopenia and a reduction in the severity of nausea and vomiting, which nonetheless, were still substantial and universal (Howell et al., 1983). However as the data in Table 9 shows, at the highest dose of thiosulfate there is a 23-32% *decrease* in mean peak cisplatin concentrations in the IP space. This undoubtedly reflects the rapid and effective distribution of thiosulfate *into* the IP compartment. In contrast, thiosulfate did not change the half-life of cisplatin in the IP space which averaged 0.85 ± 0.26 hours. In this trial, steady state plasma thiosulfate concentrations averaged 410 μg/mL following a loading dose of 4 g/m^2 and a continuous IV infusion of 2.13 $g/m^2/hr$. This translates to a mean total body clearance of 85 $mL/min/m^2$ (Howell, 1983). A follow-up pharmacokinetic study reported a slightly lower mean thiosulfate clearance value of 59 ± 52 mL/min but with nearly the same thiosulfate-mediated reduction in total platinum exposure in the peritoneal cavity (36%) and plasma (25%) [Goel et al., 1989].

It is again important to point out that all of these trials describe excellent protection from cisplatin nephrotoxicity with concurrent IV thiosulfate, a feature which no doubt relates to the concentration of thiosulfate in the kidneys (Shea et al., 1984). This suggested that possibly both agents could be administered simultaneously by the IV route. When the two agents are given concurrently by the IV route, cisplatin doses could be escalated to 225 mg/m^2 before dose-limiting nephrotoxicity occurred (Pfeifle et al., 1985). In this trial, a larger fixed dose of thiosulfate, 9.9 g/m^2, was given IV over 3 hours concurrently with escalating doses of cisplatin. The total systemic exposure to cisplatin following a 202.5 mg/m^2 dose with IV thiosulfate was roughly double that achieved by an IV cisplatin dose of 100 mg/m^2 without thiosulfate. Furthermore, there were no

Table 10. Effect of IV thiosulfate on IP cisplatin pharmacokinetics.

No. of Evaluable Courses	Thiosulfate Dose ($g/m^2/hr$)	Mean Peak Cisplatin Concentration (µg/mL)		Mean AUC (µg/mL/hr)		
		IP Cavity	Plasma	IP Cavity	Plasma	Ratio
7	0	46 ± 2.18	2.4 ± 1.4	97 ± 64.9	8.1 ± 5.1	15.2 ± 11.0
6	0.43	54 ± 12.8	2.7 ± 1.7	106.6 ± 66.0	8.3 ± 4.6	13.4 ± 5.0
5	2.13	34.6 ± 21.3	2.4 ± 2.2	74.7 ± 4.8	5.5 ± 4.8	14.7 ± 6.5

Howell et al., 1983.

alterations in the volume of distribution, or total body clearance of cisplatin when concurrent IV thiosulfate was administered. Nonetheless, concurrent IV administration of these agents is not feasible due to a major reduction in platinum's antitumor efficacy.

Overall, these findings suggest that IV sodium thiosulfate is a useful antagonist of IP cisplatin nephrotoxicity. While it is not truly selective for the kidneys, its pharmacokinetic properties help to localize most of its effects to the renal system. This clearly affords major dose-escalation of IP cisplatin which should produce greater antitumor effects in diseases localized to the peritoneal compartment. However, randomized clinical trials documenting the superiority of this approach are not yet available and therefore, this approach remains intriguing and unequivocally safe, but still experimental.

ORG 2766

ORG 2766 is a melanocortin-derived peptide which contains the similar peptide sequence as in the α-MSH and ACTH receptor binding (4-9) epitope. Several amino acid substitutions may explain this agent's resilience to degradation and unique pharmacology which involves neurotrophic activity without corticotrophic or melanotrophic effects. These actions may arise by mimicing the function of melanotropic peptides derived from the proopiomelanocortin gene whose expression has been linked to both the stress response and specifically, the response to nerve injury (DeWied, 1982). Several studies in rodents have shown that melanocortin-like peptides may accelerate peripheral nerve repair following experimental injury (Strand et al., 1986). This suggested that melanocortin-derived peptides might be useful for treating or preventing cisplatin-induced neuropathy.

Cisplatin-induced neuropathy is unique in that the damage is limited purely to sensory nerves, and in particular, those which are thickly myelinated and, even though, heavily myelinated nerves are most effected, there is no relationship to implicate cellular B_{12} levels as a mechanism for cisplatin-induced neuropathy (Trugman et al., 1985; Hamers et al., 1991). A histopathologic study in humans has shown that the dorsal root ganglia is the most sensitive structure (Gregg et al., 1992). In contrast, central structures of the brain and spinal cord are protected by the blood brain barrier from both damage and platinum accumulation. For peripheral nerves there is a linear relationship between platinum levels, toxicity, and the cumulative dose of cisplatin administered. Importantly, these intraneural platinum levels do not diminish with time suggesting that the drug is retained neuronally indefinitely in the active form. The initial symptoms of tingling and paresthesias are followed by progressively worsening sensory ataxia. This results in impeded walking and a loss of vibration perception, fine touch perception and kinesthesia. Oddly, pain and temperature sensitivity are usually spared (Elderson et al., 1989; Hamers et al., 1991; Hovesdadt et al., 1992). Unfortunately, even if cisplatin therapy is halted at the first signs of neuropathy, symptoms may continue to develop over months. And, spontaneous healing is rarely complete and is a very slow process, often extending 6 to 30 months after the last treatment course (Hamers et al., 1991; Ongerboer de Visser et al., 1985; Cersosimo , 1989).

A model of cisplatin-induced neuropathy has been developed to test the effectiveness of ORG-2766 as a neuroprotectant for cisplatin. Rodents exposed to repeated cisplatin injections develop a dose-dependent decrease in sensory-nerve conduction velocities measured after monopolar electrode stimulation of the sciatic and tibial nerves at the sciatic notch and ankle, respectively (de Koning et al., 1987). The first, m-response, results from direct stimulation of α-motor fibers and has a short latency. The second, H-response, results from the stimulation of Ia-fibers connected to α-motor neurons in the spinal cord. It has a longer latency for sensory nerve conduction velocity and is termed the "HSNCV". Initial experiments in rats showed that after 47 days, a cumulative cisplatin dose of 13 mg/kg caused significant slowing of the HSNCV to 76% of control values, whereas there was no slowing in the group given ORG-2766, 10 µg subcutaneously 4 times per week (de Koning et al., 1987). There was no effect of ORG-2766 on the motor neuron nerve conduction velocity MNCV nor on the antitumor efficacy of cisplatin against an IgM immunocytoma tumor in a different strain of inbred rats (de Koning et al., 1987). Similar preclinical results were reported by van der Hoop et al. in nude mice implanted with a FMa human mucinous adenocarcinoma and in rats given a cumulative cisplatin dose of 22 mg/kg (van der Hoop et al., 1988). A more recent electrophysiologic and histopathologic study in rats again showed that ORG-2766 prevents the decrease in HSNCV but did not prevent cisplatin-induced morphologic changes in the nucleolus and lysosomes of spinal ganglion neurons (Müller et al., 1990). These histopathologic changes included aggregation of pale-staining and fibrillar components in a few large regions of the nucleoli and increased irregularly shaped lysosomes with lipid-like inclusions surrounded by a thick electron-dense rim (Müller et al., 1990).

A pilot clinical trial has been performed with ORG-2766 compared to placebo in patients receiving cisplatin for advanced ovarian cancer. In this trial, 55 women were randomized to receive mannitol (placebo) or ORG-2766 at a low dose of 0.25 mg/m^2 or a high dose of 1 mg/m^2 intravenously twice on the day of cisplatin: at the start of a prehydration infusion, and again 24 hours later (van der Hoop et al., 1990). Patients received 6 cycles of chemotherapy containing cisplatin 75 mg/m^2 and cyclophosphamide 750 mg/m^2 every 3 weeks. The major endpoint for cisplatin-induced neuropathy was vibration perception threshold (Halonen, 1986) as standardized by Goldberg and Lindblom (1979). In addition, other neurotoxic symptoms were assessed and quantitatively scored for the 3 groups after 4 and 6 cycles of therapy had been delivered. Table 11 summarizes the findings of a major protective role for ORG-2766 in this study. In this analysis, symptoms were scored as present (1) or absent (0), and signs were scored as absent, present (I) and more severe (II). The ORG-2766 groups clearly benefitted significantly and in a dose-dependent fashion in terms of reduced signs and symptoms of cisplatin neuropathy. Likewise, vibration sensation

Table 11. Effects of the neuropeptide fragment ORG-2766 on cisplatin-induced neuropathies after 6 cylces of therapy.

Cisplatin Toxicity	Placebo (n = 22)	Low-Dose ORG-2766 (n = 17)	High-Dose ORG-2766 (n = 16)
Mean Vibration Perception in μm (SEM)	5.87 (1.97)	2.31 (0.75)	0.88 (0.17)
Number of Patients With Symptoms (%)			
Paresthesia	8 (36%)	3 (18%)	2 (13%)
Numbness	7 (32%)	2 (12%)	1 (6%)
Strength Loss	1 (5%)	0	0
Loss of Dexterity	3 (14%)	1 (6%)	0
Unsteadiness	1 (5%)	0	0
Pain	3 (14%)	2 (12%)	0
Lhermitte's Sign	2 (9%)	2 (12%)	0
Sum Score (SEM)	2.08 (0.61)	1.22 (0.45)	0.43 (0.23)
Number of Patients With Signs (%)			
Sense of Pain	5	3	0
Fine Touch	5	0	1
Sense of Vibration	5	3	2
Achilles Tendon Reflex	6	1	2
Sum Score I (SEM)	1.75 (0.55)	0.78 (0.24)	0.71 (0.21)
Sum Score II (SEM)	7.42 (1.30)	4.33 (0.33)	3.57 (0.27)

van der Hoop et al., 1990.

thresholds increased with the cumulative cisplatin dose but were halved with low dose ORG-2766. With high dose ORG-2766, mean vibration perception thresholds were not much greater than at baseline prior to cisplatin therapy. Importantly, the complete response rates were similar in 34 evaluable patients: 9/15 (60%) in the placebo group versus 5/11 (45%) in the low dose ORG group, and 5/8 (63%) in the high dose ORG-2766 group. And, no additional toxicities

Table 12. Time-course of vibration preception threshold following cisplatin and ORG-2766 therapy or placebo.

	Increase in Vibration Perception Threshold (Amplitude in μm)				
		Months After Cisplatin			
Patient Group (No.)	Before (n = 18)	1 (n = 12)	1-4 (n = 13)	4-12 (n = 16)	12-74 (n = 9)
Placebo (7)	0.6	3.7	8.1	4.8	2.9
ORG 0.25 mg (5)	0.5	2.9	14.6	3.6	0.6
ORG 1.0 mg (6)	0.5	1.1	2.5	2.0	0.8

Hovestadt et al., 1992.

were observed in either ORG-2766-treated group. A follow-up study of a subset of 18 patients in this trial showed that several ORG-treated patients did develop neuropathic signs and symptoms after the discontinuation of treatment (Hovestadt et al., 1992). Table 12 shows that while the number of symptomatic patients was smaller in the ORG-treated groups, the number of patients in each subgroup was too small to draw definitive conclusions.

Unfortunately, a second and larger double-blind trial of ORG-2766 conducted in the Netherlands was not positive (Neijt et al., 1994). A total of 131 ovarian cancer patients receiving cisplatin and cyclophosphamide were randomized to receive placebo or ORG-2766 2 mg/m^2 before and after cisplatin. Over the course of 6-8 cycles of chemotherapy, there was no difference in the increase in vibration perception threshold. Other clinical signs and symptoms of cisplatin toxicity were also similar in the two groups. Although there is conflicting data with ORG-2766, the initial pilot study results were impressive and it suggests the need for further definitive double-blind, placebo-controlled trials with this compound.

DISCUSSION

Even though the gears of science turn slowly, there has been considerable progress in identifying potentially useful modulators of platinum toxicity. Still, there have been failures. In this regard, diethyldithiocarbamate illuminates the value of randomized placebo-controlled trials as a crucible for identifying truly useful new pharmaceuticals. As the most recent study by Gandara et al. (1995) showed, DDTC lacked efficacy as a nephroprotectant for cisplatin and appeared to actually decrease the deliverable dose-intensity. This clearly relates to the more prominant side effect profile with this sulfhydryl-derivative of disulfiram. Of those agents which still hold promise, glutathione appears to be ripe for a definitive randomized, placebo-controlled trial. It has been shown to lack an effect on cisplatin's pharmacokinetic disposition and there are several pilot studies suggesting some efficacy at reducing toxicity without effecting antitumor efficacy. However, it is unclear, even in the non-randomized pilot studies, whether the degree of reduction in cisplatin-induced neuropathy and other toxicities, is large enough to warrant broader studies. Similarly, disulfiram has only received cursory initial trials. In this case, there is a suggestion that cytoprotective drug levels may not be achievable at tolerable oral doses. Thus, a careful pharmacokinetic analysis in a small population is needed before larger trials should be considered.

Two other sulfhydryl-based compounds lack even preclinical evidence for cytoprotective activity for cisplatin. N-acetylcysteine was not beneficial in a mouse model of myelosuppression and mesna appeared to reduce both toxicity and antitumor efficacy in mice. Another problematic agent is the GSH depleting drug L-buthionine sulfoximine (L-BSO). It is difficult to mechanistically explain how the depletion of GSH by BSO can protect the kidneys from cisplatin-induced nephrotoxicity in animals and at the same time, acknowledge the documented efficacy of sulfhydryl administration or actual GSH supplementation in both animals and humans receiving cisplatin. While there may be other cytoprotective mechanisms for BSO which do not involve GSH depletion in the kidney, there is certainly a worry that the loss of GSH by BSO may be deleterious in other organ systems in patients receiving cisplatin. Clearly, there is much additional preclinical work which will be needed with BSO and platinum-containing anticancer agents before any clinical trial can be contemplated. The use of BSO in patients receiving other alkylating agents, and particularly cyclophosphamide would pose additional problems since BSO significantly increases the lethality of this agent (Soble and Dorr, 1987). This disastrous effect has been shown to be due to profound GSH depletion in muscle tissues leading to severe acute cardiomyopathy in animals (Friedman, 1990).

Of the various sulfur nucleophiles, sodium thiosulfate appears to have the highest merit for use as a systemic antagonist to regionally administered cisplatin. Nephrotoxicity with IP cisplatin is significantly reduced with thiosulfate and there are studies showing that substantial cisplatin dose-intensification can be achieved. These beneficial effects stem from the predominant urinary elimination of thiosulfate thereby affording some degree of selectivity to the chemoprotective action. The limitations of thiosulfate stem from the facile equilibration of thiosulfate into- and cisplatin out of the IP compartment. This reduces IP platinum AUCs by about one-third. Nonetheless, IV thiosulfate appears to be safe and effective when used with high dose IP cisplatin. In contrast, the data on the combined IV use of both agents is not compelling. Furthermore, except for nephrotoxicity, sodium thiosulfate does not appear to substantially reduce other platinum-related toxicities, particularly neuropathy. In this regard, reduced glutathione and the aminothiol amifostine (WR-2721) appear to have consistent evidence of clinical efficacy. The limitations of studies with ORG-2766 include inconsistent observations of efficacy in pilot trials indicating the need for larger studies. In adidtion, there is a need for longer-term monitoring after ORG-2766 to see if cisplatin-induced neuropathy is prevented or merely delayed in onset.

There is now room to hope that a few effective modulators of platinum-induced nephrotoxicity and neuropathy can achieve truly selective cytoprotection. This end result will require broad Phase III confirmatory trials which should be instituted as soon as possible.

REFERENCES

Aamdal S., Fedstad O., and Pihl A., 1988, Sodium thiosulfate fails to increase the therapeutic index of intravenously administered cis-diamminedichloroplatinum (II) in mice bearing murine and human tumors, *Cancer Chemother Pharmacol.* 21:129.

Andrews P.A., and Howell S.B., 1990, Cellular pharmacology of cisplatin: perspectives on mechanisms of acquired resistance, *Cancer Cells.* 2(2):35.

Andrews P.A., Murphy M.P., and Howel S.B., 1985, Differential potentiation of alkylating and platinating agent cytotoxicity in human ovarian carcinoma cells by glutathione depletion, *Cancer Res.* 30:643.

Asta-Werke Degussa Pharma Grouppe, 1984, Uromitexan (mesna) uroprotector, basic information, Bielefeld, FRG.

Baskin S.I., Horowitz A.M., and Nealley E.W., 1992, The antidotal action of sodium nitrite and sodium thiosulfate against cyanide poisoning, *J Clin Pharmacol.* 32:368.

Berry J.M., Jacobs C., Sikic B., et al., 1990, Modification of cisplatin toxicity with diethyldithiocarbamate, *J Clin Oncol.* 8:1585.

Bodenner D.I., Dedon P.C., Keng P.C., et al., 1986, Effect of diethyldithiocarbamate (DDTC) on cis-diamminedichloroplatinum (II)-induced cytotoxicity, DNA cross-linking and gamma-glutamyl transpeptidase inhibition, *Cancer Res.* 46:2745.

Bohm S., Spatti G., De Re F., et al., 1991, A feasibility study of cisplatin administration with low-volume hydration and glutathione protection in the treatment of ovarian carcinoma, *Anticancer Res.* 11:1613.

Bonadonna G., and Karnofsky D.A., 1965, Protection studies with sodium thiosulfate against methyl bis (β-chloroethyl)amine hydrochloride (HN2) and its ethylenimonium derivative, *Clin Pharmacol Ther.* 6(1):50.

Borch R.F., and Pleasants M.E., 1979, Inhibition of cis-platinum nephrotoxicity by diethyldithiocarbamate rescue in a rat model, *Proc Natl Acad Sci USA.* 76(12):6611.

Borch R.F., Katz J.C., Lieder P.H., and Pleasants M.E., 1980, Effect of diethyldithiocarbamate rescue on tumor response to cis-platinum in a rat model, *Proc Natl Acad Sci USA.* 77(9):5441.

Burchenal J.H., Kalaher K., Dew K., et al., 1978, Studies of cross-resistance, synergistic combinations and blocking of activity of platinum derivatives, *Biochimie.* 60:961.

Cascinu S., Cordella L., Del Ferro E., et al., 1995, Neuroprotective effect of reduced glutathione on cisplatin-based chemotherapy in advanced gastric cancer: a randomized double-blind placebo-controlled trial, *J Clin Oncol.* 13(1):26.

Cavaletti G., Tredici G., Pizzini G., et al., 1992, Morphologic, morphometric and toxicologic evaluations of the effect of cisplatin alone or in combination with glutathione on the rat nervous system, *Clin Neuropathol.* 11:169.

Cersosimo R.J., 1989, Cisplatin neurotoxicity, *Cancer Treat Rep.* 16:195.

Cohen J.D., and Robins H.I., 1990, Cytotoxicity of diethyldithiocarbamate in human versus rodent cell lines, *Invest New Drugs.* 8:137.

Colombo N., Bini S., Miceli D., et al., 1995, Weekly cisplatin ± glutathione in relapsed ovarian carcinoma, *Int J Gynecol Cancer.* 5:81.

Cozzaglio L., Doci R., Colla G., et al., 1990, A feasibility study of high-dose cisplatin and 5-fluorouracil with glutathione protection in the treatment of advanced colorectal cancer, *Tumori.* 76:590.

De Re F., Bohm S., Oriana S., et al., 1993, High-dose cisplatin and cyclophosphamide with glutathione in the treatment of advanced ovarian cancer, *Ann Oncol.* 4:55.

De Wied D., and Jolles J., 1982, Neuropeptides derived from pro-opiocortin: behavioral, physiological and neurochemical effects, *Physiol Rev.* 62:976.

de Koning P., Neijt J.P., Jennekens F.G.I., and Gispen W.H., 1987, Org.2766 protects from cisplatin-induced neurotoxicity in rats, *Exper Neurol.* 97:746.

de Sauvage F.J., Hass P.E., Spencer S.D., et al., Stimulation of megakaryocytopoiesis and thrombopoiesis by the c-Mpl ligand, *Nature.* 369:533.

Dedon P.C., and Borch R.F., 1987, Characterization of the reactions of platinum antitumor agents with biologic and nonbiologic sulfur-containing nucleophiles, *Biochem Pharmacol.* 36:1955.

DeGregorio M.W., Gandara D.R., Holleran W.M., et al., 1989, High-dose cisplatin with diethyldithiocarbamate (DDTC) rescue therapy: preliminary pharmacologic observations, *Cancer Chemother Pharmacol.* 23:276.

Di Re F., Bohm S., Oriana S., et al., 1990, Efficacy and safety of high-dose cisplatin and cyclophosphamide with glutathione protection in the treatment of bulky advanced epithelial ovarian cancer. *Cancer Chemother Pharmacol.* 25:355.

Dible S.E., Siddik Z.H., Boxall F.E., et al., 1987, The effect of diethyldithiocarbamate on the haematological toxicity and antitumor activity of carboplatin, *Eur J Cancer Clin Oncol.* 23:813.

Doroshow J.H., Locker G., Ifrim I., et al., 1981, Prevention of doxorubicin cardiac toxicity in the mouse by N-acetylcysteine, *J Clin Invest.* 64:1053.

Doroshow J.H., Locker G.Y., and Myers C.E., 1980, The enzymatic defenses of the mouse heart against oxygen metabolites, *J Clin Invest.* 65:128.

Dorr R.T., and Lagel L., 1989, Interaction between cisplatin and mesna in mice. *J Cancer Res Clin Oncol.* 115:604.

Dorr R.T., Soble M., and Alberts D.S., 1988, Efficacy of sodium thiosulfate as alocal antidote to mechlorethamine skin toxicity in the mouse, *Cancer Chemother Pharmacol.* 22:299.

Elderson A., van der Hoop G.P., Haanstra W., et al., 1989, Vibration perception and thermoperception as quantitative measurements in the monitoring of cisplatin-induced neurotoxicity, *J Neurol Sci.* 93:167.

Evans R.G., Wheatley C., Engel C., et al., 1984, Modification of the bone marrow toxicity of cis-diamminedichloroplatinum(II) in mice by diethyldithiocarbamate, *Cancer Res.* 44:3686.

Filipski J., Kohn K.W., Prather R., and Bonner W.M., 1979, Thiourea reverses cross-links and restroes biological activity in DNA treated with dichlorodiaminoplatinum(II), *Science.* 204:181.

Fontanelli R., Spatti G., Raspagliesi F., et al., 1992, A preoperative single course of high-dose cisplatin and bleomycin with glutathione protection in bulky stage IB/II carcinoma of the cervix, *Ann Oncol.* 3:117.

Francis P., Markman M., Hakes T., et al., 1989, Diethyldithiocarbamate chemoprotection of carboplatin-induced hematological toxicity, *J Cancer Res Clin Oncol.* 119:360.

Freeman M.L., and Meredith M.J., 1989, The relationship between intracellular glutathione concentration and cisplatin cytotoxicity, *Proc Amer Assoc Cancer Res.* 30:460.

Friedman H.S., Colvin O.M., Aisaka K., et al., 1990, Glutathione protects cardiac and skeletal muscle from cyclophosphamide-induced toxicity, *Cancer Res.* 50:2455.

Gale G.R., Atkins L.M., and Walker, Jr., E.M., 1982, Further evaluation of diethyldithiocarbamate as an antagonist of cisplatin toxicity, *Ann Clin Lab Sci.* 12(5):345.

Gandara D.R., Nahhas W.A., Adelson M.D., et al., 1995, Randomized placebo-controlled multicenter evaluation of diethyldithiocarbamate for chemoprotection against cisplatin-induced toxicities, *J Clin Oncol.* 13(2):490.

Glover D., Glick J.H., Weiler C., et al., 1987, WR-2721 and high-dose cisplatin: an active combination in the treatment of metastatic melanoma, *J Clin Oncol.* 5:574.

Goel R., Cleary S.M., Horton C., et al., 1989, Effect of sodium thiosulfate on the pharmacokinetics and toxicity of cisplatin, *J Natl Cancer Inst.* 81:1552.

Goldberg J.M., and Lindblom U., 1979, Standardised method of determining vibratory perception thresholds for diagnosis and screening in neurological investigation, *J Neurol Neurosurg Psychiatry.* 42:793.

Gonias S.L., Oakley A.C., Walther P.J., and Pizzo S.V., 1984, Effects of diethyldithiocarbamate and nine other nucleophiles on the intersubunit protein cross-linking and inactivation of purified human α_2-macroglobulin by cis-diamminedichloroplatinum(II), *Cancer Res.* 44:5764.

Gregg R.W., Molepo J.M., Monpetit V.J.A., et al., 1992, Cisplatin neurotoxicity: the relationship between dosage, time, and platinum concentration in neurologic tissues, and morphologic evidence of toxicity, *J Clin Oncol.* 10(5):795.

Griffith O.W., and Meister A., 1979, Potent and specific inhibition of glutathione synthesis by buthionine sulfoximine (S-n-butyl homocysteine sulfoximine), J Biol Chem. 254:7558.

Hacker M.P., Ershler W.B., Newman R.A., and Gamelli R.L., 1982, Effect of disulfiram (tetraethylthiuram disulfide) and diethyldithiocarbamate on the bladder toxicity and antitumor activity of cyclophosphamide in mice, *Cancer Res.* 42:4490.

Halonen P., 1986, Quantitative vibration perception thresholds in healthy subjects of working age, *Eur J Appl Physiol.* 54:647.

Hamers F.P.T., Gispen W.H., and Neijt J.P., 1991, Neurotoxic side-effects of cisplatin, *Eur J Cancer.* 27:372.

Hamers F.P.T., Brakkee J.H., Cavalletti E., et al., 1993a, Reduced glutathione protects against cisplatin-induced neurotoxicity in rats, *Cancer Res.* 53:544.

Hamers F.P.T., Pette C., Bravenboer B., et al., 1993b, Cisplatin-induced neuropathy in mature rats: effects of the melanocortin-derived peptide ORG 2766, *Cancer Chemother Pharmacol.* 32:162.

Hamilton T.C., Winker M.A., Louis K.G., et al., 1985, Augmentation of adriamycin, melphalan and cisplatin cytotoxicity in drug-resistant and -sensitive human ovarian carcinoma cell lines by buthionine sulfoximine mediated glutathione depletion, *Biochem Pharmacol.* 34:2583.

Hovestadt A., Van der Burg M.E.L., Verbiest H.B.C., et al., 1992, The course of neuropathy after cessation of cisplatin treatment, combined with ORG 2766 or placebo, *J Neurol.* 239:143.

Howell S.B., and Taetle R., 1980, Effect of sodium thiosulfate on cis-dichlorodiammineplatinum(II) toxicity and antitumor activity in L1210 leukemia, *Cancer Treat Rep.* 64(4-5):611.

Howell S.B., Pfeifle C.E., Wung W.E., and Olshen R.A., 1982, Intraperitoneal chemotherapy with cisplatin, *Proc Am Soc Clin Oncol.* 1:27.

Howell S.B., Pfeifle C.E., Wung W.E., and Olshen R.A., 1983, Intraperitoneal cis-diamminedichloroplatinum with systemic thiosulfate protection, *Cancer Res.* 43:1426.

Leone R., Fracasso M.E., Soresi E., et al., 1992, Influence of glutathione administration on the disposition of free and total platinum in patients after administration of cisplatin, *Cancer Chemother Pharmacol.* 29:385.

Lerza R., Bogliolo G., Muzzulini C., and Pannacciulli I., 1986, Failure of N-acetylcysteine to protect against cis-dichlorodiammineplatinum(II)-induced hematopoietic toxicity in mice, *Life Sci Rev.* 38(19):1795.

Leyland-Jones B., 1988, Whither the modulation of platinum?, *J Natl Cancer Inst.* 80(18):1432.

Leyland-Jones B., Deesen P., Reidenberg M.M., et al., 1980, Amelioration of cisplatin renal and gastrointestinal toxicity by N-acytylcysteine in rats, *Clin Res.* 28:622A.

Leyland-Jones B., Morrow C., Tate S., et al., 1983, *cis*-Diamminedichloro platinum (II) nephrotoxicity and its relationship to renal γ-glutamyl transpeptidase and glutathione, *Cancer Res.* 43:6072.

Litterest C.L., Bertolero F., and Uozumi J., 1986, The role of glutathione and metallothionein in the toxicity and subcellular binding of cisplatin: In: MCBrain D.C.H., and Slater T.F. (eds) Biochemical mechanism of platinum antitumor drugs, IRL Press, Oxford p 227.

Litterst C.L., Tong S., Hirokata Y., and Siddik Z.H., 1982, Alteration in hepatic and renal levels of glutathione and activities of glutathione S-transferases from rats treated with cis-dichlorodiammine platinum-II, *Cancer Chemother Pharmacol.* 8:67.

Maines M.D., 1986, Differential effect of cis-platinum on regulation of liver and kidney haem and haemoprotein metabolism, *Biochem J.* 237:713.
Mayer R.D., and Maines M.D., 1990, Promotion of *trans*-platinum in vivo effects on renal heme and hemoprotein metabolism by D,L-buthionine-S,R-sulfoximine, *Biochem Pharmacol.* 39(10):1565.
Mayer R.D., Lee K., and Cockett A.T.K., 1987, Inhibition of cisplatin induced nephrotoxicity in rats by buthionine sulfoximine, a glutathione synthesis inhibitor, *Cancer Chemother Pharmacol.* 20:207.
Meister A., Tate S. S., and Griffith O. W., 1981, γ-Glutamyl trans-peptidase, Methods Enzymol. 77:237.
Müller L.J., van der Hoop R.G., Moorer-van Delft C.M, et al., 1990, Morphological and electrophysiological study of the effects of cisplatin and ORG • 2766 on rat spinal ganglion neurons, *Cancer Res.* 50:2437.
Neijt J., Van der burg M., Vecht C., et al., 1994, A double-blind randomised study with ORG-2766, an ACTH (4-9) analog, to prevent cisplatin neuropathy, *Proc Am Soc Clin Oncol.* 13:261.
O'Dwyer P.J., Hamilton T.C., Young R.C., et al., 1992, Depletion of glutathione in normal and malignant human cells in vivo by buthionine sulfoximine: Clinical and biochemical results, *J Natl Cancer Inst* 34:264.
Ongerboer de Visser B.W., and Tiessens G., 1985, Polyneuropathy induced by cisplatin, *Prog Exp Tumor Res.* 29:190.
Owens G., and Hatiboglu I., 1961, Clinical evaluation of sodium thiosulfate, a systemic neutralizer of nitrogen mustard, *Ann Surg.* 154:895.
Paredes J., Hong W.I., Felder T.B., et al., 1988, Prospective randomized trial of high dose cisplatin and fluorouracil infusion with or without sodium diethyldithiocarbamate in recurrent and/or metastatic squamous cell carcinoma of the head and neck, *J Clin Oncol.* 6:955.
Pfeifle C.E., Howell S.B., Felthouse R.D., et al., 1985, High-dose cisplatin with sodium thiosulfate protection, *J Clin Oncol.* 3(2):237.
Pirovano C., Balzarini A., Bohm S., et al., 1992, Peripheral neurotoxicity following high-dose cisplatin with glutathione: clinical and neurophysiological assessment, *Tumori.* 78:253.
Powell A.K., 1954, Effect of dithiocarbamates on sarcoma cells and fibrocytes in vitro, *Br J Cancer.* 8:529.
Qazi R., Chang A.Y.C., Borch R.F., et al., 1988, Phase I clinical and pharmacokinetic study of diethyldithiocarbamate as a chemoprotector from toxic effects of cisplatin, *J Natl Cancer Inst.* 80:1486.
Renoux G., and Renoux M., 1984, Diethyldithiocarbamate (DTC). A biological augmenting agent specific for T cells, In: R.L. Fenichel and M.A. Chirigos (eds) Immune Modulation Agents and Their Mechanisms, Marcel Dekker, Inc., New York, 1984, p. 7.
Rumack B.H., Peterson R.C., Koch G.G., and Amara I.A., 1981, Acetaminophen overdose: 662 cases with evaluation of oral acetylcysteine treatment, *Arch Intern Med.* 141:380.
Scheef W., Klein H.O., Brock N., et al., 1979, Controlled clinical studies with an antidote against the urotoxicity of oxazaphosphorines: preliminary results, *Cancer Treat Rep.* 63:501.
Schmalbach T.K., and Borch R.F., 1990, Mechanism of diethyldithiocarbamate modulation of murine bone marrow toxicity, *Cancer Res.* 50:6218.
Schnellmann R.G., and Mandell L.J., 1986, Multiple effects of presumed glutathione depletors on rabbit proximal tubule, *Kidney Int.* 29:858.
Shaw I.C., and Graham M.I., 1987, Mesna — a short review, *Cancer Treat Rev.* 14:67.

Shea M., Koziol J.A., and Howell S.B., 1984, Kinetics of sodium thiosulfate, a cisplatin neutralizer, *Clin Pharmacol Ther*. 35(3):419.

Slater T.F., Ahmed J., and Ibrahim S.A., 1977, Studies on the nephrotoxicity of *cis*-dichlorodiammine-platinum^{2+} and related substances, J Clin Hematol Oncol. 7:534.

Soble M.J., and Dorr R.T., 1987, Lack of enhanced myelotoxicity with buthionine sulfoximine and sulfhydryl-dependent anticancer agents in mice, *Res Commun Chem Pathol Pharmacol*. 55(2):161.

Spath A., and Tempel K., 1987, Diethyldithiocarbamate inhibits scheduled and unscheduled DNA synthesis of rat thymocytes in vitro and in vivo — dose-effect relationships and mechanisms of action, *Chem-Biol Interact*. 64:151.

Speer R.J., Ridgway H., Hall L.M., et al., 1975, Coordination complexes of platinum as antitumor agents, *Cancer Chemother Rep*. 59:629.

Speyer J.L., Green M.D., Kramer, E., et al., 1988, Protective effect of the bispiperazinedione ICRF-187 against doxorubicin-induced cardiac toxicity in women with advanced breast cancer, *N Engl J Med*., 319:745.

Speyer J.L., Green M.D., Zeleniuch-Jacquotte A., et al., 1992, ICRF-187 permits longer treatment with doxorubicin in women with breast cancer, *J Clin Oncol*. 10(1):117.

Stewart D.J., Verma S., and Maroun J.A., 1987, Phase I study of the combination of disulfiram with cisplatin, *Am J Clin Oncol*. 10(6):517.

Strand F.L., and Smith C.M., 1986, LPH, ACTH, MSH and motor systems. In: De Wied D, Gispen WH, Wimersma Greidanus TjB (eds) *Neuropeptides and Behavior*, Pergamon Press, Oxford Vol. 1, p. 245.

Sunderman F.W., 1971, The treatment of acute nickel-carbonyl poisoning with sodium diethyldithiocarbamate, *Ann Clin Res*. 3:182.

Ten Bokkel Huinink W.W., Wanders J., and Dubbelman R., 1987, High dose carboplatin in refractory ovarian cancer: a phase II trial assessing the protective role against myelosuppression by diethyldithiocarbamate, *Madrid: Eur Cancer Soc*. 214.

Trugman J., Hogenkamp H.P.C., Roelofs R., and Hrushesky W.J.M., 1985, Cisplatin neurotoxicity: failure to demonstrate vitamin B_{12} inactivation, *Cancer Treat Rep*. 69(4):453.

Uozumi J., and Litterst C.L., 1986, The effect of sodium thiosulfate on subcellular localization of platinum in rat kidney after treatment with cisplatin, *Cancer Let*. 32:279.

Uozumi J., Ishizawa M., Iwamoto Y., and Baba T., 1984, Sodium thiosulfate inhibits cis-diamminedichloroplatinum (II) activity, *Cancer Chemother Pharmacol*. 13:82.

van der Hoop R.G., de Koning P., Boven E., et al., 1988, Efficacy of the neuropeptide ORG.2766 in the prevention and treatment of cisplatin-induced neurotoxicity in rats, *Eur J Cancer Clin Oncol*. 24(4):637.

van der Hoop R.G., Vecht C.J., van der Burg M.E.L., et al., 1990, Prevention of cisplatin neurotoxicity with an ACTH(4-9) analogue in patients with ovarian cancer, *N Engl J Med*. 322:89.

Zunino F., Pratesi G., Micheloni A., et al., 1989, Protective effect of reduced glutathione against cisplatin-induced renal and systemic toxicity and its influence on the therapeutic activity of the antitumor drug, *Chem Biol Interact*. 70:89.

Zunino F., Tofanetti O., Besati A., et al., 1983, Protective effect of reduced glutathione against *cis*-dichlorodiammine Pt(II)induced nephrotoxicity and lethal toxicity, *Tumori*. 69:106.

COMBINATION THERAPY WITH CISPLATINUM AND EGF RECEPTOR BLOCKADE

John Mendelsohn, Zhen Fan, and Jose Baselga

Memorial Sloan-Kettering Cancer Center and Cornell University Medical College, New York, NY 10021

ABSTRACT

Simultaneous treatment of xenografts of human A431 vulvar squamous carcinoma cells and MDA-468 breast adenoma cells with anti-EGF receptor monoclonal antibody (MAb) plus maximal tolerated doses of cisplatinum (*cis*-DDP) results in eradication of well-established tumors that are resistant to either therapy when administered alone. Hypotheses that could explain this synergistic antitumor activity are presented. We suggest that activation of dual checkpoints by the combination therapy results in cell death.

ANTI-TUMOR ACTIVITY OF ANTI-EGF RECEPTOR Mabs

Our laboratory has produced MAbs 225 IgG1 and 528 IgG2a against the EGF receptor. They bind to the receptor with affinity comparable to the natural ligand (K_d=2nM) and compete with EGF binding. They can precipitate the receptor and block the activation of receptor tyrosine kinase by EGF or TGF-α [1-3]. It is likely that the MAbs do not react with the actual EGF binding site, but near enough to it to prevent EGF from binding, since they react with a human-specific sequence and do not recognize EGF receptors on rodent cells.

The MAbs can block EGF- and TGF-α-induced stimulation of cell growth rate. This has been demonstrated in cultures of human foreskin fibroblasts, a colon adenocarcinoma cell line, and non-malignant 184 mammary cells [2, 4, 5]. We also demonstrated that MAb-mediated inhibition of proliferation of 184 mammary cells results in growth arrest in late G_1 phase, presumably at the restriction (R) point [6].

The inhibitory effects of MAbs 528 and 225 upon malignant cells were most thoroughly characterized with A431 vulvar squamous carcinoma cells, which express both EGF receptors and TGF-α in large quantities [1, 2, 7]. The lack of an antiproliferative effect when cells were exposed to our MAb 455 provides an important control, because this antibody, which also binds to the EGF receptor, does not inhibit ligand binding and does not block activation of receptor tyrosine kinase [2]. Thus, the specificity of MAbs for the EGF

Platinum and Other Metal Coordination Compounds in Cancer Chemotherapy 2
Edited by H.M. Pinedo and J.H. Schornagel, Plenum Press, New York, 1996

receptor is not sufficient to inhibit growth; only antibodies such as 225 and 528, which have the capacity to block ligand binding and prevent receptor activation, could inhibit proliferation. This blocking activity could be reversed by addition of carefully titrated amounts of EGF or TGF-α [8].

We demonstrated that bivalent $F(ab')_2$ fragment of MAb is as effective as complete MAb in blocking receptor tyrosine kinase activation and proliferation of cultured cells; however monovalent Fab' fragment is a less effective inhibitor of cell growth [9-11]. Our studies also demonstrated that MAb225 induces dimerization of EGF receptors in intact A431 cells, without activating receptor tyrosine kinase [10, 11]. Dimerization accompanied by down-regulation of receptors accounts for the capacity of bivalent 225 MAb and 225 $F(ab')_2$ to inhibit proliferation far more effectively than monovalent 225 Fab' [11].

Studies with MAbs 225 and 528 against the EGF receptor in our laboratory and many others have examined the growth of cultured cells representing many of the common types of human cancer. In most experiments, the cells were subjected to MAb treatment in low serum-culture. Inhibition of proliferation by MAbs was observed in the following situations: skin (vulva) cancer line A431 [1, 2]; squamous lung cancer line Calu-1 [12]; colon adenocarcinoma lines SNU-C1 [13] and DiFi [14]; prostatic adenocarcinoma lines PC3, Du145, and LNCaP [15, 16]; renal carcinoma lines SKRC-4 and 29 [17]; astrocytoma lines [18]; ovarian adenocarcinoma SHIN-3 [19, 20]; head and neck cancer lines 1483, MDA 686 Ln and MDA 886 Ln [21]; cervix carcinoma lines C-4I and ME-180 [22]; and breast adenocarcinoma cell lines, MCF7, MDA-468, BT-474 ([23, 24] and unpublished observations).

In summary, many malignant cells that express high levels of EGF receptors are inhibited by MAb 225 or 528, suggesting that they have active autocrine pathways involving EGF receptor activation. Since a substantial fraction of human tumors express high levels of EGF receptors (approximately 1/3 of all epithelial malignancies, including the common ones), EGF receptor blockade may have wide application clinically.

PRECLINICAL STUDIES WITH HUMAN TUMOR XENOGRAFTS

In vivo effects of treatment with anti-EGF receptor MAbs were assayed against xenografts of human tumor cells. A treatment schedule of intraperitoneal (i.p.) injections twice weekly was selected, based on a measured MAb half-life in serum of 3 days [25]. Antitumor activity was most prominent when the malignant cells expressed high levels of EGF receptors and responded to TGF-α in culture. Administration of either 225 or 528 MAb intraperitoneally, beginning concurrent with tumor cell implantation subcutaneously, caused a dose-dependent inhibition of A431 squamous tumor cell growth [25]. Administration of 1 or 2 mg twice weekly resulted in complete suppression of tumor growth, which persisted after completion of a 3 week course of therapy. Comparable inhibition of xenograft tumor growth was observed with MDA-468 breast carcinoma [26], DiFi colon carcinoma [14], and SKRC-4 renal carcinoma [27].

An $F(ab')_2$ fragment could produce nearly comparable antitumor activity. This was extremely important, because it supported the concept that antitumor activity could be achieved by receptor blockade (a pharmacologic action), and did not require the capacity of MAb to mediate an immune reaction via its Fc fragment [9]. It should be noted that the evidence for physiologic effects of anti-EGF receptor MAbs does not rule out the possible concurrent activity of these antibodies as immune effector agents in vivo.

Treatment with MAb 225 could successfully eradicate xenografts of human tumor cells bearing EGF receptors when therapy was begun at an early stage, within 5 days of implantation subcutaneously. For most well established tumor xenografts, treatment with anti-EGF receptor MAb produced varying degrees of cytostasis, without eliminating the tumors [25, 28, 29]. This suggested to us the need for additional concurrent therapy, if antireceptor MAb therapy was to be useful against cancer in the clinical setting.

There was an early report of synergistic cytotoxicity when anti-EGF receptor MAb 108 was combined with a single dose of *cis*-DDP therapy against KB62 squamous carcinoma cell xenografts [30]. This observation stimulated us to explore the hypothesis that chemotherapy combined with growth factor receptor blockade may increase tumor cell kill in well established tumors. The drugs *cis*-DDP, doxorubicin, and Taxol were selected for initial study because of their excellent activities against many types of cancer.

ANTI-TUMOR ACTIVITY OF ANTI-EGF RECEPTOR Mab COMBINED WITH *CIS*-DDP IN CELL CULTURE

Experiments were carried out to determine the effect of *cis*-DDP, both alone and combined with 225 or 528 MAb, on the growth of A431 cell cultures [29]. Fig.1A shows the

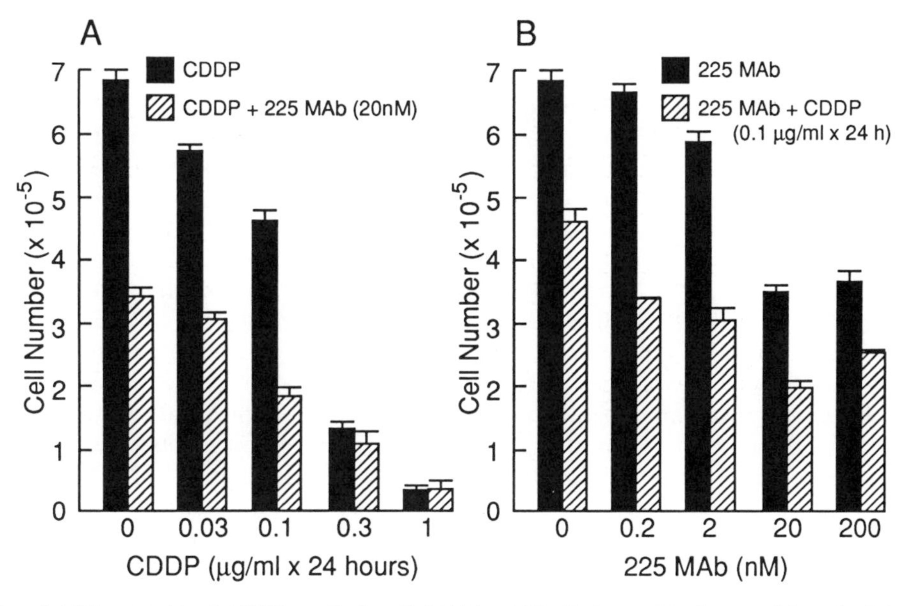

Figure 1. Additive cytotoxicity of *cis*-DDP in combination with 225 MAb on A431 cell cultures. A431 cells were seeded onto 6 well plates at 2 x 10^4 cells/well in A, A431 cells were treated with *cis*-DDP at indicated concentrations on the first day for a total of 24 h, in the continuous presence (▨) or absence (■) of 20nM 225 MAb for 6 days. In B, A431 cells were treated with indicated concentrations of 225 MAb for 6 days, plus (▨) or minus (■) 0.1 μg/ml (0.33 μM) *cis*-DDP during the first day only. The data are presented as the mean cell number of triplicates with SE bars. From [29], with permission.

response to increasing concentrations *cis*-DDP. To more closely mimic *in vivo* treatment conditions, the exposure time of A431 cells to *cis*-DDP was only for the first 24 h, and the cells were maintained in culture for an additional 5 days after removal of the drug. *cis*-DDP inhibited A431 cell proliferation in a dose-dependent fashion. The continuous presence of 20 nM 225 MAb (a saturating concentration) for 6 days produced additive inhibitory effects on cell growth. Conversely, this additive effect was also observed when the concentration of 225 MAb varied from 0.2 to 200 nM, in cultures treated for 24 h with 0.1 μg/ml (0.33 μM) *cis*-

DDP (Fig.1B). Identical results were obtained when 225 MAb was replaced by 528 MAb, which also blocks binding of EGF/TGF-α; however, no additive effect was found when 225 MAb was replaced by 455 MAb, which binds to EGF receptors but does not block binding of EGF/TGF-α (data not shown).

ANTI-TUMOR ACTIVITY AGAINST HUMAN TUMOR XENOGRAFTS OF ANTI-EGF RECEPTOR Mab COMBINED WITH *CIS*-DDP

We first established the maximum tolerated dose of *cis*-DDP which did not produce toxicity resulting in the death of the nude mice. A schedule of 150 μg/25 gm mouse, delivered i.p. and repeated and 10 days later, was found to be tolerated, but produced little antitumor activity.

We next determined the efficacy of treatment of A431 cell xenografts with MAb at varying times after tumor implantation. The dose and schedule of MAb therapy was 1 mg i.p. twice weekly, which previous studies had shown to produce receptor-saturating serum levels, as well as optimal antitumor activity when initiated on day 1 of tumor cell implantation subcutaneously. Treatment delay until day 3 achieved complete suppression of tumor growth. Treatment beginning 5 days after A431 cell implantation slowed tumor growth by 50%, while treatment beginning on day 8 had no anti-tumor effect.

To explore whether the weak antiproliferative activity of either *cis*-DDP or MAb against well established A431 cell xenografts could have additive or synergistic effects, combination treatment with concurrent *cis*-DDP and 225 MAb was administered. The therapy was started 8 days after A431 cells were inoculated and had reached a mean size of approximately 400 mm^3 (about 1 cm in diameter). The growth of A431 cell xenografts was completely suppressed, and the tumors gradually shrank. By day 32, after 8 injections of 225 MAb (1 mg/injection) and two treatments with *cis*-DDP (150 μg/ 25 g mouse x 2), tumors had been eliminated in 6 out of 7 treated mice. One mouse had a residual tumor less than 100 mm^3. All of the mice in the 225 MAb alone, *cis*-DDP alone and control groups died or had to be sacrificed due to bulky tumor burdens. However, in the combination treatment group, mice with regressed tumors were observed for over 6 months and they all remained tumor-free. The single mouse treated with combination therapy which had a small residual tumor after 32 days also survived for over 6 months with a slowly growing tumor. These results were duplicated in subsequent experiments. The treatment protocol was repeated using anti-EGF receptor MAb 528, which shares many properties with 225 MAb except that they differ in their IgG isotype. Fig.2 shows the results of this experiment. There was complete tumor regression in each of the 7 mice in the 528 MAb plus *cis*-DDP combination treatment group, and all of these mice remained tumor-free for over 6 months.

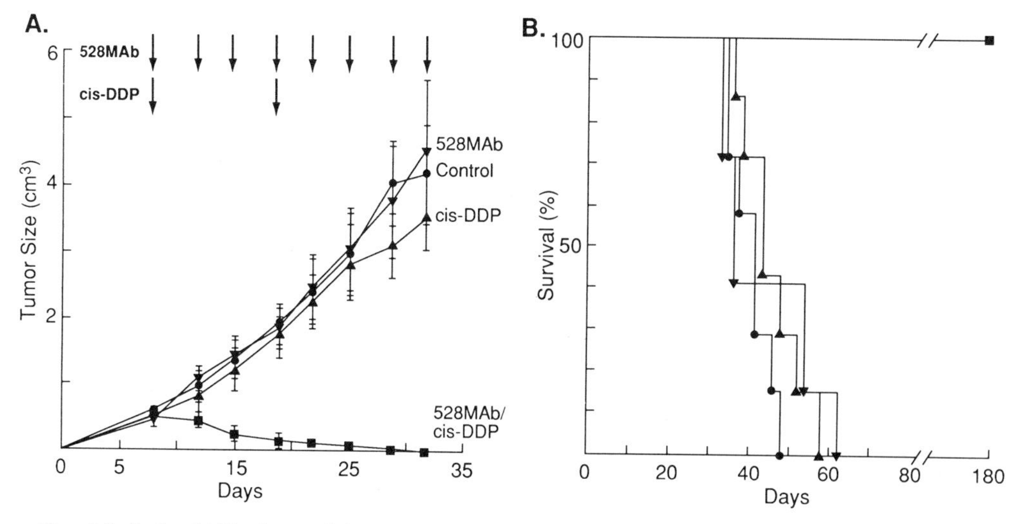

Figure 2. Eradication of A431 cell xenografts by combination treatment of 528 MAb and *cis*-DDP. A431 cells (10^7) were implanted s.c. into nude mice and allowed to grow for 8 days. In A, the mice were given i.p. injections of either phosphate-buffered saline (●); two injections of *cis*-DDP (150 μg/25 g mouse weight) on day 8 and day 18 (▲); or 528 MAb, 1 mg/mouse twice a week for 4 weeks, with (■) or without (▼) two injections of *cis*-DDP (150 μg/25 g mouse weight) on day 8 and day 18. Arrows show the timing of drug and antibody treatment. The data are expressed as the mean tumor size ± SE (7 mice per group). In B, the mice were observed for 6 months for survival. From [29], with permission.

STUDIES WITH OTHER CHEMOTHERAPEUTIC AGENTS

The synergistic antitumor activity of anti-EGF receptor MAb plus chemotherapy, which could eliminate well established xenografts of A431 cells, was not limited to these cells or to a specific drug.

Studies with xenografts of both A431 vulvar squamous carcinoma cells and MDA468 mammary adenocarcinoma cells demonstrated eradication of tumors with a combination of 225 MAb (or 528 MAb) given 1 mg i.p. twice weekly plus doxorubicin give 100 mg i.p. on two successive days (maximum tolerated dose) [28]. Treatment was initiated on day 8. By day 30, all 21 surviving mice from 3 sets of experiments were tumor-free.

In a third study, dose titration studies determined a treatment schedule in which Taxol alone only partially inhibited tumor growth. This dosing schedule of 250μg/25g mouse weight i.v. on days 1, 4, and 9 of therapy was then used in combination with MAb 225 against well established MDA-468 breast adenocarcinoma xenografts. As before, the combination of drug and MAb therapy resulted in a marked enhancement of antitumor effect [31].

In summary, we have demonstrated that combination therapy with anti-EGF receptor MAb plus one of three chemotherapeutic agents results in markedly enhanced activity against well established squamous cancer and adenocarcinoma xenografts. The antitumor activity of the combination of drug and MAb *in vivo* is far more dramatic than the additive effects observed in cell culture. Complete elimination of well established xenografts justifies use of the word "synergism" to describe the *in vivo* interaction, although we acknowledge that formal statistical proof of synergism has not been presented. The drugs that have been studied have quite specific mechanisms of action, raising the possibility that cell death resulting from combining these different chemotherapeutic agents and anti-EGF receptor MAbs may involve a common mechanism activated by drug-induced cell damage and concurrent blockade of a tyrosine kinase essential for growth.

DISCUSSION

Ongoing studies are attempting to determine the explanation for the strong antitumor response observed when MAb 225 is combined with *cis*-DDP therapy. One possibility is that *cis*-DDP is altering the activity of EGF receptors. While we found that treatment with doxorubicin upregulated EGF receptor and TGFα expression in A431 cells [28], we found no changes in cells treated with *cis*-DDP.

Another possibility is that EGF receptor blockade inhibits the capacity of cells to recover from DNA damage. Aghajanian, *et al.* showed that EGF receptor blockade of A431 cells with MAb 225 slows the rate of DNA repair of a transfected platinated plasmid, containing the CMV promoter and a luciferase reporter gene [32]. It is of interest that a previous report demonstrated a slowing of DNA repair from *cis*-DDP damage when cells were treated with a MAb against the HER-2 receptor [33].

Because similar antitumor activity was observed when anti-EGF receptor MAb treatment was combined with three drugs that have differing mechanisms of action, we are exploring, the hypothesis that a common pathway of cell death is activated by this form of combination therapy. It is well documented that chemotherapeutic agents typically kill cells by a mechanism involving programmed cell death. The problem with most epithelial tumors is that at drug doses tolerated by the patient, damage to the cell is sublethal rather than tumoricidal.

When cultured tumor cells were treated with anti-EGF receptor MAb, we found that cell growth was slowed, but not arrested. The only exception was DiFi colon adenocarcinoma cells, which undergo cell cycle arrest in G1 phase, followed by apoptosis. In these cells TGF-α/EGF is required for survival.

We postulate that when epithelial tumor cells such as A431 and MDA-468 are damaged by chemotherapeutic agents and are forced to carry out repair, TGF-α (or EGF) becomes a survival factor rather than a growth-promoting factor [34-36]. In this situation, deprivation from growth factor stimulation of receptor kinase along with damage to essential molecules by the chemotherapeutic agent results in dual signals to arrest cell proliferation at G1/S and G2/M checkpoints [37, 38]. According to our hypothesis, cells must arrest at these activated dual checkpoints or they will undergo programmed cell death. When non-malignant cells are deprived of TGF-α/EGF stimulation by incubation with 225 MAb, they appear to arrest completely in G1 phase (unlike most tumor cells we have examined), and thereby may avoid dying in the presence of chemotherapy-induced damage. However we find that tumor cells (except for the DiFi cells mentioned above) merely reduce growth rate and do not arrest when cultured with 225 MAb. When these tumor cells are damaged by chemotherapy, TGF-α/EGF may have become survival factors, and receptor blockade therefore induces cell death. Ongoing experiments are testing this hypothesis.

Phase I clinical trials with murine 225 MAb in patients with advanced squamous lung carcinoma have established the safety of treating patients with doses that produce EGF receptor saturating concentrations of antibody in the blood [39]. Effective imaging of primary lung tumors and metastases >1cm in diameter was achieved with ^{111}Indium-labeled MAb. A phase I trial has confirmed the safety of a human-chimeric 225 MAb, and we are initiating trials of multiple dose chimeric 225 plus *cis*-DDP to test the clinical efficacy of this form of combination anticancer therapy.

REFERENCES

1. T. Kawamoto, J.D. Sato, A. Le, J. Polikoff, G.H. Sato, and J. Mendelsohn, Growth stimulation of A431 cells by EGF: Identification of high affinity receptors for epidermal growth factor by an anti-receptor monoclonal antibody, *Proc Natl Acad Sci USA*. 80:1337 (1983).
2. J.D. Sato, T. Kawamoto, A.D. Le, J. Mendelsohn, J. Polikoff, and G.H. Sato, Biological effect in vitro of monoclonal antibodies to human EGF receptors, *Mol Biol Med* 1:511 (1983).
3. G.N. Gill, T. Kawamoto, C. Cochet, A. Le, J.D. Sato, H. Masui, C.L. MacLeod, and J. Mendelsohn, Monoclonal anti-epidermal growth factor receptor antibodies which are inhibitors of epidermal growth factor binding and antagonists of epidermal growth factor-stimulated tyrosine protein kinase activity, *J Biol Chem* 259:7755 (1984).
4. S.D. Markowitz, K. Molkentin, C. Gerbic, J. Jackson, T. Stellato, and J.K.V. Willson, Growth stimulation by coexpression of transforming growth factor a and epidermal growth factor-receptor in normal and adenomatous human colon epithelium, *J Clin Invest* 86:356 (1990).
5. S.E. Bates, E.M. Valverius, B.W. Ennis, D.A. Bronzert, J.P. Sheridan, M.R. Stampfer, J. Mendelsohn J, Lippman ME, Dickson RB. Expression of the TGFα/EGF Receptor Pathway in normal human breast epithelial cells, *Endocrinol* 126:596 (1990).
6. M.R. Stampfer, C.H. Pan, J. Hosoda, J. Bartholomew, J. Mendelsohn, and P. Yaswen, Blockage of EGF receptor signal transduction causes reversible arrest of normal and immortal human mammary epithelial cells with synchronous reentry into the cell cycle, *Exp Cell Res* 208:175 (1993).
7. M. Van de Vijver, R. Kumar, and J. Mendelsohn, Ligand-induced activation of A431 cell EGF receptors occurs primarily by an autocrine pathway that acts upon receptors on the surface rather than intracellularly, *J Biol Chem* 266:7503 (1991).
8. T. Kawamoto, J. Mendelsohn, A. Le, G.H. Sato, C.S. Lazar, and G.N. Gill, Relation of epidermal growth factor receptor concentration to growth of human epidermoid carcinoma A431 cells. *J Biol Chem* 259:7761 (1984).
9. Z. Fan, H. Masui, I. Atlas, and J. Mendelsohn, Blockade of epidermal growth factor (EGF) receptor function by bivalent and monovalent fragments of 225 anti-EGF receptor monoclonal antibody, *Cancer Res* 53:4322 (1993).
10. Z. Fan, J. Mendelsohn, H. Masui, and R. Kumar, Regulation of epidermal growth factor receptor in NIH3T3/HER14 cells by antireceptor monoclonal antibodies, *J Biol Chem* 268:21073 (1993).
11. Z. Fan, Y. Lu, X. Wu, J. Mendelsohn, Antibody-induced epidermal growth factor receptor dimerization mediates inhibition of autocrine proliferation of A431 squamous carcinoma cells, *J Biol Chem* 269:27595 (1994).
12. M. Reiss, E.B. Stash, V.F. Vellucci, and Z-I. Zhou, Activation of the autocrine transforming growth factor alpha pathway in human squamous carcinoma cells, *Cancer Res* 51:6254 (1992).
13. W.E.J. Karnes, J.H. Walsh, S.V. Wu, R.S. Kim, M. G. Martin, H.C. Wong, J. Mendelsohn, A.V. Gazdar, and F. Cuttitta, Autocrine stimulation of EGF receptors by TGF-alpha regulates autonomous proliferation of human colon cancer cells, *Gastro* 102:474 (1992).
14. H. Masui, B. Boman, J. Hyman, L. Castro, and J. Mendelsohn, Treatment with anti-EGF receptor monoclonal antibody causes regression of DiFi human colorectal carcinoma xenografts, *Proc of the Amer Assoc for Cancer Res USA*, 32:394 (Abs #2340) (1991).

15. D.R. Hofer, E.R. Sherwood, W.D. Bromberg, J. Mendelsohn, C. Lee, and J. M. Kozlowski , Autonomous growth of androgen-independent human prostatic carcinoma cells: Role of transforming growth factor-α. *Cancer Res* 51:2780 (1991).
16. C-J Fong, E.R. Sherwood, J. Mendelsohn, C. Lee, and J. M. Kozlowski, Epidermal growth factor receptor monoclonal antibody inhibits constitutive receptor phosphorylation, reduces autonomous growth, ans ensitizes androgen-independent prostatic carcinoma cells to tumor necrosis factor-α, *Cancer Res* 52:5887 (1992).
17. I. Atlas, J. Mendelsohn, J. Baselga, H. Masui, W.R. Fair, and R. Kumar, Growth regulation of human renal carcinoma cells: role of transforming growth factor-α, *Cancer Res* 52:3335 (1992).
18. M. Lund-Johansen, R. Bjerkvig, P.A. Humphrey, S.H. Bigner, D.D. Bigner, O-D. Laerum, Effect of epidermal growth factor on glioma cell growth, migration, and invasion in vitro, *Cancer Res* 50:6039 (1990).
19. K-I. Morishige, H. Kurachi, K. Amemiya, Y. Fujita, T. Yamamoto, A. Miyake, and O. Tanizawa, Evidence for the involvement of transforming growth factor alpha and epidermal growth factor receptor autocrine growth mechanism in primary human ovarian cancers in vitro, *Cancer Res* 51:5322 (1991).
20. K-I Morishige, H. Kurachi, K. Amemiya, H. Adachi, M. Inoue, A. Miyake, O. Tanizawa, and Y. Sakoyama, Involvement of transforming growth factor alpha/epidermal growth factor receptor autocrine growth mechanism in an ovarian cancer cell line in vitro. *Cancer Res* 51:5951 (1991).
21. E. Sturgis, P. Sacks, H. Masui, J. Mendelsohn, and S. Schantz, The effects of anti-epidermal growth factor receptor antibody 528 on the proliferation and differentiation of head and neck cancer. *Otolaryngol-Head & Neck Surgery* 111:633 (1994).
22. C.L. Brown, M. Rubin, H. Masui, and J. Mendelsohn, Growth inhibition by anti-EGF receptor monoclonal antibody in squamous cervical carcinoma cells expressing TGFα. Proc Amer Assoc *Cancer Res* 35:381(Abstr 2270) (1994).
23. C.L. Arteaga, E. Coronado, and C.K. Osborne, Blockade of the epidermal growth factor receptor inhibits transforming growth factor α-induced but not estrogen-induced growth of hormone-dependent human breast cancer, *Mol Endocr* 2:1064 (1988).
24. B.W. Ennis, E.M. Valverius, M.E. Lippman, F. Bellot, R. Kris, J. Schlessinger, H. Masui, A. Goldenberg, J. Mendelsohn, and R. B. Dickson, Monoclonal anti-EGF receptor antibodies inhibit the growth of malignant and non-malignant human mammary epithelial cells, *Mol Endocrinol* 3:1830 (1989).
25. H. Masui, T. Kawamoto, J.D. Sato, B. Wolf, G.H. Sato, and J. Mendelsohn, Growth inhibition of human tumor cells in athymic mice by anti-EGF receptor monoclonal antibodies, *Cancer Res* 44:1002 (1984).
26. J. Mendelsohn, Potential clinical applications of anti-EGF receptor monoclonal antibodies, *in:* Cancer Cells, M. Furth, and M. Greaves, eds. Cold Spring Harbor Laboratory, New York (1989).
27. I. Atlas, H. Masui, and J. Mendelsohn, In preparation.
28. J. Baselga, L. Norton, H. Masui, A. Pandiella, K. Coplan, C. Cordon-Cardo, W. Miller, and J. Mendelsohn, Anti-tumor effects of doxorubicin in combination with anti-epidermal growth factor receptor monoclonal antibodies, *J Natl Cancer Inst* 85:1327 (1993).
29. Z. Fan, J. Baselga, H. Masui, and J. Mendelsohn, Antitumor effect of anti-EGF receptor monoclonal antibodies plus cis-Diamminedichloroplatinum (cis-DDP) on well established A431 cell xenografts, *Cancer Res* 53:4637 (1993).

30. E. Aboud-Pirak, E. Hurwitz, M.E. Pirak, F. Bellot, J. Schlessinger, and M. Sela, Efficacy of antibodies to epidermal growth factor receptor against KB carcinoma in vitro and in nude mice, *J Natl Cancer Inst* 80:1605 (1988).
31. J. Baselga, L. Norton, K.Coplan, R. Shalaby, and J. Mendelsohn, Antitumor activity of placitaxel in combination with anti-growth factor receptor monoclonal antibodies in breast cancer xenografts, *Proc Amer Assoc Cancer Res* 35:380(Abstr.2262) (1994).
32. C. Aghajanian, Y. Zhuo, T.Y. Ho, C. Brown, Z. Fan, J. Baselga, J. Mendelsohn, and D.R. Spriggs, Anti-epidermal growth factor receptor monoclonal antibody tratment of A431 cells decreases cisplatin/DNA adduct repair in association with NF-KB induction. *Proc Amer. Assoc. for Cancer Res* 36(Abst.#2545):427 (1995).
33. R.J. Pietras, B. M. Fendly, V.R. Chazin, M.D. Pegram, S.B. Howell, and D. J. Slamon, Antibody to HER-2/*neu* receptor blocks DNA repair after cisplatin in human breast and ovarian cancer cells, *Oncogene* 9:1829 (1994).
34. G.I. Evan, A.H. Wyllie, C.S. Gilbert, T.D. Littlewood, H. Land, M. Brooks, C. M. Waters, L. Z. Penn, and D. C. Hancock, Induction of apoptosis in fibroblasts by c-myc protein, *Cell* 69:119 (1992).
35. C.E. Canman, T. M. Gilmer, S. B. Coutts, and M.B. Kastan, Growth factor modulation of p53-mediated growth arrest versus apoptosis, *Genes & Develop* 9:600 (1995).
36. G. T. Williams, C.A. Smith, E. Spooncer, T.M. Dexter, and D.R. Taylor, Haemopoietic colony stimulating factors promote cell survival by suppressing opoptosis, *Nature* 343:76 (1990).
37. K.W. Kohn, J. Jackman, and P.M. O'Connor, Cell cycle control and cancer chemotherapy, *J Cellular Biochem.* 54:440 (1994).
38. L. H. Hartwell, and M.B. Kastan, Cell cycle control and cancer, *Science* 266:1821 (1994).
39. C.R. Divgi, C. Welt, M. Kris, F.X. Real, S.D.J. Yeh, R. Gralla, B. Merchant, S. Schweighart, M. Unger, S. M. Larson, and J. Mendelsohn, Phase I and imaging trial of indium-111 labeled anti-EGF receptor monoclonal antibody 225 in patients with squamous cell lung carcinoma, *J Natl Can Inst* 83:97 (1991).

OXALIPLATIN : UPDATE ON A ACTIVE AND SAFE DACH PLATINUM COMPLEX

P. Soulié, E. Raymond, J.L. Misset and E. Cvitkovic

Service des Maladies Sanguines et Tumorales
Hôpital Paul Brousse
12-14 Avenue Paul Vaillant Couturier
94804 Villejuif - France

INTRODUCTION

Since the introduction of cisplatin (CDDP) in oncological chemotherapy, numerous attempts have been made to synthetize derivates that would increase the therapeutic index of this cytotoxic class, while expanding its indications. During the past decade, many platinum based therapeutic programs have been aimed at circumventing cisplatin/carboplatin resistance, as primary or acquired treatment failure represents the major limitation of the two platinum complexes currently available in the clinic [1, 2].

The substitution of the amine radicals in cisplatin by a 1,2 diaminocyclohexane ("DACH") radical yielded a family of molecules to which established tumoral cell lines selected for cisplatinum resistance remained sensitive [3]. Among them, LOHP (oxaliplatin) was selected for further development by Kidani [4], based on its activity against the CDDP resistant cell line L1210 and its good solubility in water, and was introduced to clinical trials by Mathé [5].

Recently, additional preclinical studies have confirmed the original profile of this drug, while a wealth of clinical data on the use of LOHP alone or in combination have generated very encouraging results in traditionally cisplatin/carboplatin resistant/refractory tumors (colon/pretreated ovarian cancer). Available preclinical and clinical data will be reviewed.

NEW ADVANCES IN EXPERIMENTAL STUDIES

1 - Mechanism of action

LOHP binds to repeating deoxyguanosine d(GpG) in single DNA strand in a fashion similar to cisplatin. Other DNA-adducts include intrastrand crosslinks at adenine-guanine d(ApG) sites and interstrand crosslinks with guanosines of opposing DNA strands $(dG)_2$ [6, 7, 8] DNA replication is blocked by inter and intrastrand crosslinks, resulting in cell death,

which seems not to be cell cycle phase specific. With LOHP, the formation of adducts with DNA is more rapid than with cisplatin.
This process peaks at 12 hours for cisplatin, and 24 hours for carboplatin [9, 10]. For LOHP the linkage with the d(GpG) and $(dG)_2$ sites is achieved within 15 minutes, and it appears to bind with d(ApG) sites over a 2-hour period [11].

2 - Spectrum of activity

In vitro, LOHP has shown cytotoxic activity against a wide range of murine [L1210, P 388 (leukemia)] or human [A 2780 (ovary), HT 29 (colon), MCF 7 (breast), ...] cell lines with a higher cytotoxic efficacy than cisplatin or carboplatin in the majority of the cell lines tested [4, 12]. Of note, oxaliplatin has proven to be active on most colon cancer lines in the NCI screening system, which yet distinguish it from cisplatin and carboplatin.

The NCI DISCOVER program which classifies cytotoxic agents, on the basis of their mechanism of action and resistance, has confirmed the different spectrum of activity between LOHP and CDDP and has identify the DACH compounds as a distinct cytotoxic family, differing from cisplatin and carboplatin, as well as others alkylators [12].

3 - Non cross resistance

In cells lines selected for their 20 to 100 fold resistance levels to cisplatin [L 1210 CP, A 2780 E (80) and KB CP (20)], oxaliplatin was shown active, with little or no cross resistance [13, 14] which has suggested different mechanism of resistance between these platinum compounds, arguing for their sequential or combined use in the clinical setting other than for an eventual additive dose effect.

4 - In vitro and in vivo LOHP combination with CDDP or CBDCA

Mathé was the first to study the efficacy of the simultaneous administration of LOHP and CDDP/or CBDCA in L 1210 leukemia bearing mice ; when used at optimal dosage, he obtained cure in 70% of treated animals with the CBDCA/LOHP combination, which suggested antitumoral synergism [15].

More recently, similar results were obtained in vitro against human epidermoid (KB) and ovarian (A 2780 E (80)) cells lines with LOHP and CDDP ; additive and perhaps synergistic activity was observed with the combination of both drugs [16].

These data, which relate both to the larger spectrum of activity of oxaliplatin and to the positive interaction between these different platinum compounds provide a solid basis for the inclusion of LOHP into new platinum combination.

5 - Blood distribution

Pendyala et al [13] have studied in vitro some pharmacokinetic parameter of oxaliplatin (plasma protein binding and red blood cell uptake).
When oxaliplatin was incubated in plasma, 85-88% of all measurable platinum was bound to plasma proteins within the first 5h ; when incubated in whole blood, the erythrocytes took up rapidly an average amount of 37.1 ± 2% of total platinum without significant short term efflux.

RECENT ADVANCES IN CLINICAL TRIALS

1 - Pharmacokinetics (PK)

Bastian et al [17] have determinated the oxaliplatin PK parameters in patients with normal and impaired renal function, using atomic absorption spectrophotometry (AAS).
The PK characteristics were similar to those of cisplatin, with a bicompartimental distribution for both total and ultra filtrable (UF) platinum and a linearity in the dose range studied. Fifty percent of the blood levels was plasmatic (2/3 protein bound, 1/3 UF) and 50% within the erythrocytic fraction. UF t $1/2\alpha$ was 0.3hr for patients with normal renal function (NRF) and 0.5hr in impaired renal function (IRF) patients ; total-Pt t $1/2\beta$ was 40hrs for IRF and 25hrs for NRF ; the 48hrs urinary recovery of the total amount of Pt administered was > 50% in NRF and < 30% in IRF.
The AUC and clearance of ultrafiltrable (free) platinum were found to be significantly different between patients with normal and impaired renal function (greater exposure, lower elimination in the altered renal group) even if there was no increase of toxicity checked.

2 - Phase I

While the initial work by Mathé, based on intrapatient dose escalation to a "minimally active dose" resulted in a recommendation of 67mg/sqm q 2-3 weeks, many subsequent orthodox studies have established an MTD of 200mg/sqm, and a recommended dose of 130mg/sqm q 3 weeks when administered as a short (2-4 hours) I.V. perfusion.
The continous administration over 5 days, investigated by Lévi shows the same recommended dose/cycle (130mg/sqm), while the chronomodulated infusion recommended dose is claimed as 175mg/sqm. Most of the subsequently published work by Levi with the chronomodulated infusion reports the chronomodulated administration at 135mg/sqm given over a 5 days continous infusion every 3 weeks.
The MTD limiting effect is an acute onset, cold induced and enhanced dysesthesia for the single cycle administration, which changes its characteristics and duration with cumulativity (see below). No renal impairement was initially recorded during phase I trials. The administration of oxaliplatin does not require hydration measures. It has to be dissolved and administered in 5% glucose solution, according electrolytes. Hematological toxicity, mainly thrombopenia is very seldom severe, even at the MTD levels.

3 - Clinical toxicity profile

a) Single agent LOHP

Severe nausea and vomiting, of rapid onset were common in the first studies [18, 19, 20] ; with the introduction of HT-3 antagonists, the severity of this toxicity has become very manageable, with rare instances of grade 3-4 (WHO) episodes. Diarrhea has also been reported, but seldom.

While ototoxicity is never induced by oxaliplatin, the main dose limiting toxicity is neurologic [21, 22], affecting the peripheral sensory system, being different in its pattern and characteristics to the one seen with CDDP. It consist of two types of manifestations:
- (1) acute and transient acrodysesthesia, occuring at the end of the infusion, exacerbated by cold, which appears just on the fingers and toes, sometimes perioral region and pharyngolaryngeal ; these symptoms, observed more frequently at the higher dose levels (> 100mg/sqm by cycle) wane generally in the days after during the interval of treatment but with successive cycles they acquire longer duration and persist between treatment.

- (2) distal ascending paresthesias which progressivelly increase in their intensity and duration and precede proprioception impairment (butterfingers, ataxia) ; these last debilitating symptoms are cumulative and generally appear beyond a median cumulative dose of 900 mg/sqm ; the risk of developing a moderate to severe neuropathy (gr 2-3 WHO) is 10% after 6 cycles (780mg/sqm) and > 50% after 9 cycles.
Severe neuropathies (WHO grade ≥ 2) are reversible in the majority of patients (82%) in the 3-4 months after treatment interruption, with a total disappearance of symptoms for the majority of patients in a median time of 6-8 months.

b - LOHP in combination

Llory et al have [23] reported the feasibility of the LOHP-Carboplatin combination. Thrombopenia and leucopenia were the two limiting toxicities ; these were not surprising when considering the included patient characteristics (the majority of them were heavily pretreated and/or with altered renal function). A new pharmacokinetically guided phase I study has to be carried out.

The LOHP/CDDP combination (Biplatin) has been investigated during the last two years by our group focusing on the salvage treatment of ovarian cancer and germ cell tumors [24, 25].
In this heavily pretreated population with a median prior CDDP dose of 600mg/sqm, sensory peripherical neuropathy was again the major limiting toxicity ; the risk of severe neuropathy (WHO grade > 2) was significant beyond 4 cycles, correlated with the total dose of delivered platinum. Regression of the symptoms and restoration of sensory perception was the rule within a range of 3-6 months in the majority of patients. In the future, its administration may justify the evaluation of neuroprotectors. With this combination, hematotoxicity was limited (neutropenia grade 3-4 : 41%, thrombopenia grade 3-4 : 34%) and allowed the addition of other cytotoxic drugs (anthracyclines and alkylators) at full recommended doses with G-CSF coverage.

4 - Phase II trials

The clinical development of oxaliplatin is still ongoing and the complete range of the antitumoral activity of this compound has not yet been exhaustively investigated. Clinical studies have focused on two diseases : colorectal and ovarian cancer, but significant efficacy, to be confirmed, has been demonstrated against others tumor types in phase II trials (NSCLC, HNSCC, low grade pretreated NHL). Anecdotal reports of its activity in some tumor types (melanoma, glioma) have not been confirmed, while other await evaluation (breast cancer).
An important consideration is the fact that most available single agent Phase II data have been obtained at the dose and schedule of 130mg/sqm q 3 weeks, given as a short (2-4hrs I.V. perfusion). The dose response and dose intensity relationships remain yet to be formally explored (weekly, q 2 week, continous infusion, etc...).

a - LOHP in advanced colorectal carcinoma (ACC)

The evaluation of LOHP in colorectal cancer was stimulated by its in vitro activity on human colon cancer cells lines, its demonstrated synergistic effect with the standard antitumoral agent 5-FU (tested in the L 1210 murine model) and the observation of one partial response in a patient with advanced rectal cancer during the phase I development [18].

Oxaliplatin has shown activity in 5-FU pretreated and resistant ACC when given as a single agent : three different, phase II studies [26, 27, 28] have included a total of 139 patients and documented a response rate of 10% (14 objective responses/138 evaluable patients). This modest but consistant level of activity of in ACC has never been found with CDDP or CBDCA and singles out LOHP in the platinum complex class.

Table 1 : LOHP as single agent in 5-FU refractory advanced colorectal cancer

Authors	Machover	Diaz-Rubio	Lévi
No. of patients included	58	51	30
No. of eligible patients	55	51	29
No. of evaluable patients	53	48	29
Response rate			
Evaluable patients	11.3 ± 8.5	10.4 ± 8.6	10
Included patients	10.3 ± 7.8	9.8 ± 8.1	10
Stabilizations (%)	41.5	41.6	24.1
Median survival (months)	8.5	Too early	9

In combination with the pivotal 5-FU Folinic Acid regimen, LOHP has been extensively evaluated (437 patients), administered according two different schedules : as a 2 hour infusion or as a 5 day infusion at a circadian modulated rate or at constant rate. Five prospective studies (2 multicentric, 3 monocentric) [29, 30] were carried out with this three drug combination as first or second line metastatic therapy (Table 2).

Table 2 : LOHP in association with 5-FU and folinic acid in advanced colorectal cancer

	De Gramont		Lévi			Garufi
			Phase II	Phase IIIa	Phase IIIb	25
Nb of patients	13	28	93	92	186	
LOHP infusional schedule	2h	2h	CM 5-d	FCI or CM 5-d	FCI or CM 5-d	CM 5-d .
Med dose/cycle (mg/sqm)	130	100	125	125	125	100-125
Interval/cycle (weeks)	4	2	3	3	3	2 or 3
5-FU infusional schedule	48h	48h	CM 5-d	FCI or CM 5-d	FCI or CM 5-d	CM 5-d
Med dose/cycle (mg/sqm)	4000	3500	3500	3500	3500	3500
Interval/cycle (weeks)	2	2	3	3	3	2 or 3
Nb of eligible patients	13	28	91	88	170	25
Nb of evaluable patients	11	27	84	86	170	24
Nb of responses						
CR	-	1	4	5	8	-
PR	4	12	45	34	66	7
Response rate						
evaluable patients	36	48	58	45	43.5	29
all inclusions	31	46	53	42	40	28
Stabilizations (%)	36	50	32	43	40	38
Median survival (months)	10	too early	15.5	15	17	12

FC : (Flat Continous) CM : (Chronomodulated Continous)

The observed response rate ranged from 31 to 53% and was similar in first and second line treatement ; median survival times reported vary from 15 to 19 months.
In one study [31], LOHP was added to the 5-Fluorouracil/Folinic acid infusional schedule under which tumor progression was ascertained ; with 7 partial responses and 9 stabilizations among 20 such retreated patients, the contribution of oxaliplatin in the treatment of ACC chemotherapy is clear.

Another noteworthy observation in colorectal cancer is the complete response rate (3.5-5.5%) reported in most studies reporting the 5 FUFA/LOHP combination [29] .
The high activity of this 3 drug combination has allowed to consider secondary resection of previously unresectable metastases, with prolonged survival (> 36 months) in a large proportion of such patients ; this more frequent sequential and multidisciplinary therapeutic approach in ACC is linked also to the growing perception among surgeons of a positively altered disease natural history.
There is little doubt that the introduction of oxaliplatin into ACC chemotherapy will improve its treatment results, but two ongoing controlled trials with the aim of assessing prospectively the difference between 5FU/folinic acid infusional schedules with and without oxaliplatin will further clarify the issue.

b - LOHP in ovarian cancer (OC)

CDDP remains the most active agent in OC and its use in the treatment of ovarian cancer has allowed a 20-30% 5 year disease free survival for patients presenting with advanced disease. Carboplatin has similar level efficacy and a differential toxicity profile. Paclitaxel incorporation into platinum based combinations has increased both the chances of therapeutic success and the salvage palliative treatment options ; but new drugs are required to improve the outcome of relapsing/refractory patients.

The clinical activity of oxaliplatin as single agent in pretreated ovarian cancer patients was initially described by Misset with 4 PR in 12 evaluable patients [18]. LOHP rational for further development in OC is strengtened by its in vitro non cross resistance with CDDP against human ovarian carcinoma cell lines resistant to CDDP [14].
One recent phase II experience [32] including 28 patients reported the antitumoral efficacy of single agent oxaliplatin in platinum pretreated patients with 3 partial response out of 14 platinum refractory disease (overall response rate = 30%) ; these results are being confirmed on a multicentric Phase II trial, but are already suggestive a role for LOHP in the second line therapy of OC.

LOHP has been considered by our group in combination with CDDP (Biplatin program) [33], despite their common neurotoxicity profile.
The challenge was the dose response/dose intensity concepts as supposed by Hryniuk [34] and validated by some authors [35] in ovarian cancers. Other preclinical incitations have previously been exposed (non cross resistance, additive synergistic effects).
Two regimens were tested and patients have received either Biplatin alone (called Bi = LOHP at 90-150mg/sqm with CDDP at 60-110mg/sqm) or Biplatin combined (BiC = Bi with epirubicin and ifosfamide).
In platinum pretreated ovarian cancer, the activity of Biplatin alone was remarkable (Table 3).
In platinum refractory ovarian cancer (defined according the Markman criteria) a 30% response rate has been observed whereas 75% response rate was obtained in platinum sensitive tumors [33].

Table 3 : Results of oxaliplatin/cisplatin combination (Biplatin) in pretreated ovarian cancer according to the regimen and platinum resistance status (Markman's criteria*)

PLATINUM RESISTANCE	RESPONSE				
	RC	PR	SD	PD	ORR
Potentially sensitive					
Bi N = 12	3	5	1	2	75%
BiC N = 18	5	10	1	0	73%
Primary refractory					
Bi N = 2	0	0	2	0	
BiC N = 2	1	0	1	0	
Secondary refractory					
Bi N = 10	0	3	5	2	25%
BiC N = 1	0	0	1	0	

Bi : Oxaliplatin/Cisplatin BiC : Bi + Epirubicin + Ifosfamide

c - Germ cell tumors (GCT) and choriocarcinoma (CC)

In platinum heavily pretreated patients with relapsing/refractory disease, oxaliplatin has been introduced into platinum based salvage regimens [36].
Thirteen patients (11 GCT and 2 CC) received Bi or BiC as previously defined and a 83% overall response rate was observed in this otherwise doomed population, which call for further exploration of oxaliplatin in this diseases.

d - LOHP in other indications

Several phase II trials have been reported and have shown LOHP activity in low/intermediate grade Non Hodgkin Lymphomas (NHL) (41% RR) [37], and in advanced non small cell lung carcinomas (13% RR) [38].

CONCLUSION

By its preclinical characteristics (partial non cross resistance, different spectrum of cytotoxicity) and its unusual antitumoral efficacy (against colon cancer and refractory/resistant ovarian cancer), oxaliplatin represents a new antitumoral platinum complex with a large spectrum of clinical usefulness.

REFERENCES

[1] Burchenal JH, Kalaher K, Dew K et al : Rationale for the development of platinum analogs. Cancer Treat Rep 63:1493-1498 (1979).

[2] Christian MC : The current status of new platinum analogs. Semin Oncol 19:720-733 (1992).

[3] Burchenal JH, Lokys L, Turkevich J, et al : 1,2-Diaminocyclohexane platinum derivatives of potential clinical vlaue. Recent Results Cancer Res 74:146-155 (1980).

[4] Tashiro R, Kawada Y, Sakuri Y et al : Antitumor activity of a new platinum complex, oxalato (trans-I-1,2-diaminocyclohexane) platinum (II): New experimental data. Biomed Pharmacother 43:251 (1989).

[5] Mathé G, Kidani Y, Sekiguchi M et al : Oxalato-platinum or l-OHP, a third-generation platinum complex : an experimental and clinical appraisal and preliminary comparison with cis-platinum and carboplatinum. Biomed & Pharmacother 43:237-250 (1989).

[6] Inagaki K, Kidani Y : Differences in binding of (1,2-cyclohexane) platinum (II) isomers with d(GpG). Inorg Chem. 25:1-3 (1986).

[7] Kidani Y : Oxaliplatin. Drugs Future 14 (6):529-532 (1989).

[8] Boudny V, Vrana O, Gaucheron F et al : Biophysical analysis of DNA modified by 1,2-diaminocyclohexane platinum (II) complexes. Nucleic Acids Res. 20 (2):267-272 (1992).

[9] Kohn KW : Molecular mechanisms of cross-linking by alkylating agents and platinum complexes. in : Molecular actions and targets for cancer chemotherapeutic agents, Sartorelli AC, Lazo JS, Bertino JR, eds., Academic press, New York (1981).

[10] Micetich KC, Barnes D, Erickson LC : A comparative study of the cytotoxicity and DNA-damaging effects of cis-(diammino)(1,1-cyclobutanedicarboxylaton)-platinum (II) and cis-diamminedichloroplatinum (II) on L1210 cells. Cancer Res. 45:4043-4047 (1985).

[11] Jennerwein MM, Eastman A, Khokhar A : Characterization of adducts produced in DNA by isomeric 1,2-diaminocyclohexane platinum (II) complexes. Chem Bil. Interact. 70:39-49 (1989).

[12] Myers TG, Paull KD, Fojo AT et al : Multivariate analysis of high-flux screening data using the DISCOVER computer program package: integrated analysis of activity patterns and molecular structure features of platinum complexes. Proc Am As Can Res 35:371 (1994).

[13] Pendyala L, Creaven PJ : In vitro cytotoxicity, protein binding, red blood cell partitioning, and biotransformation of oxaliplatin. Cancer Research 53: 5970-5976 (1993).

[14] Alvarez M, Ortuzar W, Rixe O et al : Cross resistance patterns of cell lines selected with platinum suggest differences in the activities and mechanisms of resistance of platinum analogues. Proc Am As Can Res 35:439 (1994).

[15] Mathé G, Chenu E, Bourut C et al : Experimental study of three platinum complexes : CDDP, CBDCA and l-OHP on L1210 leukemia. Alternate or simultaneous association of two platinum complexes. Proc Am As Can Res 30:872 (1989).

[16] Ortuzar W, Paull K, Rixe O et al : Comparison of the activity of cisplatin (CP) and oxaliplatin (OXALI) alone or in combination in parental and drug resistant sublines. Proc Am As Can Res 35:332 (1994).

[17] Massari C, Brienza S, Rotarski M : Oxaliplatin, Transplatin® comparative pharmacokinetics in normal and impaired renal function patients. Proc. Am As Can Res 35:242 (1994).

[18] Misset JL, Kidani Y, Gastiaburu J et al : Oxalatoplatinum (l-OHP): Experimental and clinical studies. in : Platinum and Other Metal Coordination Compounds in Cancer Chemotherapy, Howell SB ed., Plenum Press, New york (1991).

[19] Extra JM, Espie M, Calvo F et al : Phase I study of Oxaliplatin in patients with advanced Cancer. Cancer Chemoter Pharmacol 25:299-303 (1990).

[20] Caussanel JP, Lévi F, Brienza S et al : Phase I trial of 5-day continous venous infusion of oxaliplatin at circadian rhythm modulated rate compared with constant rate. JNCI 82:1046-1050 (1990).

[21] Brienza S, Vignoud J, Itzhaki M : Oxaliplatin (L-OHP) : Global safety in 682 patients. Proc. of 7th International Symposium on Platinum and Other Metal Coordination Compounds in Cancer Chemotherapy, Amsterdam, (1995) (Abs 140).

[22] Brienza S, Fandi A, Hugret F et al : Neurotoxicity (NTX) of long term oxaliplatin (L-OHP). Proc Am As Can Res 34:406 (1994).

[23] Llory JF, Soulie P, Cvitkovic E et al : Feasibility of high-dose Platinum delivery with combined carboplatin and Oxaliplatin. J Natl Cancer Inst 86:1098-1099 (1994).

[24] Garrino C, Cvitkovic E, Soulie P, et al : Preliminary report on the tolerance of Transplatin (L-OHP) Cisplatin association alone (Bi) or in combination with Ifosfamide and Epirubicin (Bic) in Platinum (Pt) pretreated patients. Proc Am Soc Clin Onc 13:143 (1994).

[25] Soulie P, Llory JF, Fereres M et al : Preliminary results of an active oxaliplatin (L-OHP)-CDDP association based salvage program in pretreated germ cell tumors (GCT). Proc Am Soc Clin Onc 13:250 (1994).

[26] Moreau S, Machover D, de Gramont A et al : Phase II trial of Oxaliplatin : in patients with colorectal carcinoma previously resistant to 5-FUFOL. Proc Am Soc Clin Onc 12:214 (1993) (Abs 645).

[27] Diaz-Rubio E, Marty M, Extra JM : Multicentric phase II study with Oxaliplatin (L-OHP) in 5-FU refractory patients with advanced colorectal cancer. Proc of 7th International Symposium on Platinum and Other Metal Coordination Compounds in Cancer Chemotherapy, Amsterdam, (1995) (Abs 140).

[28] Lévi F, Perpoint B, Garufi C et al : Oxaliplatin activity against metastatic colorectal cancer : a phase II study of 5-day continous venous infusion at circadian-rhythm modulated rate. Eur J Cancer 29A:1284-1293 (1993).

[29] Lévi F, Zidani R, Giachetti S et al : Chronomodulation of combined Platinum complexes and 5-Fluorouracil delivery. Proc of 7th International Symposium on Platinum and Other Metal Coordination Compounds in Cancer Chemotherapy, Amsterdam, (1995) (Abs 135).

[30] De Gramont A, Tournigand C, Louvet C et al : High-dose Folinic Acid, 5-fluorouracil 48H-infusion and Oxaliplatin in metastatic colorectal cancer. Proc of the Fifth International Congress on Anti-Cancer Chemotherapy. Paris, (1995) (Abs O269).

[31] Garufi C, Bensmaïne MA, Brienza S et al : Addition of Oxaliplatin to chronomodulated 5-FU and Folinic Acid for reversal of acquired chemoresistance in patients with advanced colorectal cancer. Proc of 7th International Symposium on Platinum and Other Metal Coordination Compounds in Cancer Chemotherapy, Amsterdam, (1995) (Abs S 125).

[32] Chollet P, Brienza S, Bensmaïne MA : Report of phase II of Oxaliplatin (L-OHP) : a new active agent in platinum pretreated ovarian cancer. Proc of 7th International Symposium on Platinum and Other Metal Coordination Compounds in Cancer Chemotherapy, Amsterdam, (1995) (Abs 137).

[33] Cvitkovic E, Bensmaïne MA, Soulié P et al : High activity of combined Oxaliplatin (LOHP) Cisplatin as salvage treatment in pretreated ovarian cancer. Proceedings of 7th International Symposium on Platinum and Other Metal Coordination Compounds in Cancer Chemotherapy, Amsterdam, (1995) (Abs S 125).

[34] Levin L, Simon R, Hryniuk W : Importance of multiagent chemotherapy regimens in ovarian carcinoma : dose intensity analysis. JNCI 85:1732-1742 (1993).

[35] Kaye SB, Lewis CR, Paul J et a : Randomized study of two doses of cisplatin with cyclophosphamide in epithelial ovarian cancer. Lancet 340: 329-33 (1992).

[36] Soulié P, Llory JF, Férérès M et al : Prliminary results of an ative oxaliplatin (LOHP)-CDDP association based salvage program in pretreated germ cell tumors (GCT). Proc Am Soc Can Oncol 13:250 (1994).

[37] Gastiaburu J, Brienza S, Rotarski M et al : Oxaliplatin (L-OHP) : a new platinum analog, active in refractory/relapsed intermediate and low grade LNH. A phase II study. ECCO 7, Jerusalem, (1993) (Abst 980).

[38] Monnet I, Brienza S, Voisin S et al : Phase II study of Oxaliplatin (L-OHP) in patients with advanced non small cell lung cancer (NSCLC) : preliminary results. Eur J Cancer 29A : S763 (1993) (Abst 901).

COMBINATION OF DIFFERENT PLATINUM COMPOUNDS:

ISSUES, STRATEGIES, AND EXPERIENCE

Franco M. Muggia, M.D.

Medical Oncology
USC-Norris Cancer Center
Los Angeles, CA 90033

INTRODUCTION

The rationale for combination of cisplatin and carboplatin is based primarily on 1) a steep dose-response in antitumor effects of platinums, and 2) the widely different dose-limiting toxicities for cisplatin and carboplatin. With the introduction of several new analogues, the rationale of lack of complete cross-resistance becomes increasingly attractive. In fact, both preclinical and clinical data have been assembled in support of the hypothesis that combinations of platinum compounds may lead to improved therapeutic effects. We review here the issues concerned with specific platinum combinations and illustrate possible strategies to test such combinations, and preliminary clinical experience.

CARBOPLATIN + CISPLATIN

The initial rationale introduced by Trump et al[1] when combining carboplatin with cisplatin included potential for less than complete cross-resistance. This last promise is unlikely. Nevertheless, the relative lack of overlapping toxicities continued to render the concept of such a combination attractive. Table I indicates the doses of both drugs in combination that were recommended for subsequent Phase II study. Although these drug combinations were empirically arrived, some preclinical work suggests an improved therapeutic index by analogous ratios of the two drugs.[4] Platinum-DNA adducts in buccal cells provided some indication of the relative contribution of each drug to staining intensity.[5] Additionally, the following points deserve emphasis: 1) dose-intensities reached do not exceed the 50 mg/m^2/week that may be achieved with cisplatin given weekly, 2) since platelet toxicity is the major hematologic finding, cytokines are unlikely at this point to facilitate dose escalation, 3) dose-limiting toxicities are both hematologic <u>and</u> non-hematologic (primarily ototoxicity), and 4) dosing on a per m^2 base is not conducive to safe dose escalation, and the AUC method of Calvert is preferable for dosing carboplatin.[6] Nevertheless, it is unlikely that this combination will allow sufficient dose-intensification to make an impact on outcome of patients with ovarian cancer or head and neck cancer - both platinum-sensitive cancers that have increased survivals and response rates with increasing exposure to platinums.

Phase II studies have been performed primarily in patients with ovarian cancer (Table 2). In previously untreated patients, results have appeared encouraging, particularly in combination with ifosfamide.[9] However, without a Phase III trial it is difficult to interpret such results. On the other hand, the schedule reported by Hardy et al[11] proved

Platinum and Other Metal Coordination Compounds in Cancer Chemotherapy 2
Edited by H.M. Pinedo and J.H. Schornagel, Plenum Press, New York, 1996

Table 1. Carboplatin + Cisplatin (Phase I Studies)

(Ref.	Series	carbo		cis
(1)	Trump A (1987)	400	q4w	50
	Trump B	280	q4w	25 x 3d
(2)	Muggia A (1991)	300 d1	q4w	100 d3
	Muggia B	480 d1	q4w	75 d3
(3)	Sessa (1991)	250	q4w	40 qw

doses are all in mg/m^2

Table 2. Carboplatin + Cisplatin (Ovarian Cancer Studies)

(Ref.)	Series	carbo	cis	Comments
(7)	Piccart	350 d1	75 d2	reduced by cycle #3
	(1987)	300 d1	100 d2	7 pCR/32 pts (22%)
(8)	Lund (1989)	300 d1	50 d2,3	8 pCR/34 pts. (24%)
(9)	(1990)	200 d1	50 d2,3	+ ifosfamide 42% pCR
(10)	(1992)			+ etoposide 32%
(11)	Hardy (1991)	AUC 11 d1	30-50 d2	Stage IV: Surv 12m
(12)	Reed (1992)	600 d1	100 d8,15,22	Salvage therapy
(13)	Bolis (1995)	375 d1	1 mg/kg* d1,8,15	Salvage

* all other doses in mg/m^2

disappointing in Stage IV patients. In the salvage setting, the combination may allow some individual dose intensification over either agent given alone in patients who already have cisplatin-induced neuropathy or carboplatin-induced persistent myelosuppression.[13]

With our gynecology group, we have employed this combination of platinums by the intraperitoneal (IP) route together with ip Floxuridine.[14] The combination was used in order to tailor platinum administration to pre-existing neurotoxicity. Six patients received 1-4 cycles of such combined platinum in doses of carboplatin (target AUC=3) plus cisplatin 30 mg/m^2 without incurring increased Gr 1 or 2 neurotoxicity. The major toxicities observed were Grade 3 and 4 neutropenia.

CARBOPLATIN + OXALIPLATIN

Oxaliplatin is an analogue undergoing clinical testing in colon cancer [15] - a reflection of different spectrum of activity. The other point of interest is the relatively little myelosuppression in doses up to 170 mg/m^2 - a property that attracted interest regarding its use in combination. The dose-limiting toxicity is neurotoxicity - but one that differs from cisplatin's, principally by its reversibility.

Preclinical activity against L1210 leukemia was reported by Mathe[16] to be associated with greater long-term survivors when combined with carboplatin. These findings persuaded French investigators to initiate clinical studies with the combination. Unfortunately, the Phase I study was carried out in patients who had considerable prior treatment including cisplatin.[17] This may have contributed to the observed frequent grade 4 thrombocytopenia (44% of cycles), even if the dose of carboplatin was lowered from 400 mg/m^2 to 300 mg/m^2 in 3 of 13 patients. This trial needs to be repeated in patients who are platinum-naive and dosing of carboplatin by AUC is recommended. The oxaliplatin dose was held fixed at 100 mg/m^2.

CISPLATIN + OXALIPLATIN

Because the neurotoxicities differ in character, the combination of these two drugs was believed feasible to explore. In Mathe's studies on L1210 leukemia, toxic deaths did not permit demonstration of therapeutic enhancement [16]. However, clinical data strongly support therapeutic advantages in the face of acquired platinum resistance[18-21].

Cvitkovic et al[19] has presented results achieved in 12 patients with refractory germ cell tumors. Misset et al[20] has presented similar data in epithelial ovarian cancer. Both studies employed alternating treatments with ifosfamide and epirubicin rendering interpretation of the results problematic. Nevertheless, the impressive results in the face of platinum resistance provide further impetus towards more extensive testing.

CISPLATIN + LOBAPLATIN

Lobaplatin is a new lipophilic analogue showing activity *in vitro* and *in vivo* [21-22] against ovarian and testis cell lines and xenografts which have been made resistant to cisplatin. This non-cross resistance coupled with absence of nephrotoxicity and neurotoxicity in clinical studies renders it attractive for combination with cisplatin.

CISPLATIN + JM 216

JM 216 has completed Phase I testing in a daily x 5 schedule[23] and longer schedules of administration are planned. This flexibility in dose-scheduling and the lack of neurotoxicity make this compound attractive for combinations with cisplatin.

DISCUSSION

Dose intensification is only modest when platinums are combined (i.e. platinum therapy). Analysis of dose-intensification with carboplatin alone suggests a plateau in efficacy as illustrated by response rates in ovarian cancer. In intrinsically less sensitive disease targets, therefore, such strategy is even less likely to result in therapeutic enhancement. Nevertheless, the lack of complete cross-resistance provides another powerful rationale for combination of new analogues such as lobaplatin or JM 216 with cisplatin.

In addition to studying with new analogues, biplatinums may be useful in overcoming some drug resistance in germ cell and ovarian cancers. Dosing of carboplatin by AUC is needed for optimal design of these studies. Studies in platinum-naive patients would also be of interest in conditions such as melanoma. Finally, the use of cytokines may allow exploration of higher dose-intensification than has been previously achieved. These studies are, therefore, awaited with interest.

REFERENCES

1. D.T. Trump, J.L. Grem, K.D. Tutsch, J.K.V. Willson, K.J. Simon, D. Alberti, B. Storer, and D.C. Tormey. Platinum analogue combination chemotherapy: cisplatin and carboplatin --a Phase I trial with pharmacokinetic assessment of the effect of cisplatin administration on carboplatin excretion. *J. Clin. Oncol.* 5:1281, (1987).

2. F.M. Muggia and I. Gill. Optimizing dose-intensity: combining carboplatin with cisplatin, *in*: "Platinum and Other Metal Coordination Compounds in Cancer Chemotherapy", S.B. Howell, ed., Plenum, New York (1991).

3. C. Sessa, A. Goldhirsch, G. Martinelli, M. Alerci, L. Imburgia, and F. Cavalli. Phase I study of the combination of monthly carboplatin and weekly cisplatin, *Ann. Oncol.* 2:123 (1991).

4. K. Kobayashi, A. Yoshimura, M. Hino, A. Gemma, K. Yoshimori, M. Shibuya, T. Takemoto, K. Hayashihara, M. Matsuzaka, S. Wasai, and H. Niitani. Combination of cisplatin and carboplatin in vitro and in clinical practice, *in*: "Cancer Chemotherapy --Challenges for the Future", Volume 8, edited by K. Kimura, H. Saito, S.K. Carter, and R.C. Bast, Jr., Excerpta Medica, Tokyo, 1992.

5. F.A. Blommaert, C. Michael, P.A.B. Terheggen, F.M. Muggia, V. Kortes, J.H. Schornagel, A.A.M. Hart, and L. den Engelse. Drug-induced DNA modification in buccal cells of cancer patients during carboplatin and cisplatin combination chemotherapy as determined by an immunocytochemical method: interindividual variations and correlation with disease response, *Cancer Res* 53:5669, (1994).

6. B. Uziely, S.C. Formenti, K. Watkins, A. Mazumder, and F.M. Muggia. Calvert's formula and high-dose carboplatin, *J. Clin. Oncol.* 12:1740, (1994)

7. M.J. Piccart, J.M. Nogaret, L. Marcelis, H. Longree, F. Ries, J.P. Kainds, P. Gobert, A.M. Domage, J.P. Sculier, G. Gompel and the Belgian Study Group for Ovarian Carcinoma. Cisplatin combined with carboplatin: a new way of intensification of platinum dose in treatment of ovarian cancer, *J. Clin. Oncol.* 2:1281 (1987).

8. B. Lund, M. Hansen, O.P. Hansen, and H.H. Hansen. High-dose platinum consisting of combined carboplatin and cisplatin in previously untreated ovarian cancer patients with residual disease, *J. Clin. Oncol.* 7:1469 (1989).

9. B. Lund, M. Hansen, O.P. Hansen, and H.H. Hansen. Combined high-dose carboplatin and cisplatin, and ifosfamide in previously untreated ovarian cancer patients with residual disease, *J. Clin. Oncol.* 8:1226 (1990).

10. B. Lund, O.P. Hansen, H.H. Hansen, and M. Hansen. Combination therapy with carboplatin/cisplatin/ifosfamide/etoposide in ovarian cancer. *Sem Oncol.* 19(1):26 (1992).

11. J.R. Hardy, E. Wiltshaw, P.R. Blake, P. Harper, M. Slevin, T.J. Perrin, and S. Tan. Cisplatin and carboplatin in combination for the treatment of stage IV ovarian carcinoma. *Ann. Oncol.* 2:131 (1991).

12. E. Reed (Personal communication).

13. G. Bolis (to be published, 1995)

14. F. Muggia, L. Muderspach, L. Roman, S. Jeffers, S. Groshen, P. Conti, S.E. Martin, and C.P. Morrow. Phase I/II study of intraperitoneal floxuridine (FUDR) with either cisplatin or carboplatin or both, *Proc. Amer. Soc. Clin. Oncol.* 14:(1995).

15. J.P. Caussavel, F. Levi, S. Brienza, et al. Phase I trial of 5-day continuous venous infusion of oxaliplatin at circadian rhythm-modulated rate compared with constant rate, *J. Natl. Cancer Inst.* 82:1046, (1990)

16. G. Mathe, E. Chenu, C. Bourut, et al. Experimental study of three platinum complexes: CDDP, CBDCA and l-OHP on L1210 leukemia. Alternate or simultaneous association of two platinum complexes, *Proc. Amer. Assoc. Cancer Res.* 30:1872 (1989).

17. J.F. Llory, P. Soulie, E. Cvitkovic, and J.L. Misset. Feasibility of high-dose platinum delivery with combined carboplatin and oxaliplatin, *J. Natl. Cancer Inst.* 86:1098 (1994).

18. P. Soulie, E. Cvitkovic, C. Garrino, M. Frereres, K. Azzouzi, J.F. Llory, M. Musset, J. Gastiaburu, and J.L. Misset. Combined oxaliplatin (l-OHP) cisplatin: an effective approach to optimize platinum treatment of malignancies, *Proc. Amer. Assoc. Cancer Res.* 35:438 (1994).

19. E. Cvitkovic, M. Frereres, P. Soulie, J.F. Llory, O. Rixe, C. Garrino, M. Marty, and J.L. Misset. High activity in pretreated germ cell tumors of combined oxaliplatin (l-OHP) cisplatin based salvage regimen, *Proc. Amer. Assoc. Cancer Res.* 35:233 (1994).

20. J.L. Misset, P. Soulie, M. Frereres, J.F. Llory, E. Raymond, M. Musset, P. Chollet, C. Jasmin, C. Garrino, and E. Cvitkovic. High efficacy of combined oxaliplatin (l-OHP) cisplatin as salvage treatment in pretreated ovarian cancer, *Proc. Amer. Assoc. Cancer Res.* 35:234 (1994).

21. C. Garrino, E. Cvitkovic, P. Soulie, M. Frereres, E. Brain, S. Brienza, M. Itzhaki, C. Jasmin, and J.L. Misset. Toxicity patterns of the oxaliplatin/cisplatin association alone or with ifosfamide/epiadriamycin to platinum pretreated patients, *Proc. Amer. Assoc. Cancer Res.* 35:244 (1994).

22. J.A.Gietema, E.G. DeVries, D.T. Sleijfer, P.H. Willemse, H.J. Guchelaar, D.R. Uges, P. Aulenbacher, R. Voegeli, N.H. Mulder. A phase I study of 1,2-diamminomethyl-cyclobutane-platinum (II)-lactate (D-19466; lobaplatin) administered daily for 5 days. *Brit. J. Cancer* 67:396 (1993).

23. M.J. McKeage, L.R. Kelland, F.E. Boxal, M.R. Valenti, M. Jones, P.M. Goddard, J. Gwynne, and K.R. Harrap. Schedule dependency of orally administered bis-acetato-amminedichloro-cyclohexylamine-platinum (IV) (JM216) in vivo, *Cancer Res.* 54:4118 (1994).

Phase I and II Studies with Lobaplatin

H.H. Fiebig

University of Freiburg
Department of Internal Medicine
Freiburg, Germany, D - 79106

INTRODUCTION

Cisplatin (Cis-diamminodichloro-platinum) (II) is widely used in cancer therapy. This agent has become the main component of regimens that cure patients with cancer of the testicle and produces high response rates in patients with small cell cancer of the lung, ovarian cancer, and bladder cancers (1-2). However, the toxicity is severe including intensive nausea and vomiting and renal toxicity. Furthermore, oto- and neurotoxicity (3) were observed especially during long term therapy. The side effects encouraged the development of analogues with a better therapeutic index. Cis-diamino-1,1-cyclobutane-dicarboxylato-platinum (Carboplatin) was selected from a large number of analogues developed by Harrap et al (4). In the clinic this compound demonstrated less emetic and no nephrotoxic effects. Myelosuppression, mainly thrombopenia, is the dose limiting toxicity of carboplatin. A cross resistance between Cisplatin and Carboplatin was shown in most tumor types. In bladder and testicular cancer Cisplatin has a higher activity than Carboplatin. In most other tumor types the spectrum of activity is similar (5).

In the search for platinum analogues with a broader spectrum of activity and a better therapeutic index Asta Medica AG initiated a program for the development of new platinum compounds. Lobaplatin, D-19466, is chemically a 1,2-diaminomethyl-cyclobutanplatinum-(II)-lactate, and showed higher or similar antitumor effects in vitro and in vivo in murine tumors compared to Cisplatin (6). In human tumor xenografts Lobaplatin was markedly more active in 3 lung and a testicular cancer in vivo, similar effects were observed in an ovarian xenograft and a stomach model (7,8). There was a lack of cross resistance with Cisplatin in platinum resistant sublines of P388 (6) and in 2 ovarian and a testicular xenograft studied in vitro (9). The preclinical toxicology studies showed no nephrotoxicity in mice and rats. Myelosuppression was the limiting toxicity in rodents (10). Based on its antitumor activity in relevant models and its favorable toxicity profile Lobaplatin was selected for clinical development. The structures of the platinum complexes are shown in figure 1.

Platinum and Other Metal Coordination Compounds in Cancer Chemotherapy 2
Edited by H.M. Pinedo and J.H. Schornagel, Plenum Press, New York, 1996

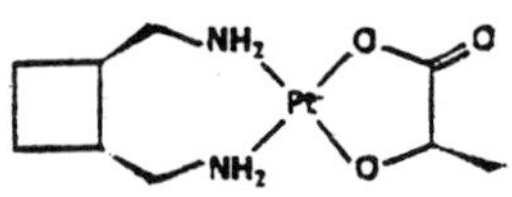

1,2-Diamminomethyl-cyclobutane-platin(II)-lactate
Lobaplatin, D-19466

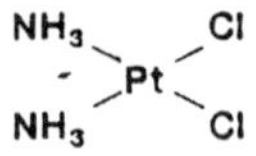

Cis-diammine-dichloroplatinum II
Cisplatin, CDDP

Fig. 1. Chemical structure of Lobaplatin and Cisplatin

Here, an overview of clinical phase I and phase II studies is given. In the phase II studies Lobaplatin was administered intravenously as a single bolus every 4 weeks without hydration if not indicated otherwise.

PHASE I STUDIES

In phase I studies 3 schedules were tested by the iv route without hydration: bolus day 1, bolus day 1 to 5 and 72 hours continuous infusion administered every 4 weeks (11-14). Thrombopenia was the dose limiting toxicity. The nadir occurred after about 2 weeks with recovery within a week. Mild leucopenia and anemia grad 1 and 2 was observed in about half of the patients. Nausea and vomiting were the most frequent nonhematological side effects. There was no evidence of renal toxicity. For phase II studies a single dose of 50 mg/m² given every 3 - 4 weeks was recommended for good risk patients.

PHASE II STUDIES

European studies

In Europe several phase II studies have been carried out by different groups. In most studies Lobaplatin was administered as a bolus iv injection of 50 mg/m2 every 3 - 4 weeks. Phase II studies in patients with advanced disease showed the following response rates:

Lobaplatin induced objective response in patients with ovarian cancer and esophageal cancer. Patients with ovarian cancer were platinum pretreated and those responding with esophageal carcinoma were chemotherapeutically naive. In head and neck cancer Lobaplatin was less effective. No relevant antitumor activity was found in lung cancer of nonsmall cell and small cell histology.

Side effects

A summary of side effects observed in phase II studies at the dose of 50 mg/m2 given iv every 3 or 4 weeks is shown in tables 3 - 5. Bone marrow suppression was the dose limiting toxicity, thrombopenia being more pronounced than leucopenia.

Table 1. Summary of phase I studies carried out with Lobaplatin.

Institution	Schedule	Patients (n)	Rec. Dose per Cycle (mg/m²)	CRCL[1] (ml/min)	Dose Limiting Toxicity
Phase I-/II-Group of the AIO (11,12)	iv bolus q4w	25	50, good risk 40, poor risk		Thrombopenia
Groningen (13)	iv bolus day 1-5 q4w	27	70 55 30	>100 81 -100 60 - 80	Thrombopenia
Groningen (14)	iv 72 hours infusion, q4w	11	50-60		Thrombopenia

[1]CRCL, creatinine clearance

Table 2. Summary of Phase II Studies of Lobaplatin performed in Europe

Tumor Type	Dose Schedule (mg/m²)	Patients (n)[1]	Response Patients (n) CR	PR	Rate (%)	Institution
Ovarian						
- Platinum pretreated	30 or 50[2], q4w	21	4	1	24	Gretema (15)
Esophageal						
- untreated	50, q3w	14	0	5	36	Schmoll (16)
- Platinum pretreated	50, q3w	4	0	0		
Head and Neck						
- untreated	50, q3w	20	1	2	15	Degardin (17)
- Platinum pretreated	50, q3w	10	0	0		
Bladder						
- untreated	Loba+MTX+Velbe, q3w	5	1	1	40	unpublished
- Platinum pretreated	50, q3w	13	0	1	8	
Lung, NSCLC						
- untreated	50, q4w	32	0	1	3	AIO, Ph.I/II (18)
Lung, SCLC						
- pretreated	40, q4w	17	0	0		AIO, Ph.I/II (19)

[1]evaluable for response
[2]CRCL < 80 ml/min: 30 mg/m², CRCL ≥ 80 ml/min: 50 mg/m²

Table 3. Phase II Studies of Lobaplatin. Summary of Hematologic Toxicity in 195 Patients

Symptoms	WHO Grade in % 0	1	2	3	4	Patients (n)
Thrombopenia	35	12	13	21	19	195
Leucopenia	39	25	24	10	2	195
Anemia	25	32	27	11	4	195

Table 4. Summary of Phase II Studies of Lobaplatin: Gastro-Intestinal Symptoms

Symptoms	WHO - Grade in %					Patients
	0	1	2	3	4	(n)
Nausea and Vomiting	35	14	38	13	0	72
Diarrhea	97	1	1	1	0	72
Appetite loss	70	12	9	7	2	57

Hematological toxicity

Thrombopenia grade 3 or 4 was observed in 40 %, leucopenia in 12 % of patients (table 3). The thrombocyte nadir occurred after about 2 weeks with recovery usually within one week. The question of cumulative thrombopenia can not be answered up to now. Anemia was less frequent, grade 3 and 4 was observed in 15 % of the patients.

Non hematological Toxicity

Overall, Lobaplatin was well tolerated (tables 4 and 5). Nausea and vomiting grade 2 or more was observed in 51 % of the patients, it was well alleviated by antiemetics. Mainly serotonin antagonists avoided or reduced intensity of nausea and vomiting in most of the patients. Nausea was evident for 1 to 3 days. Some patients complained about a loss of appetite for several days. Diarrhea was seen in 3 % (3 pats.) only. Alopecia grade 1 was found in 20 %, liver function impairment in 1 patient, a renal failure in 1 patient which was due to dehydration and not clearly drug related. There was no evidence for CNS-toxicity and ototoxicity nor for phlebitis at the injection site.

ONGOING STUDIES OF LOBAPLATIN

The clinical development of Lobaplatin is being continued intensively. Phase II studies are underway in Europe and/or in the USA (ovarian and stomach cancer) and within the EORTC in bladder cancer. Multinational randomized phase II studies in bladder cancer and in esophageal cancer are ongoing (table 6).

Table 5. Summary of Phase II Studies of Lobaplatin: Non-Hematologic Toxicity

Toxicity	WHO - Grade in %					Patients
	0	1	2	3	4	(n)
Alopecia	80	20	0	0	0	54
Dyspnea	98	0	0	0	2	57
Neuro	99	1	0	0	0	71
Renal	99.6	0	0	0	0.4[1]	260

1, 1 patient with prerenal acute renal failure due to dehydration

Table 6. Overview of ongoing Phase II Studies of Lobaplatin

Tumor Type	Modality	Number of Patients / Status	Institution
Ovarian			
- Platinum pretreated	single agent	7	USA
Gastric			
- untreated	single agent	14	USA / Europe
- pretreated	single agent	7	
Bladder			
untreated and pretreated	single agent	21	EORTC
Bladder			
untreated	Loba+MTX+Velbe vs CMV	started	Multinational
Esophageal			
untreated	single agent	7	France
pretreated	single agent	6	
Esophageal			
untreated	Loba vs LOBA + FU vs CDDP + FU	started	Multinational

A broad phase II study with Lobaplatin in several tumor types is ongoing in China. The present analysis is based on treatment of 139 patients. Promising response rates were seen in patients with mainly untreated small cell lung cancer (50%, 8/16), breast cancer (44%, 8/23), lymphoma (3/3) and chronic myelocytic leukemia.

CONCLUSIONS

Overall Lobaplatin is a promising new platinum complex with definite clinical activity in solid tumors including ovarian cancer and esophageal cancer. It is probably active in non-Hodgkin-lymphoma, breast cancer and untreated small cell lung cancer. No activity was found in non-small cell lung cancer. The favorable toxicity profile could be confirmed in more than 250 patients during clinical phase II studies. The dose limiting toxicity is thrombopenia with a nadir after about 2 weeks and recovery within a week. Lobaplatin can easily be administered in an outpatient clinic without need for hydration. The feasibility of Lobaplatin containing combination regimens has been shown, and randomized phase II studies with combination therapy have recently been started.

REFERENCES

1. L.H. Einhorn and J. Donahue. Cis-diamminedichloroplatinum, vinblastine and bleomycin combination chemotherapy in disseminated testicular cancer. Ann Inter Med 87:293 (1977).
2. E. Wiltshaw and T. Kroner. Phase II study of cis-dichlorodiammineplatinum(II) (NSC-119875, CACP) in advanced adenocarcinoma of the ovary. Cancer Treat Rep 60:55 (1976).
3. M.S. Soloway. Cis-diamminedichloroplatinum(II) (DDP) in advanced bladder cancer. J Urol 120:716 (1978).

4. R. Canetta, K. Bragman, L. Smaldone, and M. Rozenzweig. Carboplatin: current status and future prospects. Cancer Treat Rev 15:17 (1988).
5. K.R. Harrap, M. Jones, C.R. Wilkinson, H.McD. Clink, S. Sparrow, B.C.V. Mitchley, S. Clarke and A. Veasey. Antitumor, toxic and biochemical properties of cisplatin and eight other platinum complexes. In: "Cisplatin. Current status and new developments" A.W. Prestayko, S.T. Crooke and S.K. Carter edsAcademic Press New York, London, Toronto, Sydney, San Francisco, 193 (1980)
6. R. Voegeli, W. Schumacher, J. Engel, J. Respondek, and P. Hilgard. D-19466 a new cyclobutane-platinum complex with antitumor activity. J Cancer Res Clin Oncology 116:439 (1990).
7. H.H. Fiebig, D.P. Berger, K. Mross, W. Queißer, P. Aulenbacher, and P. Hilgard. Lobaplatin (D-19466): Preclinical drug profile and results of the phase I study. Ann Oncol 3 (suppl) 143 (1992).
8. H.H. Fiebig, D.P. Berger, W.A. Dengler, E. Wallbrecher, and B.R. Winterhalter. Combined in vitro/in vivo test procedure with human tumor xenografts. In: "Immunodeficient Mice in Oncology" H.H. Fiebig and D.P. Berger ed, Karger, Basel, Contrib Oncol 42:321 (1992).
9. A. Harstrick, C. Bokemeyer, M. Scharnofkse, G. Hapke, D. Reile, and H.J. Schmoll. Preclinical activity of a new platinum analogue, lobaplatin, in cisplatin-sensitive and -resistant human testicular, ovarian and gastric carcinoma cell lines. Cancer Chemoth and Pharmacol 33:43 (1993).
10. Investigator's Brochure Lobaplatin, ASTA Medica AG, Germany (1993).
11. H.H. Fiebig, K. Mross, H. Henß, D. Ludolph, F. Meyberg, P. Aulenbacher, and W. Queißer. Phase I Clinical Trial of Lobaplatin (D-19466) after intravenous bolus injection. Onkologie 17:142 (1994)
12. K. Mross, F. Meyberg, H.H. Fiebig, K. Hamm, U. Hieber, P. Aulenbacher, and D.K. Hossfeld. Pharmacokinetic and pharmacodynamic study with lobaplatin (D-19466), a new platinum complex, after bolus administration. Onkologie 15:139 (1992).
13. J.A. Gietema, H.J. Guchelaar, E.G.E. de Vries, P. Aulenbacher, D.T. Sleijfer, and N.H. Mulder. A phase I study of lobaplatin (D-19466) administered by 72 hours continuous Infusion. Anti-Cancer Drugs 4:51 (1993).
14. J.A. Gietema, E.G.E. de Vries, D.T. Sleijfer, P.H.B. Willemse, H.J. Guchelaar, D.R.A. Uges, P. Aulenbacher, R. Voegeli, and N.H. Mulder. A phase I study of 1,2-diamminomethyl-cyclobutane-platinum (II)-lactate (D-19466; lobaplatin) administered daily for 5 days. Br J Cancer 67:396 (1993).
15. J.A. Gietema, G.J. Veldhuis, H.J. Guchelaar, P.H.B. Willemse, D.R.A. Uges, A. Cats, B. Boonstra, W.T.A. van der Graaf, D.T. Sleijfer, E.G.E. de Vries, and N.H. Mulder. Phase II and pharmacokinetic study of lobaplatin in patients with relapsed ovarian cancer. BJC 71:1302 (1995).
16. H.J. Schmoll, C.H. Köhne, E. Papageorgiou, S. Luft, A. Harstrick, P. Bachmann, P. Hilgard, and U. Schubert. Single agent lobaplatin is active in patients with oesophageal squamous cell carcinoma: A phase II evaluation. Proc ASCO 14:201 (1995).
17. M. Degardin, M. De Forni, B. Chevallier, P. Kerbrat, M.A. Lentz, M. David, and H. Roché. Phase II study of Lobaplatin in head and neck cancer (HNC). Abstract 8. NCI-EORTC Symposium: 126 (1994).
18. C. Manegold, H.H. Fiebig, U. Gatzemeier, I. vov Pawel, W. Queißer, P. Bachmann, L. Edler, and P. Drings. Phase II Study of Lobaplatin in patients with advanced non small cell lung cancer. Onkologie 16, supplement 1, 13 (1993)
19. H.H. Fiebig, I. von Pawel, U. Gatzemeier, C. Manegold, L. Edler, K. Burk, P. Bachmann, W. Berdel. Phase II study of the platinum complex lobaplatin in patients with extensive small cell lung cancer after previous chemotherapy. Onkologie 16, supplement 1, 14 (1993)

PHASE I TRIALS OF ORMAPLATIN (NSC 363812)

Michaele C. Christian, M.D.

Investigational Drug Branch, Cancer Therapy Evaluation Program
Division of Cancer Treatment, National Cancer Institute
Bethesda, Maryland, USA

Tetrachloro(d,l-trans)1,2-diaminocyclohexaneplatinum (IV), ormaplatin (OP), NSC 363812, formerly called tetraplatin, is a second generation platinum (IV) compound which has been evaluated in clinical trials sponsored by the National Cancer Institute (NCI) in collaboration with Upjohn. Interest in this group of compounds was stimulated by structure-activity studies by Burchenal et al.,(1) which suggested that platinum compounds containing the 1,2-diaminocyclohexane (DACH) carrier ligand retained activity in vitro and in vivo against L1210 and P388 murine leukemia cell lines with acquired resistance to cisplatin. In addition, in preclinical antitumor testing, OP demonstrated broad spectrum activity comparable to or superior to cisplatin (CDDP) against a variety of murine and human tumor models (2,3) however, its activity in other models was variable and against a screening panel of 16 human ovarian cancer cell lines was disappointing (4).

CLINICAL TRIAL DESIGN

Between 1989 and 1994, 6 phase I trials of ormaplatin (OP) were completed. Trials were conducted on single bolus and daily for 5 days schedules because of differences in preclinical toxicity on these schedules (5,6, unpublished data on file, Division of Cancer Treatment [DCT], NCI). In addition, a day 1 and 8 schedule was pursued because of experience with cisplatin (CDDP) which suggested a potentially more tolerable toxicity profile on this schedule (7). Courses were repeated every 28 days on all schedules. The intraperitoneal route of administration was also explored in 1 trial.

These studies and the dose ranges evaluated are listed in Table 1 (8-13). For the most part these trials were similar with regard to design, patient eligibility factors and other elements. The starting doses were based on the LD_{10} (dose producing lethality in 10% of animals) values in mice and doses were escalated using a modified Fibonacci scheme. Pharmacokinetics were assessed in all trials. Important eligibility criteria included requirements for nearly normal function of a number of organ systems including: bone marrow, kidneys and liver. Patients could have only Grade 1 preexisting neurological abnormalities (loss of deep tendon reflexes) and had to have clinically normal hearing. Prehydration was commonly used, though hydration following therapy was variable. The earliest trials did not utilize prophylactic antiemetics, however they were subsequently incorporated into studies since nausea and vomiting were almost universal without pretreatment.

As can be seen from Table 1, approximately 157 patients were treated on the Phase I trials. Most trials did not proceed to a traditional maximum tolerated dose based on acute first cycle toxicity because the most troublesome toxicity observed was a debilitating neurotoxicity which appeared to be related to cumulative dose and occurred very frequently in patients who received total OP doses >150-200 mg/m^2.

Table 1 NCI Sponsored Ormaplatin Phase I Clinical Trials

institution (reference)	schedule & doses (mg/m^2)	#/# evaluable	# cycles	responses
NCI Medicine Branch (8)	daily X 1 4.3--90	35/31	75	0
University of Wisconsin (9)	daily X 1 4--123	41/39		1 lung, 1skin
University of Texas, San Antonio (10)	daily X 5 1--11.6	35/35	70	1 colon
Duke (11)	daily X 5 5--11.6	16/15	25	0
Fox Chase (12)	day 1 & 8	26/26	51	1 lung
University of California, San Diego (13)	i.p. q 28 days 10--88.4	14/13	28	0
TOTAL		**167/152**		**4 PR**

PHARMACOKINETIC DATA

OP pharmacokinetic data are presented in Table 2. Because of the extremely rapid conversion of OP to its biotransformation products, estimated to be on the order of 3 seconds in rat plasma (14), no suitable assay for parent OP was available. The pharmacokinetics of ultrafilterable and total platinum were measured in the Phase I trials. Platinum exhibited a biexponential decay. There was a linear relationship between dose and maximum plasma concentration (C_{max}) and AUC (area under the concentration x time curve) of ultrafilterable and total platinum across all of the doses and schedules evaluated.

Ultrafilterable platinum accounted for approximately 30% of the platinum administered and urinary excretion ranged from 7-32% (11,12). The pharmacokinetics of OP biotransformation products were evaluated in 1 study. $PtCl_2(DACH)$, the major active OP biotransformation product exhibited monoexponential decay with a $t_{1/2}$ of 12.9 minutes and more interpatient pk variability than ultrafilterable platinum (11). With intraperitoneal administration, a pharmacologic advantage was achieved with the ratio of peritoneal to plasma AUC of approximately 17. However, systemic exposure did occur and one patient developed neurotoxicity even with this route of administration (13).

Table 2 Ormaplatin Pharmacokinetics

schedule	dose mg/m2	Pt species	$t_{1/2\alpha}$ hour	$t_{1/2\beta}$ hour	ref
dx1	90	TP	1.12	38.9	8
	63	UP	----	13.1	9
d1,8	60.8	TP	0.16	25.8	12
	60.8	UP	0.14	14.9	
dx5	11.6	UP	0.27	----	11

schedule	dose mg/m2	Pt species	AUC mg.h/L	Cl ml/min/m2	C_{max} ng/ml	V_c L/m2	ref
dx1	90	TP	25.5	45	1332	94	8
	63	UP	----	132	490	--	9
d1,8	60.8	TP	35.6	215	----	21	12
	60.8	UP	4.9	29	----	59	
dx5	11.6	UP	----	544	227	12	11

Pt=platinum, TP=total platinum, UP=ultrafilterable platinum, $t_{1/2}$ = half-life, AUC=area under the concentration x time curve, Cl=clearance, C_{max}=maximum plasma concentration, V_c=volume of central compartment

ADVERSE EVENTS

The adverse events observed during these Phase I trials are noted in Table 3. In some cases, the actual relationship of these events to the drug therapy remains uncertain and it is likely that some of them may have been related to underlying disease. None of the trials achieved a standard MTD (maximum tolerated dose) based on acute first-cycle toxicity. Dose escalation on all schedules was terminated based on the development of a debilitating peripheral neuropathy which limited the cumulative dose which could be administered.

OP neurotoxicity was predominantly sensory, characterized by peripheral parethesias and sensory ataxia. Functional impairment was manifest primarily as difficulty walking and performing fine motor functions with the hands. There was little evidence of motor weakness. Some patients developed symptoms after completing therapy and others worsened despite cessation of OP treatment.

Recovery from the neuropathy was slow over several months and/or incomplete in the majority of symptomatic patients. In 2 patients, postural hypotension may have been related to autonomic neuropathy (12, unpublished data, DCT, NCI). Neuropathy appeared to be associated with a cumulative dose of 150-200 mg/m^2 on all schedules rather than with the size of the individual dose and was, therefore, most apparent in trials where patients received multiple courses of treatment. In one trial using the 5 day schedule, all patients who received cumulative doses $\geq$165 mg/m^2 developed neuropathy (10). Unfortunately, serial nerve conduction studies (NCS) did not detect abnormalities prior to the onset of symptoms. NCS in several patients showed prolonged latency in the sural nerve consistent with demyelination and decreased amplitude of action potentials in the median, ulnar and sural nerves consistent with axonal loss (10,12). Sural nerve biopsy in 1 symptomatic patient showed axonal loss with secondary demyelination and remyelination (10). While most of the patients on these Phase I studies had received prior chemotherapy, a number of the patients who developed significant neurotoxicity had not been treated previously with neurotoxic agents.

Other toxicities associated with OP were generally well tolerated. Nausea and vomiting were easily controlled with standard antiemetics and were often abolished altogether by the administration of prophylactic antiemetics, most commonly ondansetron. Myelosuppression was sporadic, though more common at the higher doses and appeared likely to be the acute dose-limiting toxicity, though Grade 3-4 myelosuppression was uncommon, except in heavily pretreated patients. While all cell lines were affected, thrombocytopenia and anemia were more common than neutropenia. Renal dysfunction was uncommon and usually grade 1-2. Grade 1-2 diarrhea was observed and appeared to be more frequent on the 5 daily schedule (O'Rourke). No significant ototoxicity was observed either clinically or on serial audiograms.

The DLT with intraperitoneal administration was abdominal pain. Nausea, vomiting, fever, myelosuppression, ileus and peripheral neuropathy were also seen with this route of administration.

Table 3 Ormaplatin Phase I Adverse Events

• **neurologic** **paresthesias** **peripheral neuropathy** **?autonomic neuropathy** • **nausea & vomiting** • **myelosuppression** **all cell lines** **thrombocytopenia &** **anemia > leukopenia** • **gastrointestinal** **abdominal pain** **diarrhea** **constipation** **ileus** • **hepatic** **elevated SGOT, bilirubin**	• **renal** **elevated creatinine** **elevated $\beta_2$microglobulin** • **cardiovascular** **supraventricular tachy-arrhythmias** **postural hypotension** • **metabolic** **hypomagnesemia** **hyperglycemia** • **miscellaneous** **phlebitis** **malaise/fatigue** **muscle cramps/myalgias** **altered taste** **allergic reactions** **headache**

DISCUSSION

Six Phase I trials of OP using 3 different schedules and 2 routes of administration have identified similar toxicities. The most troubling toxicity has been the sometimes debilitating sensory neuropathy which appears to occur quite commonly after cumulative doses of 150-200 mg/m^2 have been administered. This would effectively limit Phase II OP treatment to 2 cycles at otherwise reasonable doses. In addition, only 4 partial responses were observed among the 152 evaluable patients on these studies.

The neurotoxicity observed with OP will severely limit its future clinical development. It is possible that the neuropathy could be abrogated by neuroprotective agents in the future. Evidence of useful clinical activity in a platinum-refractory ovarian cancer patient population might provide sufficient incentive for such future investigation. However, the availability of other DACH-platinum compounds in the clinic must also be factored into decisions regarding the future development of OP. At present, the NCI is considering one additional limited study of OP in platinum-refractory ovarian cancer patients to explore its activity in the setting of clinical platinum resistance, since this is the clinical situation for which DACH-platinum compounds have held the greatest interest.

REFERENCES

1. Burchenal JH, Kalaher K, Dew K, et al. Studies of cross-resistance, synergistic combinations and blocking of activity of platinum derivatives. Biochimie 60 (9):961-965, 1978

2. Anderson WK, Quagliato DA, Haugwitz RD, et al. Synthesis, physical properties, and antitumor activity of tetraplatin and related tetrachloroplatinum(IV) stereoisomers of 1,2-diaminocyclohexane. Cancer Treat Rep 70(8):997-1002, 1986

3. Behrens BC, Hamilton TC, Masuda H, et al. Characterization of a cis-diamminedichloroplatinum(II)-resistant human ovarian cancer cell line and its use in evaluation of platinum analogues. Cancer Res 47(2):414-418, 1987

4. Harrap KR, Jones M, Siracky J, et al. The establishment, characterization and calibration of human ovarian carcinoma xenografts for the evaluation of novel platinum anticancer drugs. Ann Oncol 1(1):65-76, 1990

5. Smith JH, Smith MA, Litterst CL, et al. Comparative toxicity and renal distribution of the platinum analogs tetraplatin, CHIP, and cisplatin at equimolar doses in the Fischer 344 rat. Fundam Appl Toxicol 10:45-61, 1988

6. Smith JH, Smith MA, Litterst CL, et al. In vivo biochemical indices of nephrotoxicity of platinum analogs tetraplatin, CHIP, and cisplatin at equimolar doses in the Fischer 344 rat. Fundam Appl Toxicol 10:62-72, 1988

7. Gandara DR, Wold H, Perez E, et al. High-dose cisplatin in hypertonic saline; reduced toxicity of a modified dose-schedule and correlation with plasma pharmacokinetics. A Northern California Oncology Group pilot study in non-small cell lung cancer. J Clin Oncol 4:1787-1793, 1989

8. Christian MC, Kohn E, Sarosy G, et al. Phase I and pharmacologic study of ormaplatin (op)/tetraplatin. Proc Am Soc Clin Oncol 11:117, 1992

9. Tutsch KD, Arzoomanian RZ, Alberti D, et al. Phase I trial and pharmacokinetic study of ormaplatin. Proc AACR 33:536,1992

10. O'Rourke TJ, Weiss, New P, et al. Phase I trial of ormaplatin (tetraplatin, NSC 363812). Anti-Cancer Drugs 5:520-525, 1994

11. Petros WP, Chaney SG, Smith DC, et al. Pharmacokinetic and biotransformation studies of ormaplatin in conjunction with a phase I clinical trial. Cancer Chemother Pharmacol 33:347-354, 1994

12. Schilder RJ, LaCreta FP, Perez RP, et al. Phase I and pharmacokinetic study of ormaplatin (tetraplatin, NSC 363812) administered on a day 1 and day 8 schedule, Cancer Res 54(3):709-717, 1994

13. Plaxe SC, Braly PS, Freddo JL, et al. Phase I and pharmacokinetic study of intraperitoneal ormaplatin. Gynecologic Oncology 51:72-77, 1993

14. Chaney SG, Wyrick S, Till GK. In vitro biotransformations of tetrachloro(d,l-trans)1,2-diaminocyclohexaneplatinum (IV) (tetraplatin) in rat plasma. Cancer Res 50:4539, 1990

254-S, NK121 AND TRK710

Makoto Ogawa

Aichi Cancer Center
Nagoya Japan

INTRODUCTION

Cisplatin has shown to be effective againsr a wide variety of solid tumors and has contributed to cure or prolongation of survival duration in these tumors, however, both nephrotoxicity and neurotoxicity are major obstacles to use clinically.

Analogous compounds have been synthesized in order to improve these toxicities or to obtain higher antitumor activities than the parent compound.

This paper describes preclinical and clinical results of three analogous compounds developed in Japan.

254-S (Cis-Diammine Glycolato platinum)

254-S, an analog of cisplatin that was synthesized by Shionogi Pharmaceutical Company is about 10 times more water-soluble than is cisplatin. (Fig 1)

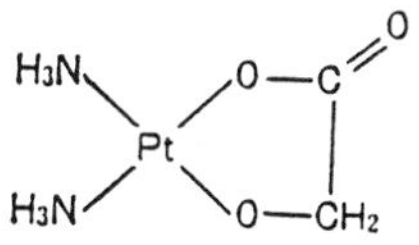

254-S

In preclinical studies[1], 254-S demonstrated comparable antitumor activity to cisplatin on transplantable murine tumors including L1210 leukemia, P388 leukemia, B16 melanoma, Colon 38 and Lewis lung carcinoma and MX-1 implanted in nude mice.

The major toxicity observed in preclinical toxicology was myelosuppression which was more profound than cisplatin, whereas nephrotoxicity and gastrointestinal toxicity were milder than cisplatin.

Platinum and Other Metal Coordination Compounds in Cancer Chemotherapy 2
Edited by H.M. Pinedo and J.H. Schornagel, Plenum Press, New York, 1996

Phase I trial

A phase I trial[2] employed a single intermittent schedule and doses were escalated from 20mg/m^2 to 120mg/m^2 by 6 steps. 254-S was dissolved 250 mL of 5% xylitol and infused over 60 minutes, and no hydration was performed.

The highest dose was a maxinum tolerated dose and the dose limiting toxicity was thrombocytopenia, reaching nadirs about 3 weeks later and requiring about 1 week for recovery. Leukopenia was dose-related and was grade 4 at 120mg/m^2. Nausea and vomiting were judged to be milder than with cisplatin, and mild elevation of BUN occured in 21%.

A recommended dose for phase II trial was decided to be 100mg/m^2 in 4-week intervals.

Phase II trial

The results obtained in phase II trials are summarized in Table 1. Inuyama[3,4] et al conducted early and late phase II trials on head and neck cancers at a dose of 100mg/m^2 at 4-week intervals. 254-S was dissolved in 300mL of 5% xylital and infused over 60 minutes with intravenous hydration according to the investigators' judgement to obtain daily urinary volume exceeding 1500mL. Of 24 patients in early phase II trial[3] there were 4 complete and 5 partial responses with an overall responrate of 37.5% and durations of responses were ranged from 71 to 141 days. Of 66 patients in late phase II trial[4] there were 7 complete and 22 partial responses with an overall response rate of 43.9% and durations of responses were ranged from 61 to 411 days, thus, the late phase II trial confirmed the result obtained in early phase II trial.

Furuse[5] et al conducted a phase II trial on lung cancer with the same administrative method as in the previous study. Of 22 patients with small cell lung cancer there were 9 partial responses with a resoinse rate of 40.9% and durations of responses were ranged from 27 to 345 days. Of 39 patients with non-small cell lung cancer, there were 8 partial responses with a response rate of 20.5% and durations of responses were ranged from 70 to 232 days.

Koyama[6] et al conducted a phase II trial on heavily pretreated patients with advanced breast cancer using the same administrative method. Of 16 patients there were 2 partial responses with a response rate of 12.5% with durations of responses lasting 78 and 95 days.

Taguchi[7] et al conducted a phase II trials on gastrointestinal tumors. Of 35 evaluable patients with esophageal cancer, there were 15 partial responses with a response rate of 42.9% and durations of responses were ranged from 29 to 282 days. Of 15 patients with advanced gastric cancer there was 1 partial response and of 13 patients with colorectal cancer there was no major reponse.

Akaza[8] et al conducted a phase II trial in genitourinary cancers. Of 13 cisplatin naive patients with testicular cancer, there were 6 complete and 6 partial responses with an overall response rate of 92.3% but two patients who had exposed prior cisplatin did not respond. Of 35 patients with urothelial cancer, there were 2 complete and 8 partial responses with an overall response rate of 28.6%, and of 16 patients with prostatic cancer, there were 3 partial responses.

Kato[9] et al conducted a phase II trial on gynecologic cancers. Of 61 patients with ovarian cancer there were 4 complete and 19 partial responses with an overall response rate of 37.7 % and durations of responses were ranged from 33 to 543 days, and of 41 patients with cervical cancer, there were 4 complete and 15 partial responses of an overall response

Table 1 : Phase II trial of 254-S

Tumor	No of Evaluable Patients	Complete Response (CR)	Partial Response (PR)	CR+PR (%)	Duration (days)	Reference
Head & Neck Cancer	24	4	5	37.5	71-141	3)
	66	7	22	43.9	61-411	4)
Small Cell Lung Cancer	22	0	9	40.9	27-345	5)
Non-Small Cell Lung Cancer	39	0	8	20.5	70-232	5)
Breast Cancer	16	0	2	12.5	78-95	6)
Esophageal Cancer	35	0	15	42.9	29-282	7)
Gastric Cancer	15	0	1	6.7	-	7)
Colorectal Cancer	13	0	0	0	-	7)
Testicular Cancer	15	6	6	80	-	8)
Urothelial Cancer	35	2	8	28.6	-	8)
Prostatic Cancer	16	0	3	18.8	-	8)
Ovarian Cancer	61	4	19	37.7	33-543	9)
Cervical Cancer	41	4	15	46.3	31-1122	10)
	40	4	9	32.5	28-352	10)

rate of 46.3% with response durations ranged from 31 to 1122 days. Noda[10] et al conducted a phase II trial on cervical cancer using a dose of 80mg/m^2 in 4-week intervals, and obtained 4 complete and 9 partical responses with an overall response rate of 32.5 % out of 40 patients. Hematologic toxicities observed in this study were judged to be milder than those in the previous study[9]. Summarizing both results, an overall response rate of 39.5% on cervical cancer was conducted.

The results obtained in phase II trials of 254-S can be summarized as follows : (1) Response rates exceeding 40 % were noted in testicular cancer, esophageal cancer, small cell lung cancer and cervical cancer therefore 254-S has comparable or even higher efficacies on these tumors as compared with cisplatin. (2) Response rates ranging from 20.5% to 37.7% were noted in head & neck cancer, non-small cell lung cancer, urothelial cancer and ovarian cancer, thus, the antitumor efficacies of 254-S on these tumors appear to be equivalent to that of cisplatin. (3) 254-S has shown modest activity on breast and prostatic cancer. (4) 254-S was inactive on gastric and colorectal cancer. (5) Nephorotoxicity was milder than with cisplatin, whereas hematologic toxicity was more profound.

NK121

NK121 is a water-soluble analog of cisplatin that was synthesized by Nippon Kayaku Company. (Fig 2)

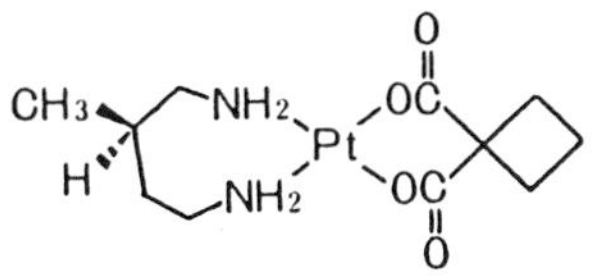

CH_3 H NH_2 Pt NH_2 OC O OC O

NK121

NK121[11] demonstrated the antitumor activity on L1210 leukemia, P388 leukemia, B16 melanoma and others which were equivalent to that of cisplatin but superior over carboplatin, in addition, NK121 was active against sublines of L1210 and P388 leukemia resistnat to cisplatin in vitro. The preclinical toxicology indicated that nephrotoxicity and gastriontestinal toxicity were milder than those of cisplatin, whereas hematologic toxicity was more profound compared to those with cisplatin.

Fukuoka[12] et al conducted a phase I trial using a single intermittent dose and doses were escalated from 40mg/m^2 to 360mg/m^2. The drug was dissolved in 250mL of 5% glucose solution and infused over 30 minutes without hydration. The dose limiting toxicity was leukopenia reaching nadirs 2 weeks later and requiring 1 week for the recovery. Nausea and vomiting occured in all cases at doses above 320mg/m^2 whereas nephrotoxicity was very mild. A recommended dose for phase II trial was judged to be 300mg/m^2 in 3-4 weeks intervals. Phase II trials were conducted in lung cancer and head and neck cancer but clinical activities on these tumors were modest amd therefore further phase II trials were discontinued.

TRK-710

TRK-710, an analog of cisplatin[13,14] that was synthesized by Basic Research Laboratory of Toray Industries. (Fig 3)

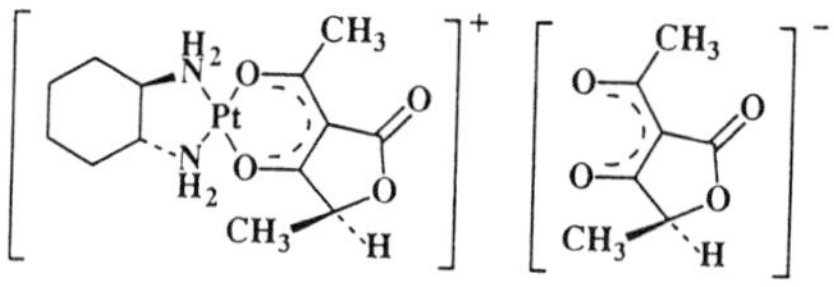

TRK-710

The antitumor activities on transplantable murine tumors including L1210 leukemia were slightly inferior to that of ciplatin but comparable to carboplatin. On the other hand, TRK-710 demonstrated significant activities on human tumor xenografts of non-small cell lung cancer and endometrial cancer which were equivalent or superior to those of cisplatin, in addition, TRK-710 showed remarkable antitumor activities on both a subline of L1210 leukemia and a human tumor xenograft of lung cancer resistnat to cisplatin. A preclinical study[15] indicated that the influx rate of TRK-710 into cisplatin-resistant cells was approximately 3 times higher than that of cisplatin and this high influx rate therefore may relate to mechanism of circumvention of the cisplatin resistance but the exact mechanism is still unclear. Major toxicities[15] observed in preclinical toxicology were neurotoxicity, hematologic toxicity and gastrointestinal toxicity which were judged to be equivalent to those with cisplatin whereas nephrotoxicity was extremely mild. A phase I trial employed a single intermittent schedule starting from a dose of 20mg/m^2 which is one-tenth of LD10 in mice, without hydration is in progress.

REFERENCES

1. Shiratori. O., H. Kasai., N. Uchida., et al : Recent Advances in Chemotherapy. Anticancer Section I. J. Ishigame ed, P635, Univ of Tokyo Press. TOKYO, 1985.
2. Ota. K., A. Wakui., H. Majima., et al : Jpn J Cancer Chemother 19 (6) : 855, 1992.

3. Inuyama. Y., H. Miyake., M. Horiuchi., et al : Jpn J Cancer Chemother 19(6) : 863, 1992.
4. Inuyama. Y., H. Miyake., M. Horiuchi., et al : Jpn J Cancer Chemother 19(6) : 871, 1992.
5. Furuse. K., M. Fukuoka., Y. Kurita., et al : Jpn J Cancer Chemother 19(6) : 879, 1992.
6. Koyama.H., M. Ogawa., Y. Kuraishi., et al : Jpn J Cancer Chemother 19(7) : 1049, 1992.
7. Taguchi. T., A. Wakui., K. Nabeya., et al : Jpn J Cancer Chemother 19(4) : 483, 1992.
8. Akaza. H., M. Togashi., Y. Nishio., et al : Cancer Chemother Pharmacol 31 : 187, 1992.
9. Kato. T., H. Nishimura., M. Yakushiji., et al : Jpn J Cancer Chemother 19(5) : 695, 1992.
10. Noda. K., M. Ikeda., M. Yakushiji., et al : Jpn J Cancer Chemother 19(6) : 885, 1992.
11. Ota. K : Jpn J Cancer Chemother 20(1) : 50, 1993.
12. Fukuoka. M., H. Nietani, K. Hasegawa., et al : Proc ASCO 8 : 62, 1989.
13. Mutoh. M., Y. Matsushima., Y. Saito., et al : Proc AACR 34 : 399, 1993.
14. Miyamoto. Y., S. Hanada., H. Hashimoto., et al : Proc AACR 35 : 437, 1994.
15. Tnoue. S. and S. Mizuno. : Proc AACR 35 : 436, 1994.

CLINICAL TRIALS OF OXALIPLATIN AND DWA2114R

T. Taguchi

Professor Emeritus
Osaka Univ.
Cancer Chemotherapy Study Society
Okada Bidg., Rm-510
1-18-12 Edobori, Nishi-ku
Osaka 550
Japan

Oxaliplatin, oxalato(trans-1,2-diaminocyclohexane)platinum(II), and DWA2114R,(-)-(R)-2-aminomethylpyrrolidine(1,1-cyclobuthanedicarboxylato) platinum(II)mono-hydrate, are platinum coordination complexes which have been identified and developed as new antitumor agents in Japan.

Oxaliplatin

A phase I clinical trial of oxaliplatin was conducted with various tumor types, i.e., histologically or cytologically confirmed solid tumors refractory to conventional therapy or for which no effective therapy existed, by multi-institution groups. Patients received a single administration of Oxaliplatin by one hour intravenous infusion without pre- or post hydration. Groups of patients (at least 3 per group) were assigned doses according to the following sequence: 20*, 40, 80, 130 and 180 mg/m² (*approximately one-fourth of the

TDL value of dogs). Of the 20 patients who entered in this study, 20 were evaluable. 19 females and 1 male were treated, with a medium age of 56 (range 42-74), and an ECOG performance status of 0-2.

Dose limiting toxicity was neurologic with a peculiar sensory neuropathy and the maximum tolerated dose was determined to be 180 mg/m². Gastrointestinal toxicity was practically constant and neither renal nor hematologic were observed (Table 1).

Early phase II clinical trials of Oxaliplatin have been conducted at a dose of 130 mg/m², administered at least 2 times, every 3 weeks for primary lung cancer, gynecological malignancies or colorectal cancer.

The response rate was 42.9% (3/7) for endometrial cancer, and 11.1% (2/18) for cervical cancer. There were no responses for ovarian or either lung and colorectal cancer.

Table 1 Clinical Toxicity

Dose (mg/m²)	No. of patients	Nausea Vomiting	Anorexia	Peripheral Neuropathy	Constipation	Pain	Others
20	3	3	1	-	-	-	-
40	3	3	2	1	1	-	-
80	3	3	1	2	-	-	1
130	6	6	1	6	-	1	1
180	5	5	1	5	-	-	1
Total (%)	20	20 (100)	6 (30)	14 (70)	1 (5.0)	1 (5.0)	3 (15)

The adverse events observed during the trials were as follows.

Thrombocytopenia or anemia of grade 3 or greater was observed in 2.7% and 1.4%, respectively, indicating only a slight effect on hematologic parameters.

As for the incidence for other clinical adverse experience with grade 3 or greater, dysesthesia (24.7%), nausea and vomiting (20.5%), and inappetence (8.2%) were observed.

Oxaliplatin is effective against corpus uteri cancer, but does not effect in ovarian, both lung and colorectal cancer.

DWA2114R

A phase I clinical trial of DWA2114R was conducted in 39 patients with various tumor types by a group of 13 institutions. Patients received a single administration of DWA2114R by 20-30 minute intravenous infusion without pre- or post hydration. Groups of patients (at least 3 per group) were assigned doses according to the following sequence: 40*, 80, 160, 240, 320, 400, 500, 600, 800 and 1000 mg/㎡(*approximately one-tenth of the LD_{10} value in mice). Of the 39 patients entered in this trial, 35 were evaluable. 15 females and 20 males were treated, with a medium age of 56 (range 22-81), and an ECOG performance status of 0-3.

The dose limiting factor was leukopenia, especially neutropenia, and the maximum tolerated dose was more than 1000 mg/㎡(25N). The major clinical toxicity was gastrointestinal. Hepatotoxicity and nephrotoxicity were mild (1). Following administration of the drug, plasma concentration of total and filterable platinum showed a triphasic and biphasic decay. Excretion into urine within 24 hours was in the range of 54.2% to 92.2% of the administrated amount of platinum (2).

Early phase Ⅱ clinical trials of DWA2114R have been conducted at a dose range of 800-1000 mg/㎡, administered at least 2 times every 3-4 weeks for 7 tumor types.

The response rate was 30% (3/10) for testicular cancer (3), 18.4% (7/38) for prostatic cancer (3), 44.1% (15/34) for ovarian cancer (4), 9.1% (2/22) for uterus cervical cancer (4), 13.3% (4/30) for non-small cell lung cancer, 15.8% (3/19) for small cell lung cancer and 20.6% (7/34) for breast cancer (5). The response rates in patients with ovarian, uterus, cervical and lung cancers were higher in patients without any prior therapy than in patients with prior therapy. Efficacy was observed in patients with testicular and ovarian cancer, who had received a Cisplatin-containing combination chemotherapy. Some prostatic cancer patients who were refractory to hormone therapy responded to DWA2114R treatment with a significant partial response. Efficacy in breast cancer, in which Cisplatin has not previously been proven useful, was also demonstrated.

The adverse events observed during the treatment courses were as follows. Leukopenia of grade 3 or greater was observed in 30.1%, which was the highest incidence. Thrombocytopenia of grade 3 or greater was observed rarely (1.9%), indicating only a slight effect of DWA2114R on platelet counts.

As for the incidence of other clinical adverse experiences with grade 3 or greater, nausea and vomiting (39.5%) occurred with the highest incidence. Anorexia (8.7%) was also observed. Diarrhea was not observed.

A late phase Ⅱ clinical trial against ovarian cancer was conducted as a randomized comparative study using the CAP regimen.

The trial involved random allocation by the controller to group A (800mg/㎡ DWA2114R, 35mg/㎡ ADM, and 500mg/㎡ CPA, day 1) or group B (50mg/㎡

CDDP, 35mg/m² ADM, and 500mg/m² CPA, day 1). The dose levels of each regimen were set by referring to the results of a phase Ⅲ trial of Carboplatin in Japan (6).

The administration of DWA2114R was carried out by 1-hour intravenous infusion without hydration, while the CDDP group was given adequate hydration. Generally, no symptomatic treatment was conducted for either dose group, the number of dosings was set as at least 2 times and the dosing interval at 3-4 weeks.

The response rates were 38.7% for the DWA2114R group and 46.7% for the Cisplatin dose group, with no statistical difference between the groups (Table 2). No significant differences were observed between the two regimens in adverse effects and laboratory findings except RBC and creatinine clearance(Table 3). However, in consideration of the fact that the DWA2114R dose group received a little hydration and a few symptomatic drugs such as diuretics, DWA2114R was considered to be a more useful drug than Cisplatin (7).

Table 2 Evaluation of Response

Group	Number of evaluation pts	Prior chemo-therapy	CR	PR	MR	NC	PD	CR+PR (%)	χ^2 test	U test
A	12	No	1	5	1	2	3	50.0	N.S.	N.S.
B	15	No	2	7	0	4	2	60.0		
A	19	Yes	3	3	1	9	3	31.6	N.S.	N.S.
B	15	Yes	1	4	2	6	2	33.3		
A	31	Total	4	8	2	11	6	38.7	N.S.	N.S.
B	30	Total	3	11	2	10	4	46.7		

DWA2114R is effective against ovarian cancer, breast cancer , and prostate cancer. The potential advantage of DWA2114R derives from the absence of neurotoxicity, nephrotoxicity, ototoxicity, and non-serious gastrointestinal toxicity, which allows for administration on an out-patient basis.

Table 3 Laboratory Findings

Test items	Group A (33 cases) Rate of occurrence	Group B (31 cases) Rate of occurrence	χ^2 test
RBC ↓	54.5% (18/33)	83.9% (26/31)	p<0.05
WBC ↓	97.0% (32/33)	96.8% (30/31)	N.S.
Plt ↓	18.2% (6/33)	32.3% (10/31)	N.S.
GOT ↑	9.1% (3/33)	12.9% (4/31)	N.S.
GPT ↑	9.1% (3/33)	19.4% (6/31)	N.S.
Ccr ↓	4.3% (1/23)	35.3% (6/17)	p<0.05

References

1)Y. Ariyoshi, A. Wakui, K. Hasegawa, et al., A Phase I Study of DWA2114R. Jpn. J. Cancer Chemother. 19(5):685-693, (1992).

2)H. Majima, and H. Kinoshita, Clinical Pharmacokinetics of (R)-(-)-Cyclobutanedicarboxylato-(2-Aminomethylpyrrolidine) Platinum(II) (DWA2114R). Platinum and Other Metal Coordination Compounds in Cancer Chemotherapy, edited by Marino Nicolini, Martinus Nijhoff Publishing, Boston/Dordrecht/Lancaster,p491-498, (1987).

3)Y. Aso, H. Yamanaka, K. Imai, et al., A Phase II Study of DWA2114R, a New Platinum, for Urogenital Cancer. Japanese Journal of Urological Surgery 5(6): 535-541, (1992).

4)T. Kato, M. Yakushiji, and H. Nishimura, A Phase II Study of DWA2114R, a New Platinum Complex, in Gynecologic Cancers. J. Jpn. Soc. Cancer Ther. 27(10):1855-1865,(1992).

5)H. Aoyama, K. Kubo, J. Uchino, et al., A Phase II Study of DWA2114R, a New Platinum Complex for Breast Cancer. Jpn. J. Cancer Chemother. 19(7):1033-1039,(1992).

6)T. Kato, H. Nishimura, T. Yamabe, et al., Phase III Study of Carboplatin for Ovarian Cancer. Jpn. J. Cancer Chemother. 15:2297-2304,(1988).

7)T. Kato, M. Yakushiji, H. Nishimura, et al. Phase III Study of DWA2114R for Ovarian Cancer. Jpn. J. Cancer Chemother. 19(9):1258-1293,(1992).

INCORPORATING ASSESSMENTS OF SEQUENCE-DEPENDENCE IN DEVELOPMENTAL STUDIES OF COMBINATION CHEMOTHERAPY REGIMENS CONTAINING NEW AGENTS AND PLATINUM COMPOUNDS

Eric K. Rowinsky

The Johns Hopkins Oncology Center
The Johns Hopkins University School of Medicine
Baltimore, Maryland USA 21287-8934

INTRODUCTION

The evaluation of combination chemotherapy regimens that incorporate new drugs has been accepted as the next rationale step in the development of new cytotoxic agents following the completion of phase I and early phase II trials. Before incorporating new agents into combination chemotherapy regimens, the new agent should have demonstrated a sufficient level of antitumor activity in relevant tumor types. In addition, adequate preclinical and clinical information about optimal scheduling should be available. New cytotoxic agents are often combined with platinum compounds, such as cisplatin and carboplatin, if the new agents possess relevant antitumor activity in ovarian, non-small and small cell lung, bladder, and germ cell carcinomas, as well as in other malignancies in which the platinum compounds are the mainstay of therapy.

Traditionally, little attention has been paid to whether drug sequencing is important during the development of new chemotherapy regimens. Instead, the sequencing of drugs has largely been empirical, with the order of drug administration generally based on: 1) practical factors such as the convenience of patients and caregivers; 2) presumed toxicologic considerations (i.e. avoiding overlapping cytopenias); or 3) hypothetical mechanistic considerations.

The haphazard sequencing of multiple chemotherapy agents that have inherent sequence-dependent cytotoxic and pharmacologic interactions may result in either optimal or suboptimal therapeutic effects. Perhaps, the best characterized example of a clinically-significant sequence-dependent interaction involves the combination of methotrexate (MTX) and 5-fluorouracil (5-FU). Optimal cytotoxicity has consistently been observed when tumor cells are treated with MTX before 5-FU. This appears to be due to the accumulation of phosphoribosylpyrophosphate (PRPP) as a consequence of MTX-induced inhibition of purine synthesis.[1-3] The accumulation of PRPP enhances the formation of 5-FU nucleotides, which results in enhanced cytotoxicity. Clinical studies have also failed to demonstrate that the activity of 5-FU is enhanced when the interval between the administration MTX and 5-FU is one hour or less. With intervals as long as 24 hours, however, there is evidence of increased

PRPP accumulation and cytotoxicity in human tumors.[4,5] Significant sequence-dependent cytotoxic effects have also been demonstrated between 5-FU and other antimetabolites. For example, the administration of cytosine arabinoside (AraC) initially for 24 hours followed by 5-FU for 24 hours results in synergism, whereas the reverse sequence of 5-FU before AraC, is antagonistic.[6] The administration of 5-FU before AraC results in the inhibition of DNA synthesis, thereby decreasing the incorporation of DNA and potentially accounting for the antagonism.

The sequence-dependent enhancement of MTX accumulation by tumor cells in the presence of the *Vinca* alkaloids has also been well described.[7,8] This effect is mediated by the blockage of MTX efflux from the cell by the *Vinca* alkaloid. MTX reaches higher steady-state intracellular levels in both acute myeloblastic and lymphoblastic leukemia cells in the presence of VCR,[9,10] but the minimal VCR concentration that is required to achieve this effect in myeloblasts (0.1 uM) is realized only momentarily *in vivo*, and even higher concentrations are needed to enhance MTX uptake in lymphoblasts. The schedule of vincristine followed by MTX, though, has not demonstrated therapeutic synergism in the L1210 murine leukemia model.[10] However, synergism is noted with the sequence of MTX followed by vincristine, but this interaction is not likely to be due the enhancement of MTX uptake.

RATIONALE FOR EVALUATING SEQUENCE-DEPENDENT EFFECTS

Mechanistic Rationale

It may not be appropriate to evaluate the potential for sequence-dependent drug interactions during the development of all new combination chemotherapy regimens. Instead, evaluations of sequence-dependence may be indicated when new regimens consist of cytotoxic agents that affect cells in specific phases of the cell cycle. Several classes of antineoplastic agents that exert optimal cytotoxicity in specific phases of the cell cycle include:

A. **Antimicrotubule agents** such as the *Vinca* alkaloids and the taxanes that principally affect cells during mitosis. However, these agents may also affect vital cellular processes in the nonmitotic phases of the cell cycle in some types of cancer cells.[11]

B. **Topoisomerase I- and II-targeting agents** that require the DNA replication machinery of the cell, which is primarily operative during S phase, to convert potentially reversible DNA lesions into cytotoxic lesions (i.e. DNA double strand breaks).

C. **Antimetabolites** such as 5-FU and AraC that principally affect cells during DNA synthesis (S phase).

Although the platinum compounds and other alkylating agents have been demonstrated to optimally induce lethal effects during specific cell cycle phases, these results have been variable, and cell cycle specific is not entirely clear.

Hypothetically, sequence-dependent interactions resulting in synergistic, additive, or antagonistic effects may occur if the particular drug either augments or inhibits cell cycle progression to the cell cycle phase that is optimal for the cytotoxicity of another agent. Such interactions may be due to the specific effects of some classes of drugs on the progression of cells through the cell cycle. For example, several classes of DNA damaging agents, such as the alkylating agents and topoisomerase I- and topoisomerase II-targeting compounds, delay cell cycle phase progression in the G_2 phase of the cell cycle, possibly due to the inhibition of cdc kinase activation.[12] In addition, both the *Vinca* alkaloids and the taxanes inhibit cell cycle progression and induce accumulation in the mitotic phases, although both classes of agents have also been demonstrated to delay cell cycle traverse during interphase (G_1 and S).

Any hypothesis regarding sequence-dependent drug interactions may be strengthened or refuted by *in vitro* studies that are directed at determining the relative cytotoxic effects of different sequence iterations, thereby further establishing the rationale for the incorporation of sequence-dependent evaluations in early clinical studies. Similar evaluations in animal models may also provide additional information about sequence-dependent toxicologic and pharmacologic interactions, which can not be obtained from *in vitro* studies.

Pharmacologic Rationale

There may be specific pharmacologic features of chemotherapy agents and elements of study design that may either increase or reduce the potential for sequence-dependent interactions. Sequence-dependency may result from metabolic interactions due to the inhibition or induction of drug metabolism. In addition, drugs may possess certain pharmacokinetic characteristics that may increase or reduce the likelihood for sequence-dependent interactions. For example, inherent sequence-dependent interactions may not be apparent *in vivo* if the agents have long elimination half-lives. This pharmacokinetic feature maximizes the temporal overlap in drug exposure between various sequence iterations, thereby abrogating inherent sequence-dependent interactions that occur *in vitro*. In contrast, the rationale to design clinical trials that incorporate assessments of sequence-dependence may be strengthened if the involved agents have short elimination half-lives, which reduce the likelihood for temporal overlap in drug exposure and increase the probability that true sequential drug exposure will occur *in vivo*. Using a similar argument, inherent sequence-dependent drug interactions may be more readily noted if prolonged administration schedules are utilized.

APPROACH TO THE ASSESSMENT OF SEQUENCE-DEPENDENCE IN PHASE I DEVELOPMENT

Overall Approach

A dual approach to the assessment of sequence-dependent drug interactions is utilized at The Johns Hopkins Oncology Center (JHOC) during the early development of combination chemotherapy regimens consisting of new cytotoxic agents. This dual approach includes:

A. **Preclinical studies** that are designed to detect substantial sequence-dependent interactions *in vitro*.

B. **Phase I and pharmacologic studies** that are designed to detect clinically relevant sequence-dependent pharmacologic and toxicologic effects.

Both preclinical and clinical studies are performed either concurrently or sequentially. Next, the final results of both types of studies are synthesized and analyzed together to determine if there is an optimal sequence that should be specifically developed in subsequent clinical trials or utilized in general oncologic practice (Figure 1). Although this study design may yield pertinent information about the magnitude of pharmacologic and toxicologic

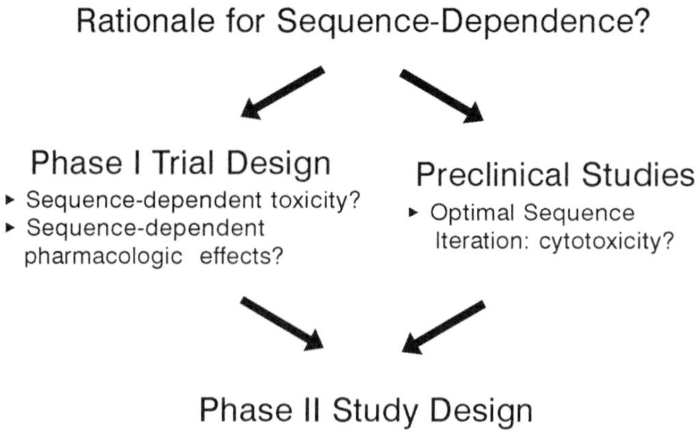

Rationale for Sequence-Dependence?

Phase I Trial Design
- Sequence-dependent toxicity?
- Sequence-dependent pharmacologic effects?

Preclinical Studies
- Optimal Sequence Iteration: cytotoxicity?

Phase II Study Design

Figure 1. The approach utilized at The Johns Hopkins Oncology Center to assess potential sequence-dependent drug interactions during the early development of combination chemotherapy regimens consisting of new cytotoxic agents.

differences between various sequence iterations, it may also be necesary to perform further randomized phase II and III studies to determine if a specific drug sequence is truly superior with respect to antitumor activity.

Preclinical Studies

The principal objective of ancillary preclinical studies is to detect significant sequence-dependent differences in cytotoxicity using relevant tumor models. Although the administration of drugs in various sequence iterations to animals bearing tumor xenografts may be superior to *in vitro* studies in simulating the clinical situation, it is much more difficult to determine if sequence-dependent differences (or the lack of such differences) in preclinical studies in animals are truly due to inherent sequence-dependent effects at the cellular level alone, sequence-dependent pharmacologic interactions alone, or a combination of both possibilities. In contrast, *in vitro* studies are more likely to detect inherent sequence-dependent cytotoxic differences in the absence of sequence-dependent pharmacologic interactions. Although it would be optimal to design *in vitro* studies to assess the magnitude of sequence-dependent synergistic interactions using classical median effect analysis,[13] it should be stressed that the evaluation of synergy is a secondary objective, and such evaluations are not required in these preclinical studies that primarily serve to assess the potential for sequence-dependent cytotoxic interactions.

Clinical Studies

The assessement of sequence-dependent toxicologic and pharmacologic effects may be incorporated into phase I developmental trials of new drug combinations. At JHOC, the following methodological approach is used.

A. **Patient Selection**: Since the major goal of the overall developmental process is to devise feasible combination chemotherapy regimens for the primary treatment of malignancies, eligibility is usually limited to subjects who have not received substantial myelosuppressive therapies. In addition, heavily-pretreated patients constitute a population that is heterogeneous with respect to hematopoietic function and bone marrow reserve, and is more likely to experience significant intersubject and intrasubject variability in myelosuppression. The following criteria are generally used at JHOC to define the "heavily-pretreated" patient with respect to prior myelosuppressive therapy:

1. More than four prior courses of combination chemotherapy containing an alkylating agent except for regimens containing low or moderate doses of cisplatin.
2. Prior treatment with mitomycin C or a nitrosourea.
3. Radiation to > 25% of marrow-bearing bones (e.g. pelvic radiation).
4. Widespread bone metastases or bone marrow involvement with bone marrow biopsies required for patients with tumor types that have a high likelihood of bone marrow involvement (e.g. lymphoma, prostate cancer).

After the maximum tolerated doses (MTD) and recommended phase II dose are defined for patients who have had minimal prior myelosuppressive therapy, a second phase of the study may be initiated for patients who have been heavily-pretreated, particularly if the new combination regimen is projected to be useful in the salvage setting.

B. **Dose Escalation** Four patients are generally treated at each dose level in which consistent dose-limiting toxicity (i.e. $\geq$ 2 dose-limiting events) does not occur. If one patient develops dose-limiting toxicity, a total of six patients are then treated at that particular dose level. The MTD is generally defined as the highest dose level in which less than one of the first six patients develops a dose-limiting event. However, at least eight total patients are usually treated at the MTD to yield ample toxicologic, pharmacologic and sequence-dependent data at a clinically relevant dose level.

Sequential dose escalation of each drug is performed in each cohort of new patients. Intrapatient dose escalation, which may obscure the detection of cumulative toxicity, is usually not permitted However, the specific dose escalation scheme employed is highly dependent on the particular agents that are being studied. It should be stressed that the dose escalation of the new agent in the combination is the primary focus. For new chemotherapy regimens containing cisplatin, cisplatin doses are generally escalated to a clinically acceptable target range of 50 to 75 mg/m^2. For carboplatin, a clinically-relevant dosing target in patients with ovarian cancer may be the achievement of an area under the plasma concentration-versus-time curve (AUC) of 5 to 7 mg/ml-min (as calculated by the Calvert formula), for untreated and previously treated patients, respectively.[14,15] In a

Patient #	Course 1	Course 2	Course 3......
1	a	b	a
2	b	a	b
3	a	b	a
4	b	a	b
.	a	b	a
.	b	a	b
n	a	b	a

Figure 2. Methodology employed at each dose level to assess sequence-dependent toxicologic and pharmacologic effects for two sequence iterations (a and b) of a drug combination.

retrospective analysis of 1028 ovarian cancer patients, the achievement of a carboplatin AUC above 5 to 7 mg/mL-min did not improve the likelihood of response but did increase myelotoxicity.[15]

C. **Evaluation of Sequence-Dependent Toxicologic Differences** Drug sequencing is alternated in each new patient entered in each dose level and in each individual subject with each successive course (Figure 2). Alternating the sequence of drug administration that is given during the first course in each patient within each dose level minimizes the confounding effects of cumulative drug interactions. In addition, alternating the sequence of drug administration with each successive course in each individual patient permits the use of each patient as their own control, thereby increasing the statistical power of the study.

D. **Evaluation of Sequence-Dependent Pharmacologic Differences** The drug sequencing scheme discussed above and depicted in Figure 2 permits an evaluation of the sequence-dependent pharmacologic effects of each drug on the other. Detailed pharmacologic studies of relevant new agents are also performed during both courses 1 and 2 (both sequences) in each individual. This design permits an assssment of the effects of each agent on the pertinent pharmacologic parameters (e.g. clearance rate, exposure (AUC), volume of distribution, peak plasma concentration) and metabolic profile of the other agent. Similar to the rationale discussed for using this particular study design to assess sequence-dependent toxicologic effects, this study design permits the use of each patient as their own control. The approach maximizes the statistical power of the study with respect to determining differences in the pharmacologic behavior between the different sequences.

E. **Data Analysis** Since drug sequencing is alternated in each individual subject treated at each dose level, as well as with each successive course in all patients, the sequence-dependent effects of the drug combination on relevant toxicologic and pharmacologic endpoints can assessed using each patient as their own control. Several statistical tests may be used to analyze these paired data. For example, the paired t-test may be used to assess toxicologic and pharmacologic differences between courses 1 and 2. If cumulative toxicity and intrasubject variability are low, mean parameter values (i.e. mean percent change in absolute neutrophil counts (ANCs) and platelet counts; mean nadir ANC and platelet count) for each sequence may be calculated for each individual patient. Next, the resultant mean values for each sequence in each individual patient may be compared. Multiple regression analysis using sequence and patient identification as variables may also be used to account for differences in the number of observations between patients, particularly if there is a propensity for cumulative toxicity and intrasubject variability, which may reduce the validity of the paired t-test using mean parameter values.

Selected Examples of Results of Combined Preclinical and Clinical Studies Incorporating Assessments of Sequence-Dependence

Paclitaxel (24 hour infusion) and Cisplatin

When paclitaxel began to be combined with other chemotherapy agents, such as cisplatin, there was a recognition of the fact that the optimal exploration of these combinations would require detailed knowledge of drug-drug interactions and possible effects of sequence of administration on both toxic and therapeutic effects. Therefore, studies were designed to

Table 1. Sequence-Dependent Effects of Paclitaxel and Cisplatin on ANCs at Various Dose Levels in Minimally Pretreated Patients With Advanced Cancers.

Dose (mg/m²) T	Dose (mg/m²) P	Sequence	ANC < 1000/μL % Courses	ANC < 1000/μL Days*	ANC < 500/μL % Courses	ANC < 500/μL Days*	Mean ANC Nadir (/μL)
110	50	P→T	77	8	77	7	672
		T→P	75	7	42	5	1291
135	50	P→T	83	8	78	8	770
		T→P	80	9	55	8	1002
110	75	P→T	97	7	62	5	400
		T→P	77	6	38	6	996
135	75	P→T	84	6	56	6	592
		T→P	83	7	60	5	670
170	75	P→T	100	10	86	8	257
		T→P	100	10	78	9	306
200	75	P→T	100	8	100	6	53
		T→P	100	10	100	7	135

T = TAXOL; P = cisplatin
*Mean duration of toxicity in days per affected course

address the possibility that sequence-dependent pharmacologic and toxicologic interactions between paclitaxel and other chemotherapy agents may occur in paclitaxel-based combination regimens. This is clearly illustrated by the results of phase I studies of the paclitaxel-cisplatin combination in which patients with minimal prior therapy received alternating sequences of these agents to determine if drug sequencing influenced the toxicity patterns and pharmacologic behavior of either agent.[16] Neutropenia was the dose-limiting toxicity of the paclitaxel-cisplatin combination without granulocyte colony-stimulating factor (G-CSF); however, the severity of neutropenia was demonstrated to be sequence-dependent. Pertient parameters pertaining to neutropenia are displayed in Table 1. Using paired t-testing of mean hematological data from individual patients, mean ANC nadirs were demonstrated to be significantly lower and the percentage of courses associated with ANCs $\leq$ 500/uL was significantly higher when patients received cisplatin before paclitaxel.

To determine if drug sequencing affected the pharmacologic disposition of paclitaxel, clearance rates for paclitaxel were calculated during courses in which paclitaxel (t) was given before cisplatin (c) [$cl_{t/c}$] and courses in which cisplatin was given before paclitaxel [$Cl_{c/t}$]. Mean paclitaxel clearance rates were significantly lower when paclitaxel followed cisplatin, 321 $\pm$ 44 mL/min/m² (range, 99 to 844 mL/min/m²) compared to 405 $\pm$ 65 mL/min/m² (range, 141 to 1097 mL/min/m²) for the alternate sequence paclitaxel followed by cisplatin ($p = 0.013$ by paired t-test). Correlation analysis of the paired clearance data revealed a linear relationship ($R = 0.93$, $p < 0.001$) as shown in Figure 3, and regression analysis demonstrated that the clearance rate values for alternate sequences were defined by the following relationship:

$$Cl_{c/t} = 0.75\ Cl_{t/c}$$

The sequence of cisplatin followed by paclitaxel induced more profound neutropenia than the reverse drug sequence in phase I studies, and was also demonstrated to be the suboptimal sequence with respect to cytotoxic activity against L1210 leukemia in concurrent *in vitro* studies compared with both the reverse sequence and simultaneous drug treatment as shown in Figure 4.[17] Similar sequence-dependent cytotoxic effects have also been demonstrated for paclitaxel using relatively long treatment durations combined with the platinum compounds in other tumor models.[18] Although the mechanism responsible for the sequence-dependent effects is not known and alkaline elution studies have demonstrated that paclitaxel does not

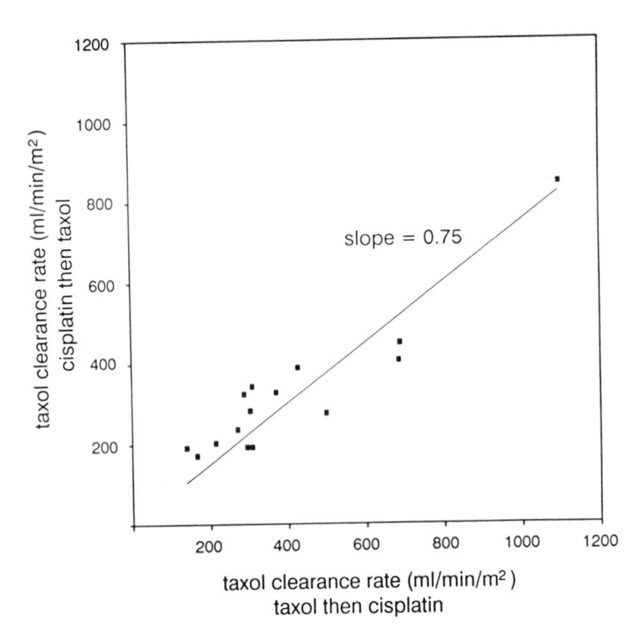

Figure 3. Paired paclitaxel clearance data for each individual patient during both paclitaxel-cisplatin sequences relationship is linear (r=0.93) with a slope of 0.75.

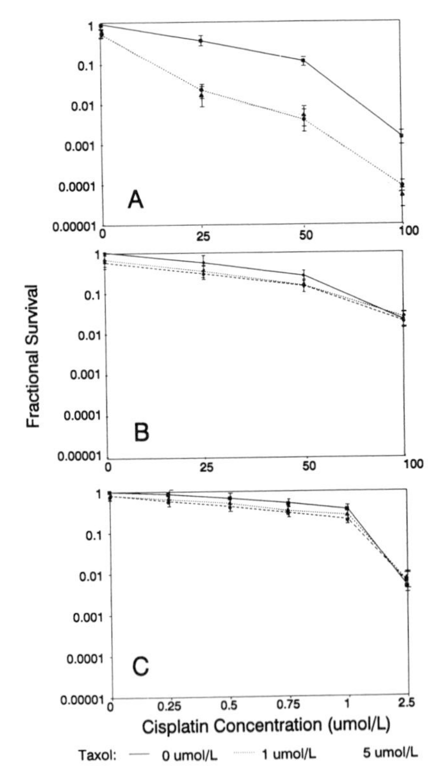

Figure 4. Clonogenic survival of L1210 leukemia cells treated with various sequence iterations of paclitaxel and cisplatin. Optimal cytotoxicity was achieved following treatment with the sequence of paclitaxel (22 hours) followed by cisplatin (30 minutes) (A) compare with the reverse sequence (B) and simultaneous treatment for 22 hours (C).

increase the magnitude of total (protein-DNA and DNA-DNA) and DNA-DNA crosslinking induced by platinum,[17] paclitaxel has been shown to inhibit the repair of cisplatin-DNA adducts.[19] Another possible mechanism for the sequence-dependent cytotoxicity *in vitro* is that treatment with cisplatin before paclitaxel is antagonistic since cisplatin inhibits cell cycle progression in the G_2 phase of the cell cycle, thereby preventing further progression into the mitotic phases, which may be the optimal period of cell sensitivity to the taxanes. Regardless of the precise mechanism, these observations formed the rationale for the selection of the sequence of paclitaxel followed by cisplatin for subsequent clinical trials of the paclitaxel-cisplatin chemotherapy combination.

Since the sequence of paclitaxel followed by cisplatin demonstrated superior cytotoxicity than the reverse sequence in preclinical studies, it might have been predicted to induce more profound toxicity in clinical trials. The mechanism for sequence-dependent interactions between paclitaxel and cisplatin in clinical trials, as well as the reason why the sequence of cisplatin followed by paclitaxel is associated with more profound neutropenia in patients, are not known. However, one potential mechanism is the modulation of cytochrome P450-dependent paclitaxel-metabolizing enzymes by cisplatin since cisplatin has been demonstrated to inhibit the activity of specific cytochrome P450 mixed function oxidases.[20] The ability to modulate cytochrome P450 mixed function oxidases is not shared by all the platinum compounds. For example, carboplatin does not appear to be capable of modulating the same P450 systems as cisplatin, which may explain, in part, why sequence-dependent toxicologic effects have not been observed with paclitaxel-carboplatin combinations.[21] However, the lack of sequence-dependent neutropenia with the combination of paclitaxel and carboplatin may also be due to the use of brief paclitaxel infusion schedules (3-hour) in developmental studies of this combination to date. The use of shorter paclitaxel infusion schedules may abrogate any inherent sequence-dependent effects because shorter paclitaxel schedules may increase the likelihood for temporal overlap in drug exposure and reduce the probability for true sequential drug exposure.

In summary, the results of these preclinical and clinical studies formed the rationale for the selection of the sequence of paclitaxel followed by cisplatin as the treatment sequence to be used in subsequent phase II/III trials of the paclitaxel-cisplatin doublet. These have included the pivotal phase III GOG study of paclitaxel plus cisplatin versus cyclophosphamide plus cisplatin in untreated patients with suboptimally-debulked ovarian epithelial neoplasms (GOG Study #111),[22] as well as in Eastern Cooperative Oncology Group (ECOG) studies of paclitaxel and cisplatin in head and neck and lung cancers.

Other Paclitaxel-Based Chemotherapy Combinations

The potential for sequence-dependent interactions has also been studied during developmental studies of paclitaxel-doxorubicin and paclitaxel-cyclophosphamide combinations for breast cancer.[23-26] In phase I studies at the M.D. Anderson Cancer Center, in which various sequence iteration of paclitaxel and doxorubicin were administered to separate groups of patients, mucositis was more prominent when paclitaxel (24-hour infusion) was administered before doxorubicin (48-hour infusion) compared to the reverse sequence.[23,24] The results of subsequent pharmacologic studies, in which patients received both sequences and acted as their own controls, indicated that the increased toxicity may be explained by a reduction in the clearance (32%) of doxorubicin when it is administered after paclitaxel.[24] Similar results have been noted in Eastern Cooperative Oncology Group pilot studies, in which patients received alternating sequences of doxorubicin (bolus injection) and paclitaxel (24-hour infusion). Based on these results and the lack of data demonstrating a superior drug sequence with respect to cytotoxicity, the sequence of doxorubicin followed by paclitaxel is being developed.

In contrast, neither sequence-dependent toxicologic or pharmacologic effects have been noted in phase I studies of doxorubicin and paclitaxel performed at both JHOC and the National Cancer Institute (Milan), in which patients received alternating sequences of both agents.[27,28] However, the major difference in these studies compared to those described previously was the administration of doxorubicin on a brief 3-hour infusion schedule, which has a greater likelihood of abrogating inherent sequence-dependent differences. Peliminary paired toxicologic and pharmacologic data from the JHOC study are shown in Figures 5-7.

In a study of the combination of paclitaxel (24-hour infusion schedule) and cyclophosphamide (brief infusion) in patients with metastatic breast cancer who are insensitive to doxorubicin, both neutropenia and thrombocytopenia have been more profound with the sequence of cyclophosphamide before paclitaxel (24-hour schedule) compared to the reverse sequence.[26] However, the mechanism for these differential effects is not clear since both pharmacologic and *in vitro* cytotoxicity studies have failed to demonstrate differences between the sequences. One possible mechanism is similar to that suggested to explain the sequence-

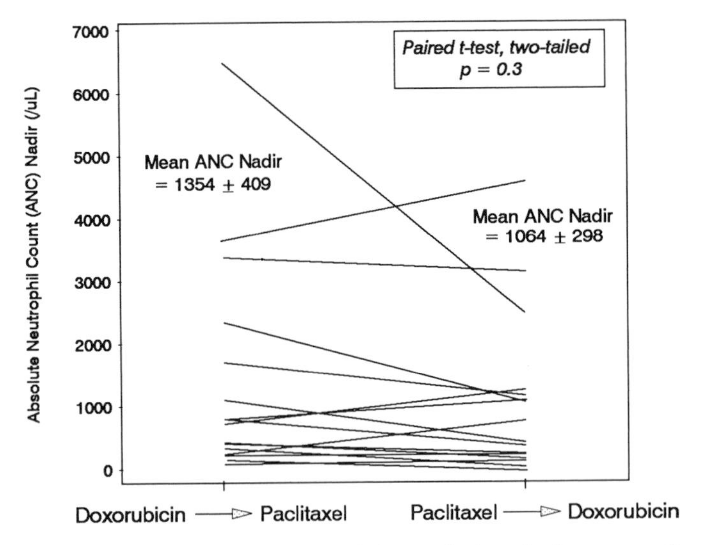

Figure 5. Preliminary results of an an ongoing phase I and pharmacologic study of sequences of paclitaxel (3-hour infusion schedule) and doxorubicin (bolus injection) at JHOC. Paired mean nadir ANCs with both treatment sequences. To date, there appears to be no sequence-dependent toxicologic differences.

The sequence of administration of paclitaxel (P) (3-hour infusion) and doxorubicin (D) does not affect paclitaxel AUC.

- **P → D 1434 +/- 315 umol-min/L**
  D → P 1331 +/- 249 umol-min/L — t-test (2-tailed) p = 0.8

- **Paired analysis of paclitaxel AUCs also demonstrates no sequence-dependence (t-test, 2-tailed, p = 0.93)**

Figure 6. Preliminary results of an an ongoing phase I and pharmacologic study of sequences of paclitaxel (3-hour infusion schedule) and doxorubicin (bolus injection) at JHOC. Paired mean steady state paclitaxel concentrations with both drug sequences. No sequence-dependent pharmacologic differences are evident.

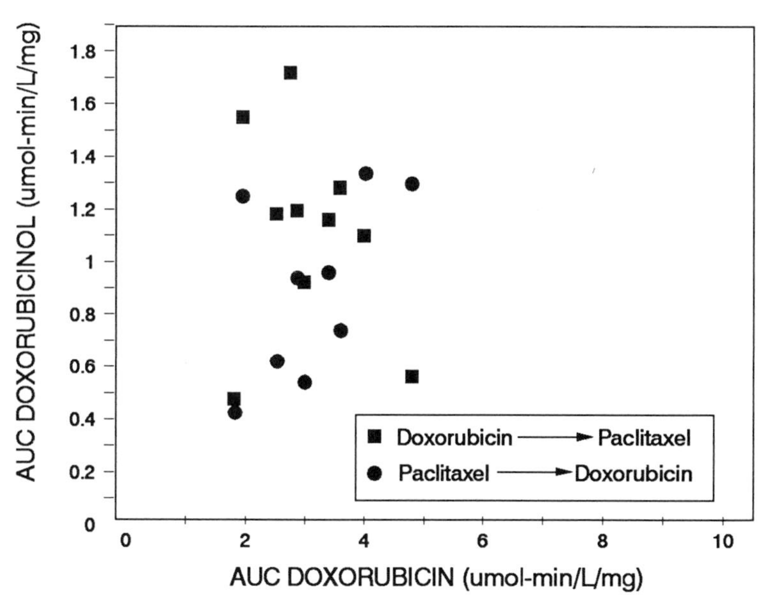

Figure 7. Preliminary results of an an ongoing phase I and pharmacologic study of sequences of paclitaxel (3-hour infusion schedule) and doxorubicin (bolus injection) at JHOC. Ratio of doxorubicinol to doxorubicin in individual patients as a function of treatment sequence. No substantial sequence-dependent differences in the ratios are evident.

dependent differences demonstrated for paclitaxel and cisplatin *in vitro*, in that the initial treatment with the alkylating agent may inhibit progression in the G_2 phase of the cell cycle, thereby preventing the optimal cytotoxic activity of paclitaxel, which occurs in mitosis.

Although these studies may be used as paradigns to assess other taxane-based chemotherapy combinations, the fact that sequence-dependent interactions have been noted only with prolonged paclitaxel infusions, and not with short (e.g. 3-hour infusion) schedules, should be stressed. Sequence-dependent effects have also not been observed with docetaxel-based combinations, in which docetaxel is administered as a 1-hour infusion.[29]

Topotecan and Cisplatin

Synergy has been noted when topoisomerse I-directed agents are combined with the alkylating agents cisplatin, 4-hydroxyperoxycyclophosphamide, and carmustine.[30-36] The addition of cisplatin to topoisomerase I inhibitors has been shown to result in additive or synergistic cytotoxicity *in vitro* and *in vivo*.[30-34] Maximal synergy between topotecan and cisplatin has been observed *in vivo* when treatment with cisplatin precedes topotecan.[33] These results might reflect an effect of topoisomerase I-directed agents on the repair of DNA damage caused by alkylating agents. Alternatively, the alkylating agents might promote unscheduled DNA synthesis, thereby providing replication forks that can interact with topo I-DNA adducts.

The rationale for combining topotecan and cisplatin is based on: 1) phase I and II studies demonstrating that topotecan is active in several tumor types in which the platinum compounds are among the most active agents, including ovarian and small cell lung cancers;

2) principal toxicities that are dissimilar (cisplatin - nephrotoxicity and neurotoxicity; topotecan- neutropenia); 3) dissimilar clearance mechanisms (topotecan - renal and hepatic; cisplatin - renal, hepatic, and metabolic); and 3) synergy between the platinum compounds and topoisomerase I-targeting agents in vitro.[30-34] In preclinical studies performed at JHOC, topotecan and cisplatin also demonstrated synergistic cytotoxocity in NIH-H82ras and A549 human non-small cell lung cancer cell lines; however, sequence-dependent cytotoxicity was not evident when cells were treated with sequence iterations of topotecan (6 hours) and cisplatin (2 hours).

In contrast to preclinical studies performed at JHOC, significant sequence-dependent toxicologic effects were evident in a phase I trial of topotecan and cisplatin at JHOC in untreated or minimally-pretreated patients with solid tumors.[37] Similar to the evaluation of cisplatin and paclitaxel as discussed previously, patients were treated with alternating sequences of cisplatin (day 1 or day 5) and topotecan (30 minute infusion daily for five consecutive days) to determine if drug sequencing influenced the toxicity patterns or pharmacologic behavior of either agent. Neutropenia and thrombocytopenia were the dose-limiting toxicies of the drug regimen and both toxicities were significantly worse when treatment with the optimal *in vitro* sequence, cisplatin before topotecan. The MTD and recommended phase II dose levels were cisplatin 50 mg/m^2 followed by topotecan 0.75 mg/m^2/day for five days every 3 weeks. In addition, the use of granulocyte colony-stimulating factor did not permit further dose escalation

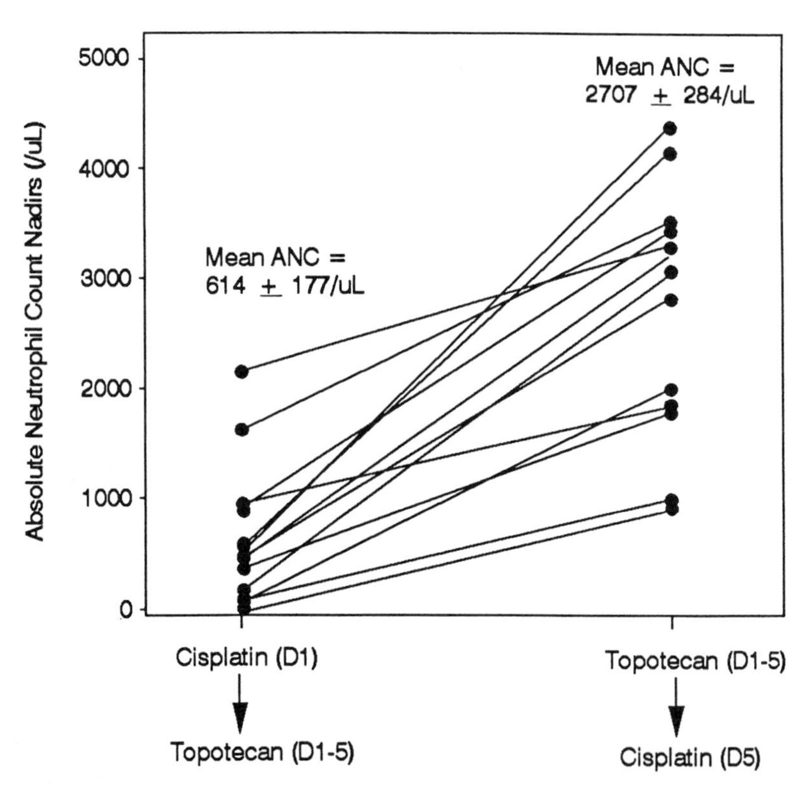

Figure 8. Preliminary analysis of paired mean nadir ANCs as a function of treatment sequence with cisplatin and topotecan revealing significantly lower mean nadir ANCs when treatment with cisplatin precedes topotecan ($p < 0.001$).

of either agent. Preliminary analyses of paired mean nadir ANCs and platelet counts as a function of drug sequence for each individual patient are shown in Figures 8 and 9. At this interim analysis, mean ANCs were 614/uL versus 2707/uL ($p < 0.001$, paired t-test [two-tailed]) and mean platelet counts were 77/uL versus 151/uL ($p < 0.001$, paired t-test [two-tailed], with the significantly lower ANC and platelet counts associated with the sequence of cisplatin followed by topotecan.

Studies assessing the effects of cisplatin on the pharmacokinetic behavior of topotecan on day 1 were also performed with both sequence iterations. The preliminary results of these studies demonstrate that cisplatin does not affect the pharmacokinetic behavior of topotecan on day 1. Both clearance and AUC values were essentially similar ($p = 0.16$, paired t-test [two-tailed]). However, topotecan AUCs were significantly higher on day 5 compared to day 1 when cisplatin was administered before topotecan ($p < 0.002$, paired t-test [two-tailed]). Therefore, the increased myelotoxicity associated with the sequence of cisplatin followed by topotecan may, in part, be due to sequence-dependent pharmacologic effects. One potential explanation for these sequence-dependent pharmacologic effects is that cisplatin may induce subclinical renal effects, which may lead to a reduction in the clearance of topotecan and increased toxicity. The temporal nature of these pharmacologic effects is currently being assessed.

These preliminary results indicate that substantial sequence-dependent toxicologic and pharmacologic interactions occur when cisplatin is combined with topotecan on a 30 minute infusion daily x 5 schedule. In contrast, to the situation with cisplatin and paclitaxel, however, subsequent clinical development of the more toxic sequence of cisplatin followed by topotecan is being recommended because it is also associated with optimal cytotoxicity in preclinical evaluations.

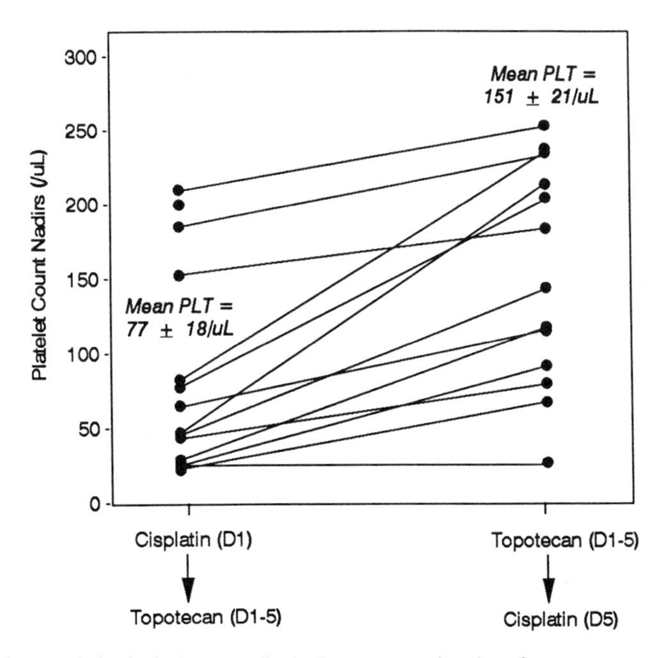

Figure 9. Preliminary analysis of paired mean nadir platelet counts as a function of treatment sequence with cisplatin and topotecan revealing significantly lower mean nadir platelet counts ($p < 0.001$).

SUMMARY

The incorporation of assessments of sequence-dependent cytotoxic, toxicologic, and pharmacologic effects during the early clinical development of new combination chemotherapy regimens, is feasible and expends minimal additional resources. As discussed in this review, such evaluations have been especially useful in optimally developing combination regimens that consist of new cytotoxic agents and platinum compounds. Although it may not be rational to perform evaluations of sequence-dependence with all new drug combinations, such studies in carefully selected cases may lead to the development of combination regimens with optimal therapeutic indices.

REFERENCES

1. E. Cadman, R. Heimer, L. Davis. Enhanced 5-fluorouracil nucleotide formation after methotrexate administration: explanation for drug synergism. Science 205:1135 (1979).

2. C. Benz, E. Cadman E. Modulation of 5-fluorouracil metabolism and cytotoxicity by antimetabolite pretreatment in human colorectal adenocarcinoma HCT-8. Cancer Res. 41:994 (1981).

3. C. Benz, T. Tillis, E. Tattelman, E. Cadman. Optimal scheduling of methotrexate and 5-fluorouracil in human breast cancer. Cancer Res 42:2081 (1982).

4. N.E. Kemeny, T. Ahmed, R.A. Michaelson, H.D. Harper, L.C. Yip. Activity of sequential low dose methotrexate and 5-fluorouracil in advanced colorectal carcinoma: attempt at correlation with tissue and blood levels of phosphoribosylpyrophosphate. J Clin Oncol 2:311 (1984).

5. C. Benz, M. DeGregorio, S. Saks, et al. Sequential infusions of methotrexate and 5-fluorouracil in advanced cancer: pharmacology, toxicity, and response. Cancer Res 45:3354 (1985).

6. J.L. Grem, C.J. Allegra. Sequence-dependent interactions of 5-fluorouracil and arabinosyl-5-azacytidine or 1-ß-D-arabinfuranosylcytosine. Biochem Pharmacol 42:409 (1991).

7. R.A. Bender, W.A. Bleyer, S.A Frisby. Alteration of methotrexate uptake in human leukemia cells by other agents. Cancer Res. 35:1305 (1975).

8. R.F. Zager, S.A. Frisby, V.T. Oliverio. The effects of antibiotics and cancer chemotherapeutic agents on the cellular transport and antitumor activity of methotrexate in L1210 murine leukemia. Cancer Res. 33:1670 (1993).

9. R.D. Warren, A.P. Nichols, R.A. Bender. The effect of vincristine on methotrexate uptake and inhibition of DNA synthesis by human lymphoblastoid cells. Cancer Res. 37: 2993 (1977).

10. R.A. Bender, A.P. Nichols, L. Norton, et al. Lack of therapeutic synergism of vincristine and methotrexate in L1210 murine leukemia in vivo. Cancer Treat. Rep. 62:997 (1978).

11. E.K. Rowinsky, R.C. Donehower, R. J. Jones, R.W. Tucker. Microtubule changes and cytotoxicity in leukemic cell lines treated with taxol. Cancer Res. 48:4093 (1988).

12. P. O'Connor, D. Ferris, G. White, et al. Relationship between cdc2 kinase activation, p34cdc2 dephosphorylation, and mitotic progression in Chinese hamster ovary cells exposed to etoposide. Cancer Res 52:1817 (1992).

13. J. Chou, P. Talalay. Quantitative analysis of dose-effect relationships: the combined effects of multiple drugs and enzyme inhibitors. Adv Enzyme Regul 22:27 (1984).

14. A.H. Calvert, D.R. Newell, L.A. Gumbrell, et al. Carboplatin dosage: Prospective evaluation of a simple formula based on renal functions. J Clin Oncol 7:1748 (1989).

15. D.I. Jodrell, M.J. Egorin, R.M. Canetta, et al. Relationships between carboplatin exposure and tumor response and toxicity in patients with ovarian cancer. J Clin Oncol 10:520, (1992).

16. E.K. Rowinsky, M. Gilbert, W.P. McGuire, et al. Sequences of taxol and cisplatin: a phase I and pharmacologic study. J Clin Oncol 9:1692 (1991).

17. E.K. Rowinsky, M. Citardi, D.A. Noe, R.C. Donehower: Sequence-dependent cytotoxicity between cisplatin and the antimicrotubule agents taxol and vincristine. J Can Res Clin Oncol 119:737 (1993).

18. R.J. Parker, M.D. Dabholkar, K-B Lee, F. Bostoick-Burton, E. Reed. Taxol effect on cisplatin sensitivity and cisplatin cellular accumulation in human ovarian cancer cells. Monograph Natl Can. Inst. 15:83, (1993).

19. E. Reed, R.J. Parker, M. Dabholkar, et al. Taxol effect on cisplatin-DNA adduct repair in human ovarian cancer cells. (Abstract) Second National Cancer Institute Workshop on Taxol and Taxus. Alexandria, VA (1992).

20. G.A. LeBlanc, S.S. Sundseth, G.F. Weber, et al. Platinum anticancer drugs modulate P-450 mRNA levels and differentially alter hepatic drug and steroid hormone metabolism in male and female rats. Cancer Res 5:540, (1992).

21. L.J.C. van Warmerdam, M.T. Huizing, S. Rodenhuis, et al: Can the Calvert formula predict the pharmacokinetics of carboplatin (C) when C is given in combination with paclitaxel (P). (Abstract) 7th International Symposium on Platinum and other metal coordination compounds in Cancer Chemotherapy. Abstract Book S67, Amsterdam, The Netherlands (March 1-4, 1995).

22. W.P. McGuire, W.J. Hoskins,, M.R. Brady, et al. A phase III trial comparing cisplatin/cytoxan (pc) and cisplatin/taxol (pc) in advanced ovarian cancer (aoc). (Abstract) Proc Am Soc Clin Oncol 12:255 (1993)

23. F.A. Holmes, D. Frye, V. Valero, et al. Phase I study of taxol and doxorubicin with G-CSF in patients without prior chemotherapy for metastatic breast cancer. (Abstract) Proc Am Soc Clin Oncol 11:600 (1992).

24. F.A. Holmes. Combination chemotherapy with Taxol (paclitaxel) in metastatic breast cancer. Ann Oncol 5(Suppl 6):S23-S27, 1994.

25. G.W. Sledge, N. Robert, J. Sparano, et al. Paclitaxel (Taxol)/doxorubicin combinations in advanced breast cancer. The Eastern Cooperative Oncology Group Experience. Sem Oncol 21(Suppl. 8):15 (1994).

26. M.J Kennedy, D. Armstrong, R. Donehower, et al. The hematologic toxicity of the taxol/cytoxan doublet is sequence-dependent. (Abstract) Proc Am Soc Clin Oncol 13:137, (1994).

27. E.K. Rowinsky (unpublished results)

28. L. Gianni, G. Straneo, F. Capri, et al: Optimal dose and sequence finding study of paclitaxel (P) by 3 h infusion with bolus doxorubicin (D) in untreated metastatic breast cancer patients (Pts). (Abstract) Proc Am Soc Clin Oncol 13:74 (1994).

29. J. Verweij, A.S.T. Planting, M.E.L. Van der Berg, et al: A phase I study of docetaxel (Taxotere) and cisplatin in patients with solid tumors. (Abstract) Proc Am Soc Clin Oncol 13:148, (1994)

30. E.J. Katz, J.S. Vick, K.M. Kling, et al. Effect of topoisomerase modulators on cisplatin cytotoxicity in human ovarian carcinoma cells. Eur J Cancer 26:724 (1990).

31. B. Drewinko, C. Green, T.L. Loo. Combination chemotherapy in vitro with cis-dichlorodiammineplatinum(II). Cancer Treat Rep 60:1619, (1976).

32. Y. Kano, K. Suzuki, M. Akutsu, et al. Effects of CPT-11 in combination with other anti-cancer agents in culture. Int J Cancer 50:604 (1992).

33. R. Johnson, F. McCabe, Y. Yu. Combination regimens with topotecan in animal tumor models. (Abstract) Ann Oncol 3(suppl 1):85, (1992).

34. K. Itoh, M. Takada, S. Kudo, et al. Synergistic effects of CPT-11 and cisplatin or etoposideon human lung cancer cell lines demonstrated by continuous infusion. (Abstract) Proc Am Assoc Cancer Res 33:259, (1992).

35. B. Drewinko, T.L. Loo, and E.J. Freireich. Combination chemotherapy in vitro. III. BCNU. Cancer Treat Rep 63:373 (1979).

36. M. Oguro. A topoisomerase I inhibitor, CPT-11: Its enigmatic antitumor activity in combination with other agents in vitro. (Abstract) In: Proceedings of the Third Conference on Topoisomerases, p 35 (1990).

37. E. Rowinsky, L. Grochow, S. Kaufmann, et al. Sequence-dependent effects of topotecan and cisplatin in a phase I and pharmacokinetic study. (Abstract) Proc Am Soc Clin Oncol 13:142, (1994).

REVIEW OF CARBOPLATIN-BASED HIGH-DOSE CHEMOTHERAPY COMBINATIONS IN THE AUTOTRANSPLANT SETTING

S. Rodenhuis, E. van der Wall, J.H. Schornagel, and J.W. Baars

Department of Medical Oncology,
The Netherlands Cancer Institute
Amsterdam, NL-1066 CX, The Netherlands

INTRODUCTION

High-dose chemotherapy is a relatively straightforward approach to overcome drug resistance. It appears to be more effective than standard chemotherapy in the salvage treatment of Non Hodgkin's Lymphomas and of relapsing germ cell cancer. Non-randomized studies have raised hope that it might improve treatment results in (other) solid tumors as well, particularly in the adjuvant treatment of high-risk breast cancer[1]. Prospective, randomized studies of high-dose therapy are currently in progress, but many more will be required to define its proper role in solid tumors.

The design of large multi-center studies of high-dose therapy obviously requires the selection of suitable high-dose regimens. In contrast with standard-chemotherapy phase III studies, however, very little data from phase II is available to estimate the relative efficacies of the available regimens. Almost all published studies are small, describe patient groups with widely varying clinical characteristics, have limited follow-up and focus on feasibility and toxicity, rather than on long-term survival.

SELECTING A HIGH-DOSE REGIMEN

In view of these difficulties, the decision to use a particular regimen must be based on known patterns of toxicity and on theoretical considerations[2]. Clearly, compounds are preferred that have a log-linear dose-response relation *in vitro* and *in vivo*, that exhibit no cross resistance to each other and that allow substantial dose escalation in the autotransplant setting. These conditions are satisfied by the alkylating agents[3]. It is reasonable to require that each of the components of the high-dose regimen should have established single-agent activity at standard dosage for the tumor type to be treated. We believe that agents associated with frequently fatal or irreversible end-organ toxicity should be avoided.

The combination of cyclophosphamide, thiotepa and carboplatin satisfies most of these requirements. Cyclophosphamide is one of the most frequently used agents because of its broad spectrum of activity and because it can be escalated to a total dose of 6 g/m^2. The dose of thiotepa can be increased to 900 mg/m^2, approximately 15 x its recently newly established maximum tolerated dose (MTD)[4]. It is also attractive because it readily crosses the blood-brain barrier and might thus be effective in eradicating CNS micrometastases. Finally, carboplatin has a broad spectrum of activity and can serve as the platinum containing drug that is essential for adequate treatment of germ cell tumors and ovarian cancer. It has also been shown to be moderately active in breast cancer[5,6]. Carboplatin can be escalated to approximately 4 times its standard single-agent dose, while the escalation factor for cisplatin is only 2.

carboplatin-etoposide based combinations

The combination of high-dose carboplatin and etoposide followed by autologous bone marrow rescue has particularly been applied in relapsing or refractory germ cell cancer. In a phase I study performed in the Indiana University program, the MTD of carboplatin was 1500 mg/m^2 in combination with etoposide 1200 mg/m^2 [7]. The dose limiting toxicity was enterocolitis; toxic deaths (21%) were observed with carboplatin doses exceeding 1500 mg/m^2. At these dose levels, hepatic toxicity developed in 8 patients (24%) with a fatal outcome in one.

The feasibility and efficacy of combination therapy with carboplatin 1500 mg/m^2 and etoposide 1200 mg/m^2 was subsequently investigated in a phase II trial in refractory germ cell cancer[8]. Patients who responded to the first high-dose chemotherapy cycle, received a second one. Five of 38 patients (13%) remained disease-free for more than 1 year. However, the toxicity of this regimen was substantial including treatment-related deaths in 5 patients (13%). All deaths occurred after the first transplantation. Toxic deaths were due to sepsis, hemorrhage or hepatic failure, the latter including one veno-occlusive disease.

In an attempt to improve its efficacy, ifosfamide has been added to the high-dose carboplatin-etoposide regimen[9]. Severe renal toxicity was observed at the first dose level (10 g/m^2), which precluded the planned further dose escalation. This is in contrast to data available from an Italian multi-center study, in which 11 patients received 12 g/m^2 ifosfamide, 1350 mg/m^2 carboplatin and 1200 mg/m^2 etoposide[10]. No renal toxicity was observed despite the simultaneous administration of the aminoglycoside antibiotic amikacine in some of these patients. Severe mucositis was the major toxicity recorded. Repeated administration of a similar combination but at somewhat lower doses has been shown to be feasible[11].

cyclophosphamide, etoposide and carboplatin (CEC)

Two phase I studies have evaluated the feasibility of a high-dose CEC chemotherapy regimen with ABMT or peripheral blood stem cell transplantation (PSCT)[12,13]. Both studies applied a dose-escalation schedule for one of the 2 alkylating agents. Dose-escalation of carboplatin resulted in acute renal failure as the dose-limiting toxicity, occurring in 2 of 14 patients who had received a dose of 1,600 mg/m^2. Both patients had previously been treated with cisplatin and/or carboplatin and both had renal function impairments with creatinine clearances of slightly above 60 ml/min prior to the administration of CEC..

In the second study, which included 30 patients with cisplatin-refractory germ cell cancer, dose-escalation of cyclophosphamide with fixed doses of carboplatin (1500 mg/m^2) and etoposide (1200 mg/m^2) was evaluated[13]. Fourteen patients who responded to the first course were retreated. The second cycle was administered 4 to 6 weeks after hematologic recovery. Two toxic deaths occurred related to myelosuppression, while other major toxicity involved the liver, with evidence of cholestasis occurring in 17 patients (57%). Veno-occlusive disease was not observed. Interestingly, the 7 patients (23%) who achieved a durable complete remission, at a median follow-up of 11.4 months (range 5.6-35.5), were those who had received 2 courses of high-dose CEC. The two patients who achieved a complete response after a single course of CEC were not retreated and subsequently relapsed.

carboplatin, thiotepa and cyclophosphamide combinations

The best known regimen with this combination is probably the CTCb regimen developed by Antman and co-workers[14]. It consists of a 4-day continuous infusion of cyclophosphamide (6 g/m^2), thiotepa (500 mg/m^2) and carboplatin (800 mg/m^2). The combination has been used extensively in breast cancer[15].

In the Solid Tumor Autologous Marrow Program (STAMP) melphalan was initially added to the combination of cyclophosphamide and thiotepa (STAMP III). This combination was considered too toxic because of severe mucositis[16]. The substitution of melphalan by carboplatin as the third alkylating agent resulted in the commonly used STAMP V (CTCb) regimen[14]. A phase II trial in 29 patients with metastatic breast cancer showed that CTCb is a regimen with acceptable morbidity and a low mortality rate[15]. This was reconfirmed in later studies, in which CTCb was preceded by high-dose melphalan[17].

In the Netherlands Cancer Institute, a similar high-dose regimen has been developed. The dose of carboplatin is however twice that of the STAMP V regimen (1600 mg/^{2}), while the dose of thiotepa is similar (480 mg/m^2) [18]. The total dose of cyclophosphamide is identical: 6000 mg/m^2. In spite of the high carboplatin dose, this CTC regimen has been shown to be well-tolerated without severe non-hematological toxicity[19], even when administered sequentially in a tandem transplantation setting[20]. Mucositis was manageable, requiring total parenteral nutrition in only a very small percentage of patients.

SINGLE AND MULTIPLE COURSES OF THE CTC REGIMEN

By February 1995, a total of eighty-three patients had received a first autologous bone-marrow and/or peripheral stem cell transplantation after CTC in the Netherlands Cancer Institute. Thirty-eight of these patients underwent high-dose therapy as part of adjuvant therapy for breast cancer (Table 1). All others had advanced malignancies and most had previously been exposed to extensive chemotherapy. In addition to bone marrow suppression, major toxicities consisted of nausea and vomiting, diarrhea, mild to moderate mucositis, fevers with or without skin rashes and reversible renal function impairments. All patients had minor and reversible elevations of liver enzymes, but no signs of veno-occlusive disease were observed. One patient died as a result of unexplained massive soft-tissue hemorrhage, followed by multi-organ failure. Notably, hearing loss and neuropathy did not occur, except in patients who had previously been exposed to high doses of cisplatin. All patients received mesna infusions and hemorrhagic cystitis was never observed.

83 First Transplantation Procedures with CTC
- 38 x high-risk breast cancer (adjuvant)
- 11 x advanced breast cancer
- 28 x germ cell cancer
- 6 x other tumors:
• 2 x ovarian cancer
• 1 x rhabdomyosarcoma
• 1 x medulloblastoma
• 1 x neuroblastoma
• 1 x gestational throphoblast
1 Toxic death (hemorrhage)
2 Patients partially reversible renal failure
No (other) irreversible toxicity

Table 1. Experience with first autologous transplantation procedures employing the CTC regimen in the Netherlands Cancer Institute.

Two further patients received CTC as a second high-dose regimen after failing high-dose therapy with BCNU, etoposide, cytarabine and melphalan (BEAM) because of malignant lymphoma. The first of these patients achieved a complete remission, but at the cost of symptomatic hearing loss. The second patient died of septicemia caused by *Streptococci.*

	Number of Patients	Toxic Deaths	Major Toxicity
CTC-1	83	1 (hemorrhage)	
CTC-1 (2nd transplant)	2	1 (sepsis)	
CTC-2	28	1 (sepsis)	VOD (1x) hemorrhagic cystitis (3x)
CTC-3	11	3 (sepsis, VOD, HUS)	VOD (3x) HUS (2x)

Table 2. Major toxicities of 124 courses of CTC with autotransplantation. Abbreviations: VOD: veno-occlusive disease; HUS: hemolytic-uremic syndrome.

A total of 28 second CTC courses were administered. Second courses were started on day 29 after the previous stem cell reinfusion and were usually administered in the same dose as the previous course. The rate of bone marrow recovery and the toxicity were similar to that of the first course of CTC, with two exceptions: hemorrhagic cystitis was observed in three patients despite the administration of mesna, and one patient developed (non-lethal) veno-occlusive disease. The hemorrhagic cystitis, which occurred in three of the patients with refractory germ cell cancer, was prolonged in all three, and had not resolved by the time that they died of disease progression.

Eleven patients received a total of three CTC courses. The third course was planned on day 29 after the second stem cell reinfusion, but only three patients were able to receive the chemotherapy at the planned time and in the planned dose. Three of the eleven patients died of toxicity (Table 2).

CONCLUSION

A single course of CTC in the dose and schedule developed in our institute is feasible and relatively safe. It lacks irreversible organ toxicity such as pneumonitis or veno-occlusive disease. Based on its toxicity profile, CTC may be considered suitable for adjuvant chemotherapy in breast cancer, as is the CTCb regimen. CTC is the high-dose regimen in two ongoing randomized adjuvant chemotherapy studies in the Netherlands, which had recruited over 200 patients by February 1995. Whether or not the higher carboplatin dose of CTC as compared to CTCb is advantageous in this setting is still unknown.

Repeated courses of CTC are apparently associated with significant morbidity in a small percentage of patients (14% in this study) and even lethal toxicity in one. Because of the very high cumulative dose of alkylating agents, including carboplatin, the double CTC regimen is theoretically attractive in germ cell cancer and in ovarian cancer. Preliminary experience with the approach is very encouraging in relapsing germ cell cancer (but not in refractory disease)[20] and a confirmatory multi-center study is currently in progress in the Netherlands.

Triple CTC is clearly too toxic in many patients and leads to a high frequency of both veno-occlusive disease and a delayed hemolytic uremic syndrome, which resembles the one associated with mitomycin-C. This excess toxicity illustrates the novel situation in oncology that bone marrow toxicity is no longer dose-limiting but that cumulative drug dosages define the limits of dose-intensive therapy.

REFERENCES

1. Peters WP, Ross M, Vredenburgh JJ, Meisenberg B, Marks LB, Winer E, Kurtzberg J, Bast RC Jr, Jones R, Shpall E, Wu K, Rosner G, Gilbert C, Mathias B, Coniglio D, Petros W, Henderson IC, Norton L, Weiss RB, Budman D, Hurd D. High-dose chemotherapy and autologous bone marrow support as consolidation after standard-dose adjuvant therapy for high-risk primary breast cancer. J Clin Oncol 1993; 11: 1132-43
2. Van der Wall E, Beijnen JH, Rodenhuis S. High-dose chemotherapy regimens for solid tumors: a review. Cancer Treatm Rev 1995; in press

3. Frei III, E, Antman K, Teicher B, Eder P, Schnipper L. Bone marrow autotransplantation for solid tumors - prospects. J Clin Oncol 1989; 7: 515-26
4. O'Dwyer PJ, LaCreta F, Engstrom PF, Peter R, Tartaglia L, Cole D, Litwin S, DeVito J, Poplack D, DeLap RJ, Comis RL. Phase I pharmacokinetic reevaluation of ThioTEPA. Cancer Res 1991; 51: 3171-76
5. Martin M, Diaz-Rubio E, Casado A, Santabárbara P, Manuel López Vega J, Adrover E, Lenaz L: Carboplatin: an active drug in metastatic breast cancer. J Clin Oncol 1992; 10: 433-7
6. O'Brien MER, Talbot DC, Smith IE. Carboplatin in the treatment of advanced breast cancer: a phase II study using a phamacokinetically guided dose schedule. J Clin Oncol 1993; 11: 2112-17
7. Nichols CR, Tricot G, Williams SD, Van Besien K, Loehrer PJ, Roth BJ, Akard L, Hoffman R, Goulet R, Wolff SN, Giannone L, Greer J, Einhorn LH, Jansen J. Dose-intensive chemotherapy in refractory germ cell cancer-a phase I/II trial of high-dose carboplatin and etoposide with autologous bone marrow transplantation. J Clin Oncol 1989; 7: 932-9
8. Nichols CR, Andersen J, Lazarus HM, Fisher H, Greer J, Stadtmauer EA, Loehrer PJ, Trump DL. High-dose carboplatin and etoposide with autologous bone marrow transplantation in refractory germ cell cancer: an Eastern Cooperative Oncology Group Protocol. J Clin Oncol 1992; 10: 558-63
9. Broun ER, Nichols CR, Tricot G, Loehrer PJ, Williams SD, Einhorn LH. High-dose carboplatin/VP-16 plus ifosfamide with autologous bone marrow support in the treatment of refractory germ cell tumors. Bone Marrow Transplant 1991; 7: 53-6
10. Rosti G, Albertazzi L, Salvioni R, Pizzocaro G, Cetto GL, Bassetto MA, Marangolo M. High-dose chemotherapy supported with autologous bone marrow transplantation (ABMT) in germ cell tumors: a phase two study. Ann Oncol 1992; 3: 809-12
11. Lotz JP, Machover D, Malassagne B, Hingh B, Donsimoni R, Gumus Y, Gerota J, Lam Y, Tulliez M, Marsiglia H, Mauban S, Izrael V. Phase I-II study of two consecutive courses of high-dose epipodophyllotoxin, ifosfamide, and carboplatin with autologous bone marrow transplantation for treatment of adult patients with solid tumors. J Clin Oncol 1991; 9: 1860-70
12. Shea TC, Storniolo AM, Mason JR, Newton B, Mullen M, Taetle R, Green MR. A dose-escalation study of carboplatin/cyclophosphamide/etoposide along with autologous bone marrow or peripheral blood stem cell rescue. Semin Oncol 1992; 19(Suppl 2): 139-44
13. Motzer RJ, Gulati SC, Tong WP, Menendez-Botet C, Lyn P, Mazumdar M, Vlamis V, Lin S, Bosl G. Phase I trial with pharmacokinetic analyses of high-dose carboplatin, etoposide, and cyclophosphamide with autologous bone marrow transplantation in patients with refractory germ cell tumors. Cancer Res 1993; 53: 3730-35
14. Eder JP, Elias A, Shea TC, Schryber SM, Teicher BA, Hunt M, Burke J, Siegel R, Schnipper LE, Frei E III, Antman K. A phase I-II study of cyclophosphamide, thiotepa and carboplatin with autologous bone marrow transplantation in solid tumor patients. J Clin Oncol 1990; 8: 1239-45
15. Antman K, Ayash L, Elias A, Wheeler C, Hunt M, Eder JP, Teicher BA, Critchlow J, Bibbo J, Schnipper LE, Frei E III. A phase II study of high-dose cyclophosphamide, thiotepa and carboplatin with autologous marrow support in women with measurable advanced breast cancer responding to standard dose

therapy. J Clin Oncol 1992; 10: 102-10
16. Eder JP, Antman K, Elias A, Shea TC, Teicher B, Henner WD, Schryber SM, Holden S, Finberg R, Chritchlow J, Flaherty M, Mick R, Schnipper LE, Frei E III. Cyclophosphamide and thiotepa with autologous bone marrow transplantation in patients with solid tumors. J Natl Cancer Inst 1988; 80: 1221-6
17. Ayash LJ, Elias A, Wheeler C, Reich E, Schwartz G, Mazanet R, Tepler I, Warren D, Lynch C, Gonin R, Schnipper L, Frei E III, Antman K. Double dose-intensive chemotherapy with autologous marrow and peripheral-blood progenitor-cell support for metastatic breast cancer: a feasibility study. J Clin Oncol 1994; 12: 37-44
18. Rodenhuis S, Baars JW, Schornagel JH, Vlasveld LT, Mandjes I, Pinedo HM, Richel DJ. Feasibility and toxicity study of a high-dose chemotherapy regimen for autotransplantation incorporating carboplatin, thiotepa and cyclophosphamide. Ann Oncol 1992; 3: 855-60
19. Van der Wall E, Nooijen WJ, Baars JW, Holtkamp MJ, Schornagel JH, Richel DJ, Rutgers EJT, Slaper-Cortenbach ICM, Van der Schoot CE, Rodenhuis S. High-dose carboplatin, thiotepa and cyclophosphamide (CTC) with peripheral blood stem cell support in the adjuvant therapy of high-risk breast cancer: a practical approach. Br J Cancer 1995; in press
20. Rodenhuis S, Van der Wall E, Ten Bokkel Huinink WW, Schornagel JH, Richel DJ, Vlasveld LT. Pilot study of a high-dose carboplatin-based salvage strategy for relapsing or refractory germ cell cancer. Cancer Invest 1995; in press

TREATMENT INTENSIFICATION IN GERM-CELL TUMOURS

Jean-Pierre DROZ[(1)] - Pierre BIRON[(1)] - Stéphane CULINE[(2)] - Andrew KRAMAR[(3)]

(1) Department of Medical Oncology
Centre Léon Bérard, 28 rue Laënnec, 69373 LYON cedex 08 - France
(2) Centre Val d'Aurelle Paul-Lamarque, 326 rue des Apothicaires, Parc Euromédecine,
34298 MONTPELLIER cedex 5 - France
(3) Department of Biostatistics
Institut Gustave Roussy, rue Camille Desmoulins, 94805 VILLEJUIF cedex - France

INTRODUCTION

Germ-cell tumours are rare diseases. However they occur in young patients and their prognosis was unfavourable before the introduction of cisplatin in chemotherapy regimens. Cisplatin based chemotherapy followed by the surgical removal of residual disease is the standard treatment of these tumours (1). Different prognostic classifications or models allow to assign patients in good-risk or poor-risk groups (2).
The standard chemotherapy regimen in the good-risk patients group is either three cycles of a combination of cisplatin, etoposide and bleomycin (BEP) or four cycles of combination of etoposide and cisplatin (EP) (1).

The standard treatment of patients of the poor-risk group is four cycles of BEP (1). The continuously non-evolutive disease (NED) rate of patients with good-risk and poor-risk characteristics is 90-95 % and 50-70 % respectively (1). A small proportion of patients fail to be cured because either they have uncomplete response to first-line chemotherapy or they experience relapse after achievement of a complete remission. The standard salvage regimen of chemotherapy is a combination of ifosfamide, cisplatin and either vinblastine or etoposide (3). However the long-term NED rate is rather low in this group of patients : it is 20 % with a range of 0 to 50 % depending on prognostic factors (4). From the beginning of the modern era of chemotherapy in germ-cell tumours to the present time, the study of dose-intensive fixation was one of the most important areas of research.

Platinum and Other Metal Coordination Compounds in Cancer Chemotherapy 2
Edited by H.M. Pinedo and J.H. Schornagel, Plenum Press, New York, 1996

Dose-intensity is defined according to Hryniuk (5). The dose-intensity is the total dose of the drug (expressed per square meter) given during the whole chemotherapy protocol duration and divided by the number of weeks of treatment. The mean dose-intensity is the sum of the dose-intensities of each drug divided by the number of drugs of the protocol. The relative dose-intensity (RDI) of a protocol is the dose-intensity of this protocol divided by the dose-intensity of a standard protocol. There are different means to increase the dose-intensity :

- To increase the dose of each drug in the protocol but with the same time interval between each cycle.
- To decrease the time interval between each cycle of chemotherapy with a lower dose of each drug at each cycle.

These two different possibilities allow to deliver a higher dosage of drugs during a certain time. However it is also possible to increase the pic-dose intensity that is to increase by three to five fold the dose of all drugs, one time or two times. It needs the support of hematopoietic stem-cells. We will review the different means to increase a dose-intensity or the pic-dose in germ-cell tumours.

I / ROLE OF DOSE-INTENSITY IN CONVENTIONAL CHEMOTHERAPY OF GERM-CELL TUMOURS

The first experience was published by Stoter who studied combination of cisplatin, vinblastine and bleomycin (PVB) where cisplatin and bleomycin were given at fixed doses (cisplatin 20 mg/m^2/day for five days and bleomycin 30 mg IV each week) (6).

Patients were randomized to receive either vinblastine 0,3 mg/kg or vinblastine 0,4 mg/kg each cycle (at three weeks interval). This study failed to show an increasing response rate parallel to increasing dose-intensity of vinblastine. The same demonstration was published by Einhorn in 1980 (7) in a three arm phase III randomized trial studying a fixed combination of cisplatin 20 mg/m^2/day for five days and bleomycin 30 mg/week and either vinblastine 0,3 mg/kg or vinblastine 0,4 mg/kg or a combination of vinblastine 0,2 mg/kg and adriamycin 50 mg/m^2 every three weeks. There were no difference in complete remission rates and long term NED rates in the three arms. However there was an increasing toxicity with the combination of vinblastine/doxorubicin. Restrospective results published by Samuels were not confirmed by these two protocols (8).

Different trials studied the role of cisplatin dose-intensity. The first trial was published by the SWOG Group and consisted on fixed dose of vinblastine 12 mg/m^2 plus bleomycin
15 mg/m^2/week and cisplatin administered every four weeks (9). The dose of cisplatin was either 15 mg/m^2/day for five days or 120 mg/m^2. The dose-intensity of the former dosage was 18,7 mg/m^2/week and 30 mg/m^2/week for the latter. This increasing cisplatin dose-intensity translated in higher complete remission (CR) and NED rates. The CR and NED rates were 43 % and 57 % with the 18,7 mg/m^2/week dosage and 63 % and 80 % with the 30 mg/m^2/week dosage respectivelly.

Further studies were performed with higher cisplatin dose-intensity. The first was performed by Ozols at the NCI (10). This investigator studied cisplatin dosage of 20 mg/m^2/day for five days versus 40 mg/m^2/day for five days. However the increased cisplatin dose-intensity was not the only change in the protocol. Patients received also in the higher dose-intensity arm a combination of vinblastine and etoposide instead of vinblastine alone. The results of this protocol showed higher CR and NED rates with a higher dose of cisplatin combined with etoposide.

It was not demonstrated whether these results were related to the increased cisplatin dose-intensity or to the introduction of etoposide in the protocol. The question was resolved by the intergroup study which compared a combination of fixed doses of etoposide and bleomycin with either conventional cisplatin dose-intensity (33 mg/m²/week) or double-dose of cisplatin (66 mg/m²/week) (11). There was neither a difference of complete remission rates, nor a difference in long term NED rates between the two arms. However the latter protocol was more toxic than the former.

A compilation of cisplatin dose-intensity in different protocols published in the literature led Loehrer to conclude that there was a relationship between effect and dose-intensity with suboptimal chemotherapy regimens (that is cisplatin dose-intensity of less than 30 mg/m²/week) but there was no evidence of better results with higher dose-intensity than 30 mg/m²/week (12).

There are no data on the role of dose-intensity of etoposide and ifosfamide.

II/ DOSE INTENSIFICATION OF CONVENTIONAL CHEMOTHERAPY

1 - Accelerated chemotherapy without hematopoietic growth factor support

Different protocols of accelerated chemotherapy with a one week interval between cycles where recently published following the first publication of Wettlaufer (13). Investigators at the Royal Marsden Hospital published in 1989 the combination of BOP/BEP (14). The first combination chemotherapy consisted of bleomycin 15 mg IV weeks 1, 3 and bleomycin 15 mg/day in continuous infusion for five days weeks 2, 3, vincristine 2 mg total dose, weeks 1, 4, cisplatin 20 mg/m²/day days 1-5, weeks 1, 3 and cisplatin 1 mg/kg, weeks 2, 4. This four week protocol was followed by the BEP standard protocol. Twenty seven patients were studied and there were eleven clinical complete responses, nine pathological complete responses and five patients with partial response and normalized tumour markers. In total, there were twenty two long term NED patients.

The same group studied and published in 1994 a combination of chemotherapy with a same kind of schedule which incorporated carboplatin (15). Patients received bleomycin 15 mg IV per week, weeks 1, 3, 5, 6 and 15 mg/day in continuous infusion for five days on weeks 2, 4, vincristine 2 mg total dose every week for six weeks, cisplatin 100 mg/m² week 1, 3 and combination of cisplatin 40 mg/m² and carboplatin with a projected AUC of three on weeks 2, 4. Twenty one patients were treated.

There were eleven clinical complete responses and ten pathological complete responses. Nineteen patients were long term NED patients.

Other similar protocols were described : they combine either vinblastine, bleomycin combination and the BOP regimen (16) or cisplatin and oxaliplatin (17).

Table 1 shows the different dose-intensities in these protocols and compares the complete remission rates in each study. A protocol was initiated at the Indiana University which combined the VIP protocol and the administration of vinblastine and bleomycin every week during cycles intervals (18).

2 - Dose intensification with hematopoietic growth factor support

Investigators at the Hannover University developed a combination of cisplatin, etoposide and ifosfamide with increasing doses of each drug (19). The higher dose level consisted of (total dose) cisplatin 150 mg/m², etoposide 1000 mg/m² and ifosfamide 8 g/m², that is a relative dose-intensity of 1,6 when compared to the standard VIP regimen. At the higher dose level all patients experienced aplasia with a mean duration of eight days, neutropenic fever in 69 % of patients. The planned/achieved relative dose-intensity

Table 1. Relative dose-intensity of accelerated regimens of chemotherapy

	PVB	BOP	C-BOP
Bleomycin	30 mg/week	45 mg/week	45 mg/week
Cisplatin	33 mg/m²/week	70 mg/m²/week	95 mg/m²/week
RDI	1	1.81	2.19
CR	60 %	92 %	100 %

Abbreviations : see the text

was 82 %. The actually given relative dose-intensity (RDI) was 1.31.

A phase II randomized trial of accelerated BOP/VIP regimen of chemotherapy versus standard BEP in poor-risk patients failed to demonstrate survival advantage with the experimental arm (20).

III / DOSE INTENSIFICATION WITH HEMATOPOIETIC STEM-CELL SUPPORT

The different hematopoietic stem-cell supports have been tested during the last years. The most standard procedure was autologous bone marrow transplantation until the late eighties. Nowadays, patients receive peripheral blood stem-cell support in almost every case. The first experience with high dose chemotherapy and autologous bone marrow transplantation were reported from 1985 to 1988 (21). Patients received high dose treatment with etoposide, cyclophosphamide and cisplatin. In the late 1980, when carboplatin was available, patients were treated by high dose carboplatin-based chemotherapy (22). Several drugs are not convenient for dose intensification : bleomycin, vinblastine and partly cisplatin.

However experience with double dose of cisplatin allowed to increase significantly the cisplatin dosage. Conversely, increasing dose of etoposide, ifosfamide, cyclophosphamide with bone marrow suppression was easily overpassed by autologous bone marrow support (21).

The first patients were treated in feasability and phase I trials and subsequent experiences were allowed in patients in the salvage setting. Only few experiences tested high dose chemotherapy in the first line treatment setting.

1 - Phase I trial with high dose chemotherapy and autologous bone marrow support

Seven studies concerned phase I trials of high dose chemotherapy. One experience at Indiana University studied increasing doses of etoposide and carboplatin (22). Treatment was planned for two cycles of high dose treatment chemotherapy. Investigators at the Memorial Hospital and at the Institut Gustave Roussy studied the combination of carboplatin cyclophosphamide and etoposide. At the Memorial Hospital (23) the dose of cyclophosphamide was increased with fixed dose of etoposide and carboplatin. Conversely, at the Institut Gustave Roussy, the dose of carboplatin was increased and etoposide and cyclophosphamide were given at a constant dose (24). The similar regimen of chemotherapy regimen was tested at the Vienna University (25). However, ifosfamide was tested instead of cyclophosphamide at the Indiana University and at the Berlin University (26). Ifosfamide was demonstrated to increase the risks of renal and encephalitic toxicities. A trial was performed at the Tenon Hospital to use a tandem schedule of chemotherapy with bone marrow support (27). Patients received two cycles of high dose-VP16, etoposide, ifosfamide and carboplatin (27-28). These different experiences led to the following conclusions :

- First, it was demonstrated that high dose chemotherapy with different dosages of drugs at each cycle was tolerable even when tandem cycles were performed.
- The second conclusion was that carboplatin based chemotherapy regimens were manageable and that the use of cisplatin was obsolete in this setting.

However none of these studies allowed to define the standard regimen and to demonstrate whether the addition of an oxazaphosphorine derivate was useful.

Recent studies showed that increased dosages could be performed with the use of peripheral blood stem-cell support (28).

2 - Experiences with salvage high dose chemotherapy treatment

The review of high dose chemotherapy with bone marrow support in the literature is shown in table 2. Patients are stratified according to whether they were responder to the salvage conventional chemotherapy or not. Results are expressed by the number of patients who were NED after one year of follow up. These patients are assumed to be cured. It is observed that sixty two out of 101 patients are long term NED in the responder patients group and that only nineteen patients out of 163 may be cured in the non-responder patients group. Patients are considered as non-responder or refractory when they experience increasing tumour markers levels during cisplatin treatment or less than one month after the last cycle of this chemotherapy. It is concluded that high dose chemotherapy seems to have no effect in patients with refractory disease but may lead to more than 50 % long term NED rates in patients who are still responders to conventional salvage treatment.

An analysis of prognostic factors for the long term NED status of patients treated by high dose chemotherapy shows that the disease status (refractory or non refractory disease), the nature of drugs (Oxazaphosphorine or not) and drug dosages are important prognostic (29). It seems that patients who were refractory to conventional cisplatin doses may be sensitive to high doses of etoposide and carboplatin. Conversely, patients who are responder to chemotherapy may benefit from increasing dose of etoposide and oxazaphosphorine but not of carboplatin.

An european study is testing the role of high dose chemotherapy consolidation in the salvage setting. A multivariate analysis of prognostic factors for favourable outcome

Table 2. Long-term NED rate in responder and refractory patients in the salavage setting : review of the literature (21-28).

Author	Protocol	Responder		Refractory	
		Nb	NED	Nb	NED
Droz	PEC	14	7	18	1
Biron	VIC	8	4		
Brown	EcP x 2				
Nichols	EcP x 2	18	4	50	9
Motzer	cPEC	4	0	10	2
Pico	cPEC	3	1	10	2
Barnett	cPEC	11	9	4	0
Linkesch	cPEC	27	9	13	0
Siegert	cPEI	45	23	23	1
Rosti	cPE(I)	11	5	17	0
Lotz	cPEI x 2			18	4
		141	62 (44 %)	163	19 (11.6 %)

Abbreviations :
P = cisplatin, cP = carboplatin, V/E = etoposide, I = ifosfamide, C = cyclophosphamide

showed that different factors are important: tumour origine, initial tumour marker level at the start of salvage treatment, presence of pulmonary metastases and response to first line chemotherapy.
The on-going randomized trial tests four cycles of VIP versus three cycles of VIP and high dose chemotherapy with carboPEC (carboplatin, etoposide and cyclophosphamide).

IV / HIGH DOSE CONSOLIDATION CHEMOTHERAPY IN FIRST LINE TREATMENT

Only several studies have tested the role of pic-dose-intensity in consolidation of chemotherapy in poor-risk non semenimatous germ-cell tumours. Three phase II studies were reported in the literature.
Two concerned double dose cisplatin-based chemotherapy regimens. One was performed at the Institut Gustave Roussy in a group of twenty nine patients with poor-risk germ-cell tumours (30). The complete remission rate was 61 % and the long term NED rate was 45 %. These results compared favorably with previous results obtained by cisplatin-based conventional chemotherapy regimens. However the former protocols did not include etoposide in the chemotherapy association. Another phase II study was performed at the Centre Léon Bérard in Lyon and showed that 14 out of 16 patients attained long term NED status (31). All these patients had poor-risk characteristics. The third study was performed at the Memorial Hospital (32) : patients received two cycles of vinblastine, actinomycin, bleomycin, cyclophosphamide and cisplatin (VAB6) chemotherapy and two cycles of high dose etoposide-carboplatin with autologous bone marrow transplantation. There was a 65 % complete remission rate and a 40 % long term NED rate. These results compared favourably with results published previously with VAB6 protocol.
However VAB6 does not include etoposide and these results do not demonstrate the role of high dose chemotherapy because it might not be excluded that etoposide could induce these good results.

A randomized phase III study was published by a French group. It compared the PVeBV (double dose cisplatin, etoposide, vinblastine and bleomycin) regimen as described at the NCI to the former protocol tested in phase II (two cycles of modified PVeBV regimen followed by high-dose double-dose-cisplatin based chemotherapy regimen plus autologous bone marrow transplantation) (33). This trial failed to show any difference between the two arms. It was concluded that pic-dose-intensity with the PEC regimen did not induce higher long term NED rates in patients with poor-risk nonseminomatous germ-cell tumours. However a trial is on-going in the USA which tests, in the same patient population, two cycles of BEP followed by two cycles of high-dose etoposide-carboplatin with stem-cell support versus the standard four cycles of BEP. This protocol finally will answer the question to know whether dose intensification is active in the treatment of poor-risk germ-cell tumour patients.

V / VERY HIGH DOSE INTENSIVE CHEMOTHERAPY PROTOCOL

The use of peripheral blood stem-cell (PBSC) support allows to test repeated very high dose chemotherapy cycles. Only one experience has been published until now (19). Investigators at the Hannover University treated patients with a first cycle of PEI (cisplatin, etoposide and ifosfamide) plus GM-CSF, then the collection of PBSC and three cycles of escalated dose PEI plus PBSC plus GM-CSF. Drug dosages of etoposide, cisplatin and ifosfamide in this protocol were the same as the higher dose level tested with GM-CSf only. The use of PBSC allowed to decrease hematological toxicity. It is observed that the median duration of aplasia is less than four days and that the median duration of thrombocytopenia of less than 25000/mm^3 is four to five days whatever the cycle of chemotherapy. However, this protocol was only tested in the phase II setting. It needs to be investigated in a phase III randomized trial versus standard VIP regimen.

CONCLUSION

There is some evidences that there is a dose-response relationship in the chemotherapy of germ-cell tumours. However this relationship is evident only when suboptimal and optimal treatments are considered. Moreover no dose relationship effect with cisplatin was demonstrated with dosages higher than 100 mg/m^2 every three weeks. Dose-intensity could be increased by the use of hematopoietic growth factors or peripheral blood stem-cell support.
One other possibility is to study pic-dose-intensity as consolidation of conventional treatment. Until now there is no demonstration of a role of high dose chemotherapy consolidation treatment of first line chemotherapy of poor-risk patients group. There is some evidence for a role of dose-intensification consolidation treatment in the salvage setting. A European randomized trial is currenthly testing this hypothesis. Dose-intensity in the treatment of germ-cell tumours warrants further investigations.

Aknowledgements :
This work was partly sponsored by a grant PHRC 94 of the Ministry of Health.
The authors thank Mrs Sandrine Siau for her skilfull help in the preparation of the manuscript.

REFERENCES

1- LH EINHORN
Treatment of testicular cancer : a new and improved model.
J Clin Oncol 8 : 1777-1781, 1990.

2- JP DROZ, A KRAMAR, A REY.
Prognostic factors in metastatic disease.
Sem Oncol 19 : 181-189, 1992.

3- PJ LOEHRER, R LAUER, BJ ROTH, SD WILLIAMS, LA KALASINSKI, LH EINHORN.
Salvage therapy in recurrent germ cell cancer : ifosfamide and cisplatin plus either vinblastine or etoposide.
Ann Int Med 109 : 540-546, 1988.

4- JP DROZ, A KRAMAR, C NICHOLS, H SCHMOLL, A AUPERIN, A HARSTRICK, L EINHORN.
Second line chemotherapy with ifosfamide, cisplatin and either etoposide or vinblastine in recurrent germ cell cancer : assignment of prognostic groups. (abstract)
Proc Am Soc Clin Oncol 12 : 229, 1993.

5- W HRYNIUK, H BUSH.
The importance of dose intensity in chemotherapy of metastatic breast cancer.
J Clin Oncol 2 : 1281-1288, 1984.

6- G STOTER, Dth SLEIJFER, WW TEN BOKKEL HUININK, SB KAYE, WG JONES, AT VAN OOSTEROM, CPJ VENDRIK, P SPAANDER, M DE PAUW, R SYLVESTER.
High-dose versus low-dose vinblastine in cisplatin-vinblastine-bleomycin combination

chemotherapy of non-seminomatous testicular cancer : a randomized study of the EORTC genitourinary tract cancer cooperative group.
J Clin Oncol 4 : 1199-1206, 1986.

7- LH EINHORN, SD WILLIAMS.
Chemotherapy of disseminated testicular cancer.
Cancer 46 : 1339-1344, 1980.

8- ML SAMUELS, VJ LANZOTTI, PY HOLOYE, LE BOYLE, TL SMITH, DE JOHNSON.
Combination chemotherapy in germinal cell tumors.
Cancer Treat Rev 3 : 185-204, 1976.

9- MK SAMSON, SE RIVKIN, SE JONES, JJ COSTANZI, AF LOBUGLIO, RL STEPHENS, EA GEHAN, GD CUMMINGS.
Dose-response and dose-survival advantage for high versus low-dose cisplatin combined with vinblastine and bleomycin in disseminated testicular cancer.
Cancer 53 : 1029-1035, 1984.

10- RF OZOLS, DC IHDE, WM LINEHAN, J JACOB, Y OSTCHEGA, RC YOUNG.
A randomized trial of standard chemotherapy versus a high-dose chemotherapy regimen in the treatment of poor prognosis non-seminomatous germ-cell tumors.
J Clin Oncol 6 : 1031-1040, 1988.

11- CR NICHOLS, SD WILLIAMS, PJ LOEHRER, FA GRECO, ED CRAWFORD, J WEETLAUFER, ME MILLER, A BARTOLUCCI, L SCHACTER, LH EINHORN.
Randomized study of cisplatin dose intensity in poor-risk germ cell tumors : a southeastern cancer study group and southwest oncology group protocol.
J Clin oncol 9 : 1163-1172, 1991.

12- PJ LOEHRER, SD WILLIAMS, LH EINHORN.
Testicular cancer : the quest continues.
J Natl Cancer Inst 80 : 1373-1382, 1988.

13- JN WETTLAUFER, AS FEINER, WA ROBINSON.
Vincristine,cisplatin and bleomycin with surgery in the management of advanced metastatic non-seminomatous testis tumours.
Cancer 53 : 203-209, 1984.

14- A HORWICH, M BRADA, J NICHOLLS, G JAY, WF HENDRY, DP DEARNALEY, MJ PECKHAM.
Intensive induction chemotherapy for poor risk non-seminomatous germ cell tumours.
Eur J Cancer Clin Oncol 25 : 177-184, 1989.

15- A HORWICH, DP DEARNALEY, A NORMAN, J NICOLLS, WF HENDRY.
Accelerated chemotherapy for poor prognosis germ cell tumours.
Eur J Cancer 30A : 1607-1611, 1994.

16- JP DROZ, S CULINE, B BUI, R DELVA, A CATY, M MOUSSEAU, MH FILIPPI, B MINIER.
Shortly recycled, intensive alternating, bleomycin containing chemotherapy in heavily pretreated germ cell tumors : an effective strategy. (abstract)
Ann Oncol 5 : 68, 1994.

17- E CVITKOVIC (personal communication)

18- C BLANKE, P LOEHRER, L EINHORN, C NICHOLS.
A phase II study of VP-16 plus ifosfamide plus cisplatin plus vinblastine plus bleomycin (VIP/VB) with filgrastim for advanced stage testicular cancer. (abstract)
Proc Am Soc Clin Oncol 13 : 234, 1994.

19- C BOKEMEYER, HJ SCHMOLL.
Treatment of advanced germ cell tumours by dose intensified chemotherapy with hematopoietic growth factors or peripheral blood stem cells.
Eur Urol 23 : 223-230, 1993.

20- SB KAYE, GM MEAD, S FOSSA, M CULLEN, R DE WIT, I BODROGI, C VAN GROENINGEN, R SYLVESTER, S STENNING, K VERMEYLEN, E LALLEMAND, P DE MULDER.
An MRC/EORTC randomised trial in poor prognosis metastatic teratoma, comparing BEP with BOP-VIP. (abstract)
Proc Am Soc Clin Oncol 14 : 246, 1995.

21- JP DROZ, JL PICO, A KRAMAR.
Role of autologous bone marrow transplantation in germ-cell cancer.
Urol Clin North Am 20 : 161-171, 1993.

22- CR NICHOLS, G TRICOT, SD WILLIAMS, K Van BESIEN, PJ LOEHRER, BJ ROTH, L AKARD, R HOFFMAN, R GOULET, SN WOLFF, L GIANNONE, J GREER, LH EINHORN, J JANSEN.
Dose-intensive chemotherapy in refractory germ cell cancer - A phase I/II trial of high-dose carboplatin and etoposide with autologous bone marrow transplantation.
J Clin Oncol 7 : 932-939, 1989.

23- R MOTZER, SC GULATI, WP TONG, C MENENDEZ-BOTET, P LYN, M MAZUMBAR, V VLAMIS, S LIN, G BOSL.
Phase I trial with pharmacokinetic analysis of high-dose carboplatin, etoposide and cyclophosphamide with autologous bone marrow transplantation in patients with refractory germ cell tumors.
Cancer 53 : 3730-3735, 1993.

24- A IBRAHIM, E ZAMBON, JH BOURHIS, M OSTRONOFF, F BEAUJEAN, P VIENS, C LHOMME, M CHAZARD, D MARANINCHI, M HAYAT, JP DROZ, JL PICO.
High-dose chemotherapy with etoposide, cyclophosphamide and escalating dose of carboplatin followed by autologous bone marrow transplantation in cancer patients. A pilot study.
Eur J Cancer 29A : 1398-1403, 1993.

25- W LINKESCH, HT GREINIX, P HOCKER, M KRAINER, A WAGNER.
Longterm follow up of phase I/II trial of ultra-high carboplatin, VP16, cyclophosphamide with ABMT in refractory or relapsed NSGCT. (abstract)
Proc Am Soc Clin Oncol 12 : 232, 1993.

26- W SIEGERT, J BEYER, I STROHSCHEER, H BAURMANN, H OETTLE, J ZINGSEM, R ZIMMERMANN, C BOKEMEYER, HJ SCHMOLL, D HUHN.
High-dose treatment with carboplatin, etoposide and ifosfamide followed by autologous

stem-cell transplantation in relapsed or refractory germ cell cancer : a phase I/II study.
J Clin Oncol 12 : 1223-1231, 1994.

27- JP LOTZ, D MACHOVER, B MALASSAGNE, B HINGH, R DONSIMONI, Y GUMUS, J GEROTA, Y LAM, M TULLIEZ, H MARSIGLIA, S MAUBAN, V IZRAEL.
Phase I-II study of two consecutive courses of high-dose epipodophyllotoxin, ifosfamide and carboplatin with autologous bone marrow transplantation for treatment of adult patients with solid tumors.
J Clin Oncol 9 : 1860-1870, 1991.

28- JP LOTZ, T ANDRE, R DONSIMONI, C FIRMIN, C BOULEUC, H BONNAK, Z MERAD, A ESTESO, J GEROTA, V IZRAEL.
High dose chemotherapy with ifosfamide, carboplatin and etoposide combined with autologous bone marrow transplantation for the treatment of poor-prognosis germ cell tumors and metastatic trophoblastic disease in adults.
Cancer 75 : 874-885, 1995.

29- JP DROZ, A KRAMAR, JL PICO.
Prediction of long-term response after high-dose chemotherapy with autologous bone marrow transplantation in the salvage treatment of non-seminomatous germ cell tumours.
Eur J Cancer 29A : 818-821, 1993.

30- JP DROZ, JL PICO, M GHOSN, A KRAMAR, A REY, M OSTRONOFF, D BAUME.
A phase II trial of early intensive chemotherapy with autologous bone marrow transplantation in the treatment of poor prognosis non seminomatous germ cell tumors.
Bull Cancer 79 : 497-507, 1992.

31- P BIRON, D BERTON, C TERRET, JP DROZ.
Long term results of high-dose chemotherapy (HDCT) and autologous bone-marrow transplantation (ABMT) in poor-risk and relapsed germ-cell tumors (GCT). (abstract)
Proc Am Soc Clin Oncol 14 : 320, 1995.

32- RJ MOTZER, M MAZUMDAR, SC GULATI, DF BAJORIN, P LYN, V VLAMIS, GJ BOSL.
Phase II trial of high-dose carboplatin and etoposide with autologous bone marrow transplantation in first-line therapy for patients with poor-risk germ cell tumors.
J Natl Cancer Inst 85 : 1828-1835, 1993.

33- JP DROZ, JL PICO, P BIRON, P KERBRAT, H CURE, JF HERON,C CHEVREAU, B CHEVALLIER, P FARGEOT, J BOUZY, A KRAMAR.
No evidence of a benefit of early intensified chemotherapy (HDCT) with autologous bone marrow transplantation (ABMT) in first line treatment of poor risk non seminomatous germ cell tumors (NSGCT) : preliminary results of a randomized trial. (abstract)
Proc Am Soc Clin Oncol 11 : 197, 1992.

IL-1α ENHANCEMENT OF PLATINUM-MEDIATED ANTI-TUMOR ACTIVITY

Candace S. Johnson,[1,2] Ming-Jei Chang,[1] Wei-Dong Yu,[1]
Ruth A. Modzelewski,[1] Derick M. Russell,[1] Theodore F. Logan,[3]
Daniel R. Vlock,[4] Leonard M. Reyno,[5] Merrill J. Egorin,[5]
Kadir Erkmen,[5] and Philip Furmanski[6]

[1]Department of Otolaryngology
[2]Department of Pharmacology
[3]Department of Medicine
University of Pittsburgh School of Medicine
University of Pittsburgh Cancer Institute
Pittsburgh, PA 15213
[4]Brigham and Women's Hospital
Boston, MA 02115
[5]Department of Medicine, Division of Hematology-Oncology
University of Maryland School of Medicine
University of Maryland Cancer Center
Baltimore, MD 21201
[6]Department of Biology
New York University
New York, NY 10003

INTRODUCTION

Cytokines or biological response modifiers have been utilized in combination with conventional chemotherapeutic agents to increase cytotoxic drug-mediated efficacy and to reduce the associated toxic side effects. Interleukin-1 (IL-1) is a multifunctional cytokine, primarily produced by monocyte/macrophages[1-4], that plays a central role in the activation of T and B cells[3,4], potentiates hematopoiesis both in vitro and in vivo[5-9], and induces the synthesis of other cytokines and regulatory molecules[3,10,11]. Studies in our laboratories have demonstrated that IL-1α induces acute tumor hemorrhagic necrosis, changes in tumor blood flow, an increase of tumor clonogenic cell kill and enhancement of cytotoxic drug-mediated anti-tumor effects in murine tumor model systems[12-20]. We focus our discussion here on IL-1's activities in the treatment of solid tumors either alone or in combination with the platinum agents, cisplatin or carboplatin.

IL-1

IL-1 is primarily responsible for the host response to infection and inflammation[1-4] and has been shown to potentiate hematopoiesis[5-9], effect vascular endothelium[21-23], enhance bone resorption[24,25] and muscle protein degradation[26], stimulate hepatic acute phase proteins[27], enhance proliferation of some cells[11,28,29] with anti-proliferative effects on others[30,31], enhance the synthesis of other cytokines[3,10,11] and stimulate T and B cell function[3,4]. The two biochemically distinct types of IL-1 isofocus at a pI of 5 (IL-1α) and 7 (IL-1β). Although they share only 22% homology, IL-1α and β share target cells and biologic activities[32,33]. There are three IL-1 receptors: IL-1RI, which is found on nearly all cells; IL-1RII, which is also found on many cells but primarily on neutrophils, monocytes and B-lymphocytes; and the T1/ST2/*Fit*-1 receptor[34]. While all three are derived from a common receptor gene, T1/ST2/*Fit*-1 does not appear to bind human IL-1. A natural IL-1 inhibitor termed IL-1 receptor antagonist (IL-1ra) has been isolated which specifically inhibits both IL-1α and IL-1β[35-38]. IL-1ra binds to IL-1 receptors with affinities similar to those of IL-1α and IL-1β but does not transduce a signal.

IL-1 AND HEMATOPOIESIS

IL-1 has been shown to have a profound effect on hematopoiesis[5-9]. IL-1 induces increased levels of colony stimulating factors (CSF) which results in a significant peripheral neutrophilia[6,7]. IL-1 can act in concert with CSF to stimulate primitive hematopoietic progenitors[7,8,39] and initiate cycling of myeloid progenitor cells in the marrow[40]. In addition, IL-1 has been shown to enhance the generation of megakaryocytic and early erythroid progenitor(s). Neta et al[41] have demonstrated that a single injection of IL-1 acts as a radioprotector and increases survival of lethally irradiated mice. IL-1 also enhances myeloid recovery following 5-fluorouracil in normal and tumor-bearing mice[42,43]. IL-1 has been demonstrated to protect marrow progenitors from the toxic effects of 4-hydroperoxy-cyclophosphamide[44]. In addition, the combination of IL-1 with either GM-CSF or G-CSF produced greater protection than either agent alone and allowed dose-escalation of the cytotoxic drug[45].

IL-1 EFFECT ON VASCULAR ENDOTHELIUM

Recent studies indicate that many endothelial cell activities can be regulated by IL-1 and TNF through modulation of cell surface molecules. IL-1 enhances endothelial cell procoagulant activity through induction of tissue factor production[21], enhanced prostacyclin production[46], and increased synthesis of plasminogen activator inhibitors[23,47]. IL-1 and TNF also cause endothelial cells to become markedly adhesive for neutrophils, monocytes and lymphocytes[22,48,49], thus providing an important early step in the establishment of leukocyte infiltration. These vascular effects culminate in hypotension, diffuse intravascular coagulopathy and vascular leakage.

Acute hemorrhagic necrosis of tumors is characterized by decreased pH, hypoxia and tumor cell death, and tumor microvascular injury as evident by hemorrhage, congestion, thrombus formation and circulatory blockage[50-54]. The classical mediator of this effect is TNF, which is induced in vivo in response to bacterial lipopolysaccharide (LPS)[55] although the IL-1 effects appear not to be mediated by TNF[13]. Interferon-α/ß (IFN-α/ß)[56], IL-1ß[57], and IL-1α[12] have also been implicated in inducing hemorrhagic necrosis in Friend erythroleukemia cell tumors and murine fibrosarcomas. Hemorrhagic necrosis occurs within hours without a strict requirement of T cell function[55,58], whereas complete clinical regression of solid tumors occurs days later and involves a competent host immune component.

ANTI-TUMOR ACTIVITIES OF IL-1

While IL-1 is generally considered not to mediate direct tumor cytotoxic activity, cytotoxicity has been observed for a number of human cell lines, including melanoma, breast and ovarian carcinoma[59-61]. North et al[62] have demonstrated that IL-1ß induced regression of murine tumors requires intact T cell function with dependence on tumor immunogenicity. IL-1ß has also been reported to inhibit the growth of B16 melanoma in C57BL/6 mice[63]. Nakamura et al[64] have observed that IL-1α inhibits the growth of several tumors and that indomethacin enhances the inhibition[65].

Based on the effect of IL-1 on hematopoiesis, studies were originally initiated to examine the ability of IL-1α to increase the therapeutic efficacy cytotoxic agents. We have observed that syngeneic MCA induced fibrosarcomas in NIH/PLCR inbred mice become significantly hemorrhagic 4 hours after a single injection of IL-1α[12]. Using several tumor systems (including the RIF-1 and Panc02 tumor systems in C3H/HeJ and C57BL/6, respectively), this hemorrhage has been characterized and quantitated and its kinetics determined in tumor and normal tissues by the ^{59}Fe-labeled RBC dilution method. Acute hemorrhage is observed as early as 3 hours and is maximal at 6-12 hours post ip injection in tumor tissue with no effect observed in skin (pinna) or muscle (femoralis). Extracellular water volume is significantly higher in IL-1α treated RIF-1 tumors when compared to untreated controls. A marked reduction in tumor blood flow is also observed as early as 30 minutes after IL-1 treatment. In vitro, IL-1α is not directly cytotoxic to RIF-1 tumor cells, however, in vivo, IL-1α significantly decreases the clonogenicity of RIF-1 tumors by 55% at 24 hours post. An increase in tumor cell proliferation is observed at 24 hours and by 48 hours, a rapid regrowth has occurred. Despite this acute anti-tumor activity, no substantial regrowth delay is observed in RIF-1 tumors after a single treatment with IL-1α. In addition, no significant differences in anti-tumor activity are observed when animals are treated with equivalent D10 unit doses of IL-1ß. The effects of IL-1α on RIF-1 tumors have also been characterized by ^{31}P-NMR spectroscopy, which has demonstrated an increase in inorganic phosphates and a decrease in high energy phosphates with correlated changes in tumor blood flow[14].

TNF-induced hemorrhagic necrosis of solid tumors occurs within hours without a strict requirement for T cell function[55]. We have shown that a single ip dose of IL-1 also induces hemorrhagic necrosis of RIF-1 tumors in athymic BALB/c nude mice, as measured by μl packed RBC/g tumor tissue[12]. As described with intact mice, no effects are observed in normal tissues. Therefore, if other mediators are involved in the process of hemorrhagic necrosis induced by IL-1α, our data suggest that T cells are not candidate producers.

Studies have also been undertaken to determine the role of adrenal hormones in IL-1α's acute hemorrhagic response[16]. When animals are adrenalectomized, IL-1α induced hemorrhage is increased, however, toxicity is severe. When dexamethasone is given before or at the same time as IL-1α, no hemorrhage is observed. Further studies have demonstrated that ketoconazole, a potent inhibitor of adrenal hormone biosynthesis, inhibits the IL-1α induced increase in plasma corticosterone, increases IL-1α mediated hemorrhagic necrosis and potentiates clonogenic cell killing[15]. Therefore, the anti-tumor activity of IL-1 appears to be reduced in vivo by IL-1 mediated effects on adrenal hormone release.

The primary mediator of hemorrhagic necrosis is TNF. IL-1 has been shown to induce TNF production or release by mononuclear cells in vitro[3,10]. The cytologic and histologic appearance of IL-1α induced hemorrhagic necrosis is identical to that observed with TNF. Therefore, we have further characterized IL-1α induced hemorrhagic necrosis and examined whether these acute effects are mediated through TNF[13]. To determine whether IL-1's effects are mediated through TNF, tumor-bearing animals are pre-treated with 250μg of monoclonal anti-murine TNF. This dose of antibody has been shown to abrogate TNF or LPS induced hemorrhagic necrosis[55]. Anti-TNF antibody has no effect on the IL-1 induced increase in packed RBC volumes for tumor, whereas the antibody abrogates hemorrhage induced by TNF or LPS. The amount of TNF found in the sera or tumors of animals treated with LPS is more than 20-fold higher than in mice treated with IL-1α, and LPS induces similar degrees of hemorrhagic necrosis. A significantly hemorrhagic dose of IL-1α induces no detectable TNF ($<$50pg/g) in tumors. Thus, IL-1 has profound effects on solid tumors that are quite similar, histologically and physiologically, to those of TNF, but that appear to be independent of the induction of or mediation by TNF.

ANTI-TUMOR ACTIVITIES OF IL-1 AND PLATINUM ANALOGUES

We have demonstrated that IL-1α can potentiate the anti-tumor activity of mitomycin C and porfiromycin as measured by an increase in clonogenic cell kill and tumor regrowth delay[17]. Based on these studies, we have examined the effect of IL-1α on cisplatin (cDDP)-mediated anti-tumor activities[18]. In vitro, IL-1α has no direct cytotoxic effect on RIF-1 tumor cells[12] and does not significantly enhance cDDP-mediated tumor cell kill. To examine in vivo efficacy, RIF-1 tumor bearing C3H/HeJ mice (14 days post implant) are treated concurrently with single ip injections of IL-1α and/or cDDP at various doses. Increased clonogenic tumor cell kill with IL-1α/cDDP is dose-dependent with significant enhancement by IL-1α even at the lowest doses tested (2mg/kg and 6μg/kg for cDDP and IL-1α, respectively). An increase in clonogenic tumor cell kill with IL-1α and cDDP does not correlate with an increase in tumor hemorrhage. Using median dose-effect analysis, this interaction is determined to be strongly synergistic. When treated animals are monitored for long-term anti-tumor effects, the combination of IL-1α with cDDP significantly increases tumor regrowth delay and decreases fractional tumor volume when compared to cDDP alone. These results demonstrate that IL-1α synergistically enhances cDDP-mediated in vivo anti-tumor activity.

In addition, the ability of IL-1α to enhance carboplatin (CBDCA)-mediated anti-tumor activity has also been examined[20]. Similar to cDDP, IL-1α has no effect on the dose response of CBDCA directed cytotoxicity against RIF-1. IL-1α also has no effect on CBDCA-mediated cell kill even at high IL-1α concentrations (10^5u/ml) nor when cells are treated at various intervals with IL-1α either before or after CBDCA.

To determine the in vivo effect of IL-1α and CBDCA, 14 day RIF-1 tumor bearing mice are treated on day 0 with concurrent ip administration of a standard dose of IL-1α (480μg/kg) and/or varying doses of CBDCA. After 24 hr, animals are killed, tumors removed, and clonogenic tumor cells enumerated. IL-1α significantly enhances CBDCA-mediated tumor cell kill (decrease in surviving fraction) even at low doses (not cytotoxic) or CBDCA alone.

To examine the optimum schedule for administration of platinum (cDDP or CBDCA) and IL-1α, time sequence studies have been performed where cDDP or CBDCA is always given to tumor-bearing mice at time 0 and IL-1α given either at time 0, before or after platinum at various times. Animals are killed 24 hr after platinum and the surviving fraction of tumor cells determined by excision clonogenic cell assay. The potentiation of CBDCA-mediated tumor cell kill is highly time dependent, with maximum anti-tumor activity observed when IL-1α is administered 4-12 hours before CBDCA. In contrast, administration of IL-1α from 24 hours before or as late as 6 hours after cDDP results in the same anti-tumor activity as concurrent administration of the two agents. When IL-1α is injected 3-7 days before either CBDCA or cDDP, enhancement of tumor cell kill is not observed and surviving fraction is significantly higher when compared to either CBDCA or cDDP alone.

Table 1. Potential of IL-1 in combination with platinum analogues.

Stimulation of late/early stage hematopoietic progenitor cells
Protection of bone marrow cells from insult
Ability to induce changes in cell cycle status
Effect on endothelial cell function
Production of other cytokines and regulatory molecules
Induction of acute tumor hemorrhagic necrosis
Activation of immune function

IL-1α has been shown to have a profound effect on endothelial cells[21-23,46-49]. We have shown significant acute effects on tumor blood flow following IL-1α injection as well as capillary damage, hypoxia and reduced tumor pH in vivo[12,14,15]. One possible explanation for the enhanced anti-tumor activity observed with the combination of platinum plus IL-1α could be that IL-1α is allowing more drug to get to the tumor thereby causing an increase in cell kill. To determine platinum drug levels, RIF-1, 14 tumor-bearing mice are treated with either cDDP or CBDCA at time 0 and IL-1α at various times before and after cytotoxic drug. At 24 hr post-treatment (a time when maximum clonogenic cell kill is observed), tumor and normal tissues are removed, frozen and platinum analyzed by atomic absorption spectro-scopy. Tumor and normal tissue platinum content are significantly increased by IL-1α in animals treated with CBDCA at time points which demonstrate IL-1 en-hancement of clonogenic tumor cell kill. In contrast, significant differences are observed in those treated with cDDP. These results suggest differences may be partially due to IL-1α induced alteration of CBDCA pharmacokinetics.

CLINICAL STUDIES

IL-1 has been in phase I clinical trials mainly as a bone marrow-sparing or stimulating agent in conjunction with chemotherapeutic drugs. In the phase I setting, the administration of IL-1 alone is reasonably well tolerated with hypotension the dose-limiting toxicity[66,67]. In addition, an increase in platelet counts is also observed 6 days after IL-1 initiation, with a sustained elevation for 24 days. When IL-1 is combined with 5-FU in gastrointestinal adenocarcinoma patients, a dose-dependent neutrophil leukocytosis is observed the day of IL-1 administration followed by an increase in platelets 14 days later[68]. Fewer days of neutropenia are observed after 5FU alone, although this difference is not reported to be statistically significant. IL-1 has also been shown to decrease the duration of thrombocytopenia following administration of CBDCA[69]). Recently, IL-1β has been administrated to patients with acute myelogenous leukemia undergoing autologous bone marrow transplant[70]. Despite somewhat moderate toxicity, patients receiving IL-1β have a reduced incidence of infection and enhanced survival when compared to historic controls.

We have initiated a phase I trial to determine the toxicity, maximum tolerated dose (MTD), pharmacokinetics and pharmacodynamics of IL-1α and CBDCA[71]. Patients have received CBDCA at $400\mu g/m^2$ over 30 min followed 90 min later by IL-1α over 2 hr every 28 days. Groups of patients have been given escalating doses of IL-1α starting at $0.4\mu g/m^2$. Patients (three/dose tier) have been entered at each dose and are currently at the $30\mu g/m^2$ dose tier level. Toxicities include fever, chills, anemia and hypotension. Patients have tolerated higher doses of IL-1α than may have been predicted previously. At the higher doses of IL-1α, a platelet sparing effect has been observed. While no tumor responses have been noted, time to progression has been significantly increased in a number of patients. To date, the MTD has not been reached and these studies continue.

Another trial that has been initiated and is currently ongoing involves a phase I trial of IL-1α and CBDCA where IL-1 was administered 4 hours prior to CBDCA, a time when maximum cell kill is observed[20] with effects on tumor blood flow evaluated by positron emission tomography (PET)[72]. The dose of IL-1α has been escalated, as described above, with a starting dose of 1μg/m^2 given iv over 2 hours, followed at 4 hours after IL-1 initiation by CBDCA at 400μmg/m^2 iv over 30 min every 28 days. PET scanning by ^{15}O water has been performed to assess tumor blood flow both before and after IL-1. To allow for blood flow measurements, eligible patients (three/dose tier) must have pulmonary metastatic disease within 10cm of the left atrium. Patients are also monitored for changes in granulocyte kinetics, integrin expression, fibrinolytic activity, nitric oxide production and carboplatin pharmacokinetics. This schedule has been well tolerated and to date 10 patients have been treated and have demonstrated a decrease in tumor blood flow, 2 hours after IL-1α administration.

SUMMARY

The potential for IL-1, a cytokine with numerous regulatory and biologic functions, in combination with chemotherapeutic drugs, especially the platinum agents, is yet to be realized (Table 1). In pre-clinical studies, IL-1α clearly enhances the anti-tumor activity of both cDDP and CBDCA in murine tumor model systems. The ability of IL-1α to enhance CBDCA-mediated tumor cell kill may be explained partially by an IL-1 induced alteration of CBDCA pharmacokinetics. This pharmacokinetic difference, however, can not explain IL-1α enhancement of cDDP anti-tumor activity. Because of the lack of a direct effect of IL-1α and platinum in vitro, the mechanism(s) involved must be indirect and involve other cells present in the tumor microenvironment (Figure 1). IL-1 has significant effects on fibroblasts, macrophages, endothelial cells, as well as T and B cells. Further studies are in

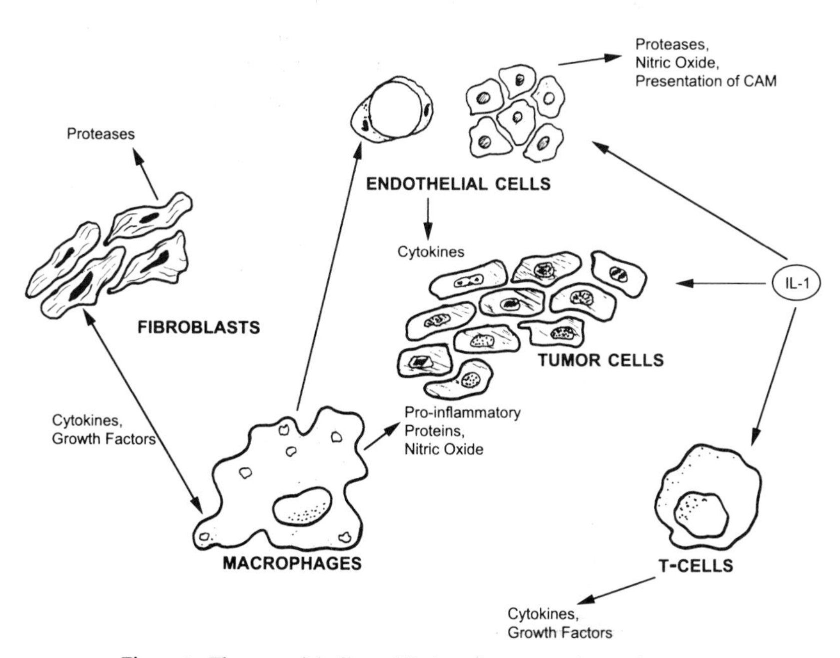

Figure 1. The potential effect of IL-1 on the tumor microenvironment.

progress to dissect out the mechanisms for IL-1 enhancement of platinum-mediated tumor cell kill. The diverse effects of IL-1 make it a promising cytokine in the design of novel therapeutic approaches for solid malignancies.

ACKNOWLEDGMENTS

This work was supported by Pubic Health Service Grants CA48077 and CA56756 from the National Cancer Institute, NIH, Department of Health and Human Services; the Mary Hillman Jennings Foundation and the John R. McCune Charitable Trust Foundation. We thank Andrea Piacentini for preparation of the manuscript.

REFERENCES

1. I. Gery, R.K. Gershon, and B.H. Waksman, Potentiation of the T-lymphocyte response to mitogens: I. The responding cell, J. Exp. Med. 136:128 (1972).
2. C.A. Dinarello, Interleukin-1, Rev. Infect. Dis. 6:51 (1984).
3. J. Le, and J. Vilcek, Biology of disease: tumor necrosis factor and interleukin 1. Cytokines with multiple overlapping biological activities, Lab. Invest. 56:234 (1987).
4. K.A. Smith, L.B. Lachman, J.J. Oppenheim, and M.J. Favata, The functional relationship of the interleukins, J. Exp. Med. 151:1551 (1980).
5. C.S. Johnson, D.J. Keckler, M.I. Topper, P.G. Braunschweiger, and P. Furmanski, In vivo hematopoietic effects of recombinant interleukin-1α in mice: stimulation of granulocytic, monocytic, megakaryocytic, and early erythroid progenitors, suppression of late-stage erythropoiesis, and reversal of erythroid suppression with erythropoietin, Blood 73:678 (1989).
6. L.C. Stork, U.M. Peterson, C.H. Rundus, and W.A. Robinson, Interleukin-1 enhances murine granulopoiesis in vivo, Exp. Hematol. 16:163 (1988).
7. M.A.S. Moore, and D.J. Warren, Synergy of interleukin 1 and granulocyte colony-stimulating factor: in vivo stimulation of stem-cell recovery and hematopoietic regeneration following 5-fluorouracil treatment of mice, Proc. Natl. Acad. Sci. 84:7134 (1987).
8. D.Y. Mochizuki, J.R. Eisenman, P.J. Conlon, A.D. Larsen, and R.J. Tushinski, Interleukin 1 regulates hematopoietic activity, a role previously ascribed to hemopoietin 1, Proc. Natl. Acad. Sci. 84:5267 (1987).
9. J.R. Zucali, H.E. Broxmeyer, C.A. Dinarello, M.A. Gross, and R.S. Weiner, Regulation of early human hematopoietic (BFU-E and CFU-GEMM) progenitor cells in vitro by interleukin-1-induced fibroblast-conditioned medium, Blood 69:33 (1987).
10. R. Philip, and L.B. Epstein, Tumor necrosis factor as immunomodulator and mediator of monocyte cytotoxicity induced by itself, gamma-interferon and interleukin-1, Nature 323:86 (1986).

11. J.J. Oppenheim, E.J. Kovacs, K. Matsushima, and S.K. Durum, There is more than one interleukin 1, Immunol. Today 7:45 (1986).
12. P.G. Braunschweiger, C.S. Johnson, N. Kumar, V. Rod and P. Furmanski, Antitumor effects of recombinant human interleukin-1α in RIF-1 and Panc02 solid tumors, Cancer Res. 48:6011 (1988).
13. C.S. Johnson, M.-J. Chang, P.G. Braunschweiger and P. Furmanski, Acute hemorrhagic necrosis of tumor induced by interleukin-1α: effects independent of tumor necrosis factor, J. Natl. Cancer Inst. 12:842 (1990).
14. I. Constantinidis, P.G. Braunschweiger, J.P. Wehrle, N. Kumar, C.S. Johnson, P. Furmanski and J.D. Glickson, P-nuclear magnetic resonance studies of the effect of recombinant human interleukin-1α on the bioenergetics of RIF-1 tumors, Cancer Res. 49:6379 (1989).
15. P.G. Braunschweiger, N. Kumar, I. Constantinidis, J.P. Wehrle, J.D. Glickson, C.S. Johnson and P. Furmanski, Potentiation of interleukin 1α mediated antitumor effects by ketoconazole, Cancer Res. 50:4709 (1990).
16. P.G. Braunschweiger, C.S. Johnson, N. Kumar, V. Ord and P. Furmanski, The effect of adrenalectomy and dexamethasone on interleukin-1α induced responses in RIF-1 tumors, Br. J. Cancer 61:9 (1990).
17. P.G. Braunschweiger, S.A. Jones, C.S. Johnson and P. Furmanski, Potentiation of mitomycin C and porfiromycin antitumor activity in solid tumor models by recombinant human interleukin 1, Cancer Res. 51:5454 (1991).
18. C.S. Johnson, M.J. Chang, W.D. Yu, R.A. Modzelewski, J.R. Grandis, D.R. Vlock and P. Furmanski, Synergistic enhancement by interleukin-1 α of cisplatin-mediated antitumor activity in RIF-1 tumor-bearing C3H/HeJ mice, Cancer Chemother. Pharmacol. 32:339 (1993).
19. C.S. Johnson, Interleukin-1: therapeutic potential for solid tumors, Cancer Invest. 11:600 (1993).
20. M.J. Chang, W.D. Yu, L.M. Reyno, R.A. Modzelewski, M.J. Egorin, K. Erkmen, D.R. Vlock, P. Furmanski and C.S. Johnson, Potentiation by interleukin 1α of cisplatin and carboplatin antitumor activity: schedule-dependent and pharmacokinetic effects in the RIF-1 tumor model, Cancer Res. 54:5380 (1994).
21. M.P. Bevilacqua, J.S. Pober, G.R. Majeau, W. Fiers, R.S. Cotran, and M.A. Gimbron, Recombinant tumor necrosis factor induces precoagulant activity in cultured human vascular endothelium: characterization and comparison with the actions of interleukin 1, Proc. Natl. Acad. Sci. 83:4533 (1986).
22. T.H. Pohlman, K.A. Stanness, P.G. Beatty, H.D. Ochs, and J.M. Harlan, An endothelial cell surface factor(s) induced in vitro by lipopolysaccharide, interleukin 1, and tumor necrosis factor-alpha increases neutrophil adherence by a CDw 18-dependent mechanism, J. Immunol. 136:45 (1986).
23. R.L. Nachman, K.A. Hajjar, R.L. Silverstein, and C.A. Dinarello, Interleukin 1 induces endothelial cell synthesis of plasminogen activator inhibitor, J. Exp. Med. 163:1595 (1986).

24. M. Gowen, and G.R. Mundy, Actions of recombinant interleukin 1, interleukin 2, and interferon-gamma on bone resorption in vitro, J. Immunol. 136:2478 (1986).
25. J.E. Horton, L.G. Raisz, H.A. Simmons, J.J. Oppenheim, and S.E. Mergenhagen, Bone resorbing activity in supernatant fluid from cultured human peripheral blood leukocytes, Science 177:793 (1972).
26. U. Barocos, H.P. Rodemann, C.A. Dinarello, and A.L. Goldberg, Stimulation of muscle protein degradation and prostaglandin E2 release by leukocytic pyrogen (interleukin-1). A mechanisms for the increased degradation of muscle proteins during fever, N. Eng. J. Med. 308:553 (1983).
27. G. Ramadori, J.D. Sipe, C.A. Dinarello, S.B. Mizel, and H.R. Colten, Pretranslational modulation of acute phase hepatic protein synthesis by murine recombinant interleukin 1 (IL-1) and purified human IL-1, J. Exp. Med. 162:930 (1985).
28. J.A. Schmidt, S.B. Mizel, D. Cohen, and I. Green, Interleukin 1, a potential regulator of fibroblast proliferation, J. Immunol. 128:2177 (1982).
29. M. Gowen, D.D. Wood, and R.G. Russell, Stimulation of the proliferation of human bone cells in vitro by human monocyte products with interleukin-1 activity, J. Clin. Invest. 75:1223 (1985).
30. U. Ruggiero, and C. Baglioni, Synergistic anti-proliferative activity of interleukin 1 and tumor necrosis factor, J. Immunol. 138:661 (1987).
31. F. Cozzolino, M. Torcia, D. Aldinucci, M. Ziche, F. Almerigogna, D. Bani, and D.M. Stern, Interleukin 1 is an autocrine regulator of human endothelial cell growth, Proc. Natl. Acad. Sci. 87:6487 (1990).
32. C.J. March, B. Mosley, A. Larsen, D.P. Cerretti, G. Braedt, V. Price, S. Gillis, C.S. Henney, S.R. Kronheim, K. Grapstein, P.J. Conlon, T.P. Hopp and D. Cosman, Cloning, sequence and expression of two distinct human interleukin-1 complementary DNAs, Nature 315:641 (1985).
33. E.V. Gaffney and S.C. Tsai, Lymphocyte-activating and growth-inhibitory activities for several sources of native and recombinant interleukin 1, Cancer Res. 46:3834 (1986).
34. C.A. Dinarello, The interleukin-1 family: 10 years of discovery, FASEB 8:1314 (1994).
35. C.H. Hannum, C.J. Wilcox, W.P. Arend, F.G. Joslin, D.J. Dripps, P.L. Heimdal, L.G. Armes, A. Sommer, S.P. Eisenberg and R.C. Thompson, Interleukin-1 receptor antagonist activity of a human interleukin-1 inhibitor, Nature 343:336 (1990).
36. D.B. Carter, M.R. Deibel, C.J. Dunn, C.S.C. Tomich, et al, Purification, cloning, expression and biological characterization of an interleukin-1 receptor antagonist protein, Nature 344:633 (1990).
37. C.A. Dinarello, Interleukin-1 and interleukin-1 antagonism, Blood 77:1627 (1991).
38. D.J. Dripps, B.J. Brandhuber, R.C. Thompson, S.P. Eisenberg, Interleukin-1 (IL-1) receptor antagonist binds to the 80-kDa IL-1 receptor but does not initiate IL-1 signal transduction, J. Biol. Chem. 266:10331 (1991).

39. J.R. Zucali, C.A. Dinarello, D.J. Oblon, M.A. Gross, L. Anderson, and R.S. Weiner, Interleukin 1 stimulates fibroblasts to produce granulocyte-macrophage colony-stimulating activity and prostaglandin E2, J. Clin. Invest. 77:1857 (1986).
40. R. Neta, M.B. Sztein, J.J. Oppenheim, S. Gillis, and S.D. Douches, The in vivo effects of interleukin 1: I. bone marrow cells are induced by cycle after administration of interleukin 1, J. Immunol. 139:1861 (1987).
41. R. Neta, S. Douches, and J.J. Oppenheim, Interleukin-1 is a radioprotector, J. Immunol. 136:2483 (1986).
42. T.R. Bradley, N. Williams, A.B. Kriegler, J. Fawcett, and G.S. Hodgson, In vivo effects of interleukin-1α on regenerating mouse bone marrow myeloid colony-forming cells after treatment with 5-fluorouracil, Leukemia 3:893 (1989).
43. M.A.S. Moore, R.L. Stolfi, and D.S. Martin, Hematologic effects of interleukin-1β, granulocyte colony-stimulating factor, and granulocyte-macrophage colony-stimulating factor in tumor-bearing mice treated with fluorouracil, J. Natl. Cancer Inst. 82:1031 (1990).
44. J. Moreb and J.R. Zucali, Role of interleukin-1 in 4-hydroperoxy-cyclophosphamide toxicity to bone marrow progenitor cells: a review, Biotherapy 1:273 (1989).
45. M.A.S. Moore, R.L. Stolfi and D.S. Martin, Hematologic effects of interleukin-1β, granulocyte colony-stimulating factor, and granulocyte-macrophage colony-stimulating factor in tumor-bearing mice treated with fluorouracil, J. Natl. Cancer Inst. 82:1031 (1990).
46. V. Rossi, F. Breviario, P. Ghezzi, E. Dejana, and A. Mantovani, Prostacyclin synthesis induced in vascular cells by interleukin-1, Science 229:174 (1985).
47. J.J. Emeis, and T. Kooistra, Interleukin 1 and lipopolysaccharide induce an inhibitor of tissue-type plasminogen activator in vivo and in cultured endothelial cells, J. Exp. Med. 163:1260 (1986).
48. M.P. Bevilacqua, J.S. Pober, M.E. Wheeler, R.S. Cotran, and M.A. Gimbrone, Interleukin 1 acts on cultured human vascular endothelium to increase the adhesion of polymorphonuclear leukocytes, monocytes, and related leukocyte cell lines, J. Clin. Invest. 76:2003 (1985).
49. A.M. Lamas, C.M. Mulroney, and R.P. Schleimer, Studies on the adhesive interaction between purified human eosinophils and cultured vascular endothelial cells, J. Immunol. 140:1500 (1988).
50. E.A. Carswell, L.J. Old, R.L. Kassel, S. Green, N. Fiore, and B. Williamson, An endotoxin-induced serum factor that causes necrosis of tumors, Proc. Natl. Acad. Sci. 72:3666 (1975).
51. L.J. Old, Tumor necrosis factor (TNF), Science 230:630 (1985).
52. K. Tracey, B. Beutler, S. Lowry, J. Merryweather, S. Wolpe, I. Milsark, R. Hariri, T. Fahey, A. Zentella, J. Albert, G. Shired, and A. Cerami, Shock and tissue injury induced by recombinant human cachectin, Science 234:470 (1987).

53. P.P. Nawroth, and D.M. Stern, Modulation of endothelial cell hemostatic properties by tumor necrosis factor, J. Exp. Med. 163:740 (1986).
54. P.P. Nawroth, D. Haudley, G. Matsulda, R. DeWall, H. Gerlach, D. Blohn, and D.M. Stern, Tumor necrosis factor/cachectin-induced intravascular fibrin formation in meth A fibrosarcomas, J. Exp. Med. 168:637 (1988).
55. R.J. North, and E.A. Havell, The antitumor function of tumor necrosis factor (TNF) II. Analysis of the role of endogenous TNF in endotoxin-induced hemorrhagic necrosis and regression of an established sarcoma, J. Exp. Med. 167:1086 (1988).
56. H.F. Dvorak, and I. Gresser, Microvascular injury in pathogenesis of interferon-induced necrosis of subcutaneous tumors in mice, J. Natl. Cancer Inst. 81:497 (1989).
57. F. Belardelli, E. Proietti, J. Ciolli, P. Sestili, G. Carpinelli, M. Divito, A. Ferretti, D. Woodrow, D. Boraschi, and F. Podo, Interleukin-1 beta induces tumor necrosis and early morphologic and metabolic changes in transplantable mouse tumors. Similarities with the anti-tumor effects of tumor necrosis factor alpha or beta, Int. J. Cancer 44:116 (1989).
58. E.A. Havell, W. Fiers, and R.J. North, The antitumor function of tumor necrosis factor (TNF), I. Therapeutic action of TNF against an established murine sarcoma is indirect, immunologically dependent, and limited by severe toxicity, J. Exp. Med. 167:1067 (1988).
59. P.L. Kilian, K.L. Kaffka, D.A. Biondi, J.M. Lipman, W.R. Benjamin, D. Feldman and C.A. Campen, Antiproliferative effect of interleukin-1 on human ovarian carcinoma cell line, Cancer Res. 51:1823 (1991).
60. K. Onozaki, K. Matsushima, B.B. Aggarwal, and J.J. Oppenheim, Human interleukin-1 is a cytocidal factor for several tumor-cell lines, J. Immunol. 135:3962 (1985).
61. L.B. Lachman, C.A. Dinarello, N.D. Llansa, I.J. Fidler, Natural and recombinant human interleukin-1β is cytotoxic for human melanoma cells, J. Immunol. 136:3098 (1986).
62. R.J. North, R.H. Neubauer, J.J.H. Huang, R.C. Newton, and S.E. Loveless, Interleukin 1-induced, T cell-mediated regression of immunogenic murine tumors, J. Exp. Med. 168:2031 (1988).
63. M. Pezzella, M.E. Neville, J.J. Huang, In vivo inhibition of tumor growth of B16 melanoma by recombinant interleukin-1β: I. Tumor inhibition parallels lymphocyte activating factor activity of interleukin-1β proteins, Cytokine 2:357 (1990).
64. S. Nakamura, K. Nakata, S. Kashimoto, H. Yoshida, and M. Yamada, Antitumor effect of recombinant human interleukin 1 alpha against murine syngeneic tumors, Jpn. J. Cancer Res. 77:767 (1988).
65. K. Nakata, S. Kashimoto, H. Yoshida, T. Oku, and S. Nakamura, Augmented anti-tumor effect of recombinant human interleukin-1α by indomethacin, Cancer Res. 48:584 (1988).
66. A. Tewari, W.C. Buhles Jr. and H.F. Starnes Jr, Preliminary report: effect of interleukin-1 on platelet counts, Lancet 336:712 (1990).

67. J.W. Smith, W.J. Urba, B.D. Curti, et al, The toxic and hematologic effects of interleukin-1α administered in a phase I trial to patients with advanced malignancies, J. Clin. Oncol. 10:1141 (1992).
68. J. Crown, A. Jakubowski, N. Kemeny, et al, A phase I trial of recombinant human interleukin-1β alone and in combination with myelosuppressive doses of 5-fluorouracil in patients with gastrointestinal cancer, Blood 78:1420 (1991).
69. J.W. Smith II, D.L. Longo, W.G. Alvord, J.E. Janik, et al, The effects of treatment with interleukin-1α on platelet recovery after high-dose carboplatin, N. Engl. J. Med. 328:756
70. J. Nemunaitis, F.R. Appelbaum, K. Lilleby, W.C. Buhles, C. Rosenfeld, Z.R. Zeigler, R.K. Shadduck, J.W. Singer, W. Meyer and C.D. Buckner, Phase I study of recombinant interleukin-1β in patients undergoing autologous bone marrow transplant for acute myelogenous leukemia, Blood 83:3473 (1994).
71. D.R. Vlock, C.S. Johnson, M.-J. Chang, L.M. Reyno, K. Erkmen, M.J. Egorin, T. Logan and C. McCauley, Phase I trial of interleukin-1 alpha (IL-1) and carboplatin (CBDCA), Proc. Amer. Assoc. Cancer Res. 34:296 (1993).
72. T.F. Logan, H. Bishop, M.A. Mintun, Y. Choi, D. Sashin, M.A. Virji, T. Billiar, D.L. Trump, D. Smith, J.M. Kirkwood, D.R. Vlock, M.-J. Chang, M.J. Egorin and C.S. Johnson, Phase I trial of interleukin-1α and carboplatin in patients with metastatic disease to the lung: effects on tumor blood flow evaluated by positron emission tomography, Proc. Amer. Assoc. Cancer Res. 35:198 (1994).

OVERVIEW OF TUMOR-INHIBITING NON-PLATINUM COMPOUNDS

Bernhard K. Keppler* and Ellen A. Vogel

Anorganisch-Chemisches Institut der Universität Heidelberg,
Im Neuenheimer Feld 270, D-69120 Heidelberg, Germany

INTRODUCTION

Cancer mortality in the western world is still on the increase today. In the last few decades, platinum and other metal coordination compounds have been the subject of numerous investigations in the field of cancer chemotherapy. Cisplatin and its derivatives feature prominently here (Figure 1) [1,2,3,4]. One of the reasons why tumor-inhibiting non-platinum compounds are receiving increasing attention is the fact that cisplatin and other platinum complexes have a relatively limited spectrum of indication. Cisplatin shows its best activity in testicular carcinomas and has good activity in ovarian carcinomas, tumors of the head and neck and bladder tumors[2,3,4]. It is not, or only insufficiently, active against the tumors that account for the major share of cancer mortality today, such as tumors of the lung and the gastrointestine. The synthesis of new compounds which are active in these platinum-resistant tumors must therefore be one of the aims in inorganic chemistry.

There are three ways to develop new tumor-inhibiting metal complexes: synthesis of classical and non-classical derivatives of cisplatin, synthesis of tumor-inhibiting non-platinum complexes, and synthesis of platinum complexes linked to carrier systems which are able to accumulate the cytotoxic drug in certain organs and tissues[5].

This contribution will give an overview of the state of development of tumor-inhibiting non-platinum compounds.

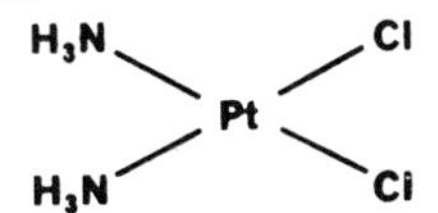

Figure 1. Structure of cisplatin.

Platinum and Other Metal Coordination Compounds in Cancer Chemotherapy 2
Edited by H.M. Pinedo and J.H. Schornagel, Plenum Press, New York, 1996

NON-PLATINUM COMPOUNDS IN CLINICAL STUDIES

In contrast to the tremendous number of compounds which have been investigated in the preclinical stage, only very few compounds have reached the clinical stage of development. These include gallium salts like gallium chloride or gallium nitrate[6,7,8], germanium compounds like carboxyethylgermaniumsesquioxide (germanium-132) and spirogermanium[9,10,11,12], and two titanium compounds, budotitane[13,14] and titanocene dichloride[15] (Figure 2).

Gallium salts
e.g. $Ga(NO_3)_3$
Galliumnitrate

$[(GeCH_2CH_2COOH)_2O_3]_n$
Germanium-132
Carboxyethylgermaniumsesquioxide

CH_3CH_2, Ge, CH_3CH_2, $N-CH_2CH_2CH_2N(CH_3)_2 \cdot 2\,HCl$

Spirogermanium
N-(3-Dimethylaminopropyl)-2-aza-8,8-diethyl-8-germa-spiro-4,5-decanedihydrochloride

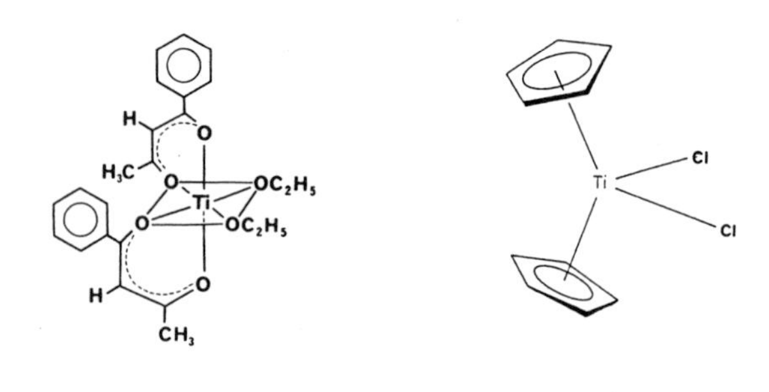

Budotitane (INN)
cis-Diethoxybis(1-phenylbutane-1,-3-dionato)titanium(IV)

Titanocene dichloride
Bis(cyclopentadienyl)-titanium(IV)dichloride

Figure 2. Non-platinum complexes in clinical trials.

In preclinical studies, the gallium salts had shown systemic activity mainly against the subcutaneously transplanted Walker 256 carcinosarcoma. ^{67}Gallium salts are in use as radiopharmaceutical agents for the scintigraphic detection of lymphomas, bone tumors and metastases. In clinical studies, the gallium salts showed nephrotoxicity and gastrointestinal toxicity. They are active against Hodgkin and non-Hodgkin lymphomas and against tumor-induced hypercalcaemias. This spectrum of indication is not very promising, but recently it was found that there is a synergistic effect between gallium and cisplatin in the treatment of lung cancer patients[16,17]. These results have revived interest in gallium complexes as anticancer agents.

It is relatively difficult to reach sufficient gallium plasma levels using gallium chloride, which is rapidly excreted. In this connection, we have synthesized a number of new gallium derivatives with the aim to obtain compounds which show a higher accumulation in plasma and organ tissues. The pharmacokinetic investigations showed that the compound tris(8-chinolinolato)gallium(III), KP 46, is the most promising representative as it reaches the highest plasma and tissue levels[18] (Figure 3).

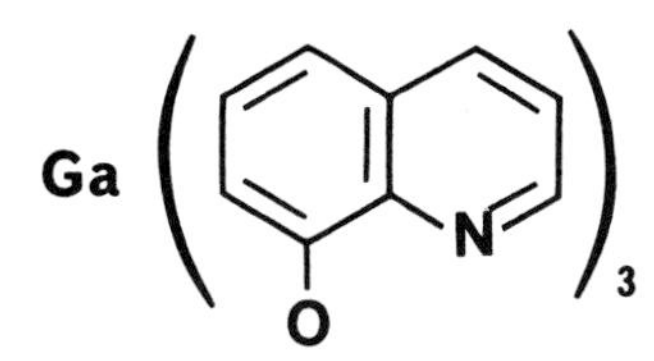

Figure 3. Structure of tris(8-chinolinolato)gallium(III), KP 46.

The two germanium compounds, germanium-132 and spirogermanium, are active in a number of preclinical tumor models and have been tested in clinical trials, but the tumor-inhibiting activity that was found turned out to be not sufficient for further clinical use[9,10,11,12].

The two titanium compounds which are under clinical studies are titanocene dichloride and budotitane. The development of budotitane is well described in the literature[13,14,19,20]. The antitumor activity of this compound strongly depends on the phenyl rings in the outer sphere of the molecule. If these phenyl rings are replaced by methyl groups, the activity totally disappears. This structure-activity relation was confirmed in more than 200 bis(β-diketonato) complexes.

These bis(β-diketonato) complexes can occur as *cis*- and *trans*- isomers. The *cis*- configurated isomers are more stable. If the β-diketonato ligands of the *cis*- isomers are substituted asymmetrically, there are, in addition, three *cis*- isomers, which each form enantiomers (Figure 4). This could be confirmed in NMR spectroscopic investigations, which show altogether four signals of the methyl groups or the methino ring protons of the three *cis*- isomers.

The percentage of the different isomers in solution could be calculated in two-dimensional NMR studies. The *cis-cis-cis* isomer accounts for about 60 %, and the other two isomers with the *trans*- standing phenyl rings and the neighbouring phenyl rings account for about 20 % each. This could also be confirmed by molecular mechanics investigations[21].

When budotitane is applied in humans, it always takes the form of this mixture of isomers because the energy difference between the isomers is only about 40 kJ/mole.

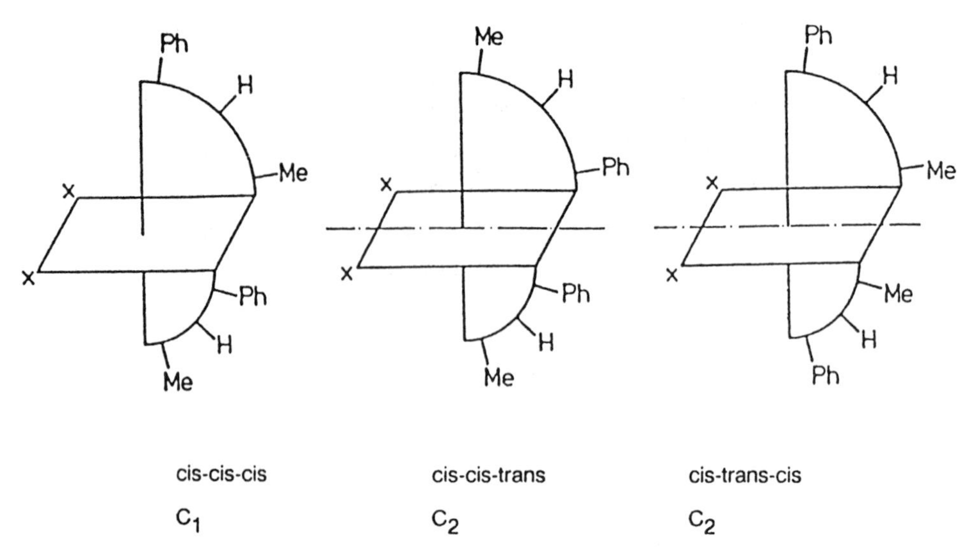

Figure 4. Number of isomers within the *cis-* form of asymmetrically substituted bis(β-diketonato) metal complexes. This example: a benzoylacetonato complex.

So far it has not been possible to crystallize the active drug budotitane in a quality that would permit X-ray crystallographic investigations. Figure 5 shows the X-ray crystal structure of one derivative with two additional phenyl groups in the molecule. Three of these four phenyl rings are nearly coplanar to the metal enolate ring system, and so it can be assumed that in the case of budotitane there is also coplanarity between the phenyl rings in the metal enolate system.

Budotitane has been extensively investigated in numerous preclinical studies[19,20]. The clinical studies with this compound started with a single-dose phase Ia study. Here, doses of 1, 2, 4, 6, 9, 14 and 21 mg/kg were applied in three patients per dose level. The maximum tolerated dose ranges between 14 and 21 mg/kg.

Side-effects could first be detected at the 9 mg/kg level. The patients showed an impairment of the sense of taste, which was reversible within 24 hours. An increase in liver enzymes could be observed at a dose of 14 mg/kg, and dose-limiting nephrotoxicity was seen at 21 mg/kg. There was no hematological toxicity during this study.

The titanium levels in the plasma reached about 2 to 4 μg/g at the 14 mg/kg level and up to 12 μg/g at the 21 mg/kg level. Significant titanium levels could still be detected seven days after the infusion.

The treatment with budotitane requires a chronic application to achieve the best effects, so a phase Ib study was started with a four-week application of this compound. This phase Ib study could be completed only recently. Budotitane was administered as a 2 to 3

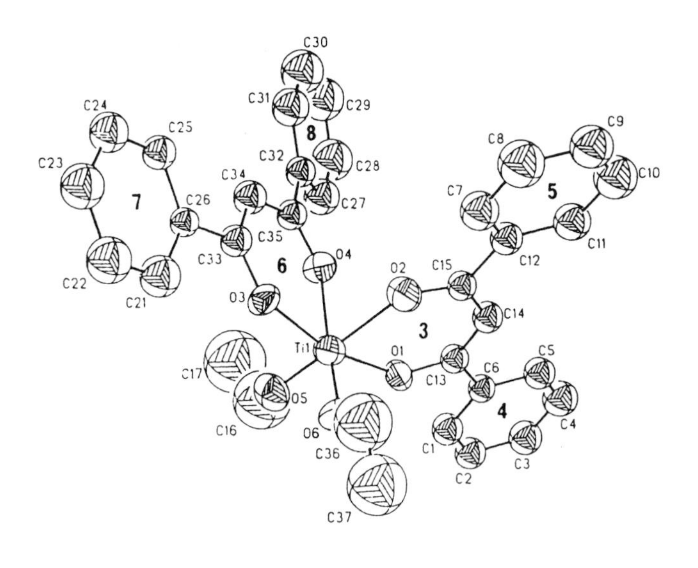

Selected bond distances (Å) and angles (°):

Ti(1)-O(1)	1.999 (9)	O(1)-Ti(1)-O(2)	82.6 (4)
Ti(1)-O(2)	2.057 (10)	O(2)-Ti(1)-O(4)	82.3 (4)
Ti(1)-O(3)	1.996 (9)	O(3)-Ti(1)-O(4)	81.9 (4)
Ti(1)-O(4)	2.074 (11)	O(3)-Ti(1)-O(5)	98.2 (4)
Ti(1)-O(5)	1.792 (11)	O(5)-Ti(1)-O(6)	98.4 (5)
Ti(1)-O(6)	1.793 (12)	O(1)-Ti(1)-O(6)	100.1 (5)

The two phenyl rings 4 and 5 are coplanar to the metal enolate ring 3, the phenyl ring 7 is coplanar to plane 6, and plane 8 is drawn out.

Figure 5. X-ray crystal structure of *cis*-Diethoxybis(1,3-diphenylpropane-1,3-dionato)titanium(IV), $Ti(bzbz)_2(OEt)_2$.

hour's infusion in 100 ml of isotonic mannitol solution intravenously twice per week over four weeks. The dose levels were 100, 120, 150, 180, and 230 mg/m^2. Three to five patients were treated at each dose level. There was no leukopenia, no thrombocytopenia, and no nephrotoxicity. Side-effects included an infrequent elevation of bilirubin and alkaline phosphatase. 15 out of 18 patients reported loss of taste on the day of infusion, which resolved within 24 hours. At 230 mg/m^2, two out of four patients developed polytope ventricular extrasystoles and non-sustained ventricular tachycardia. The most important results of this study are that the MTD is at 230 mg/m^2 and that the recommended dose for clinical phase II studies is 180 mg/m^2 twice per week over four weeks[22].

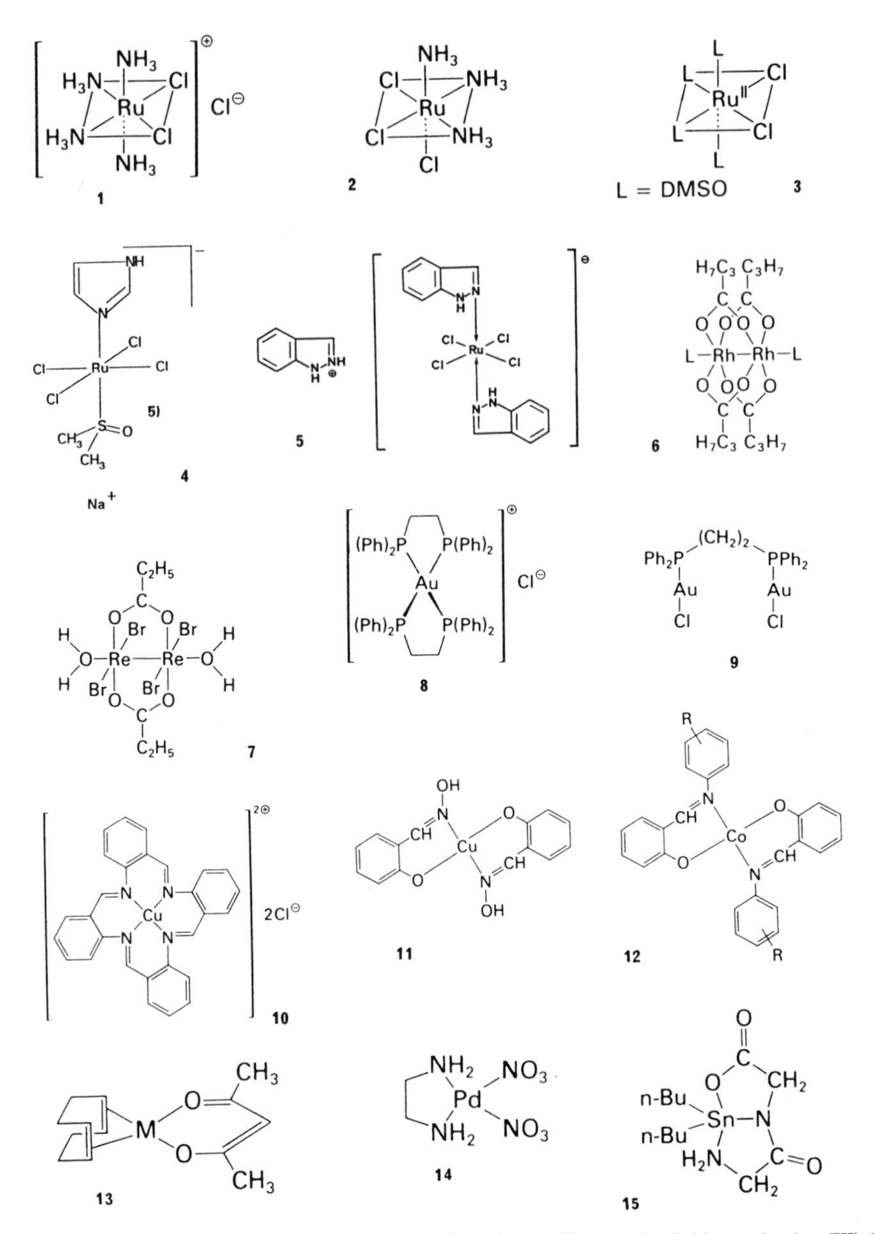

Figure 6. Non-platinum compounds in preclinical studies. 1 = *cis*-Tetramminedichlororuthenium(III)chloride, 2 = *fac*-Trisamminetrichlororuthenium(III), 3 = *cis*-Dichlorotetrakis(dimethylsulfoxide)ruthenium(II), 4 = Sodium *trans*-(dimethylsulfoxide)(imidazole)tetrachlororuthenate(III), 5 = *trans*-Indazolium-tetrachlorobis-(indazole)ruthenate(III), 6 = Tetra-μ-butyratodirhodium(II), 7 = Bis-μ-propionatodiaquatetrabromorhenium(II), 8 = Bis[1,2-bis(diphenylphosphino)ethane]gold(I)chloride, 9 = 1,2-Bis(diphenylphosphino)ethane-bis[gold(I)chloride], 10 = [Tetrabenzo(b,f,j,n)(1,5,9,13)tetrazacyclohexadecinecopper(II)]dichloride, 11 = Bis(1,2-hydroxybenzaldoximato)copper(II) complexes, 12 = Bis(2-hydroxybenzanilinato]cobalt(II) complexes, 13 = Pentane-2,4-dionatocycloocta-1,5-diene complexes of rhodium(I) and iridium(I), 14 = 1,2-Diaminoethanedinitratopalladium(II), 15 = Di-n-butyltin(IV)glycylglycinate.

This has been an overview of the most important data on non-platinum compounds in clinical trials. In the following, the non-platinum complexes under preclinical studies will be discussed.

NON-PLATINUM COMPOUNDS IN PRECLINICAL STUDIES

There is a tremendous number of metal complexes which have been investigated in one or the other model *in vivo* or *in vitro*. The most important compounds are summarized in Figure 6. They are different ruthenium compounds, bridged carboxylato complexes of

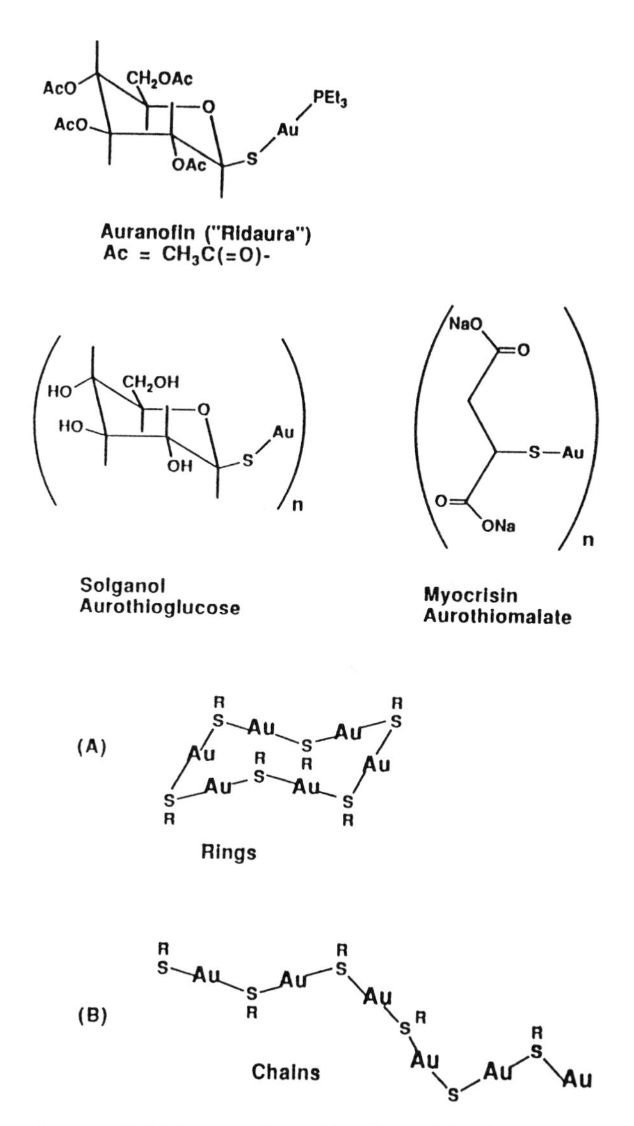

Figure 7. Structures of some gold thiolate complexes clinically used for the treatment of rheumatoid arthritis[25].

rhodium and rhenium, gold compounds, various copper complexes, Schiff base complexes of cobalt, cyclooctadien complexes, palladium complexes similar to platinum, and tin complexes.

Gold is known for its use in the treatment of primary chronic polyarthritis[23]. Auranofin is a gold complex which can be given orally because it is well absorbed from the gastrointestine[23,24]. This compound and its derivatives were also investigated for their antitumor activity. The experiments with the gold thiolates and phosphine gold thiolates shown in Figure 7 suggest that the phosphine ligand is necessary for cytotoxic activity, but all these compounds were absolutely inactive *in vivo*[25]. The diphosphine complexes shown in Figure 8 exhibit antitumor activity against a wider range of experimental tumors. Examples of the ligands used are shown at the bottom of Figure 8[26].

Bridged Linear Digold Diphosphine

Annular Digold Bisdiphosphine

Tetrahedral Gold Bisdiphosphine

General structures of cytotoxic gold(I) diphosphine complexes.

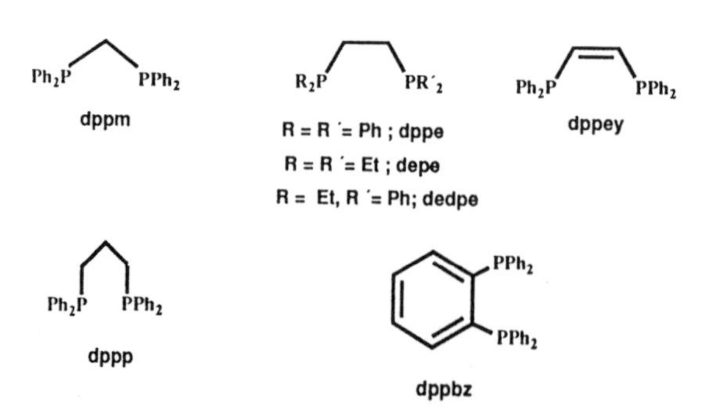

Figure 8. General structures of cytotoxic gold(I) diphosphine complexes.

The most important complexes are the gold(dppe) compounds[23,24]. However, their antitumor activity decreases when the drug is given intravenously instead of intraperitoneally. Besides, the activity is relatively poor in subcutaneously growing tumors. Another disadvantage of these compounds is the unexpectedly high cardiac, hepatic and vascular toxicity. The compounds are able to uncouple the mitochondrial oxidative phosphorylation by increasing the permeability of the mitochondrial membrane to cations, with collapse of the mitochondrial membrane potential[25].

There are still many areas of gold chemistry that are poorly explored and hence there is much scope for the discovery of novel gold antitumor agents.

In the field of tumor-inhibiting tin compounds[27], recent investigations have centered for example on novel triphenyltin carboxylates[28]. The structure is shown in Figure 9. The compounds reached interesting, low, ID values compared to clinically established antitumor drugs. The antitumor activity of some di-n-butyltin-difluorobenzoates was also investigated. They also exhibit interesting activity[29].

Compound	Cell lines	
	MCF-7	WiDr
2,3-$F_2C_6H_3CO_2SnPh3$	17	24
2,4,5-$(CH_3O)_3C_6H_2CO_2SnPh_3$	16	15
2-SCH_3-3-$C_5NH_3CO_2SnPh_3$	16	18
Cisplatin	850	624
Etoposide	187	62
Doxorubicin	63	31
Mitomycin C	3	1

Inhibition doses ID_{50} in ng/mL against two human tumour cell lines, MCF-7 and WiDr, obtained for three compounds in comparison with clinically established antitumour drugs[28].

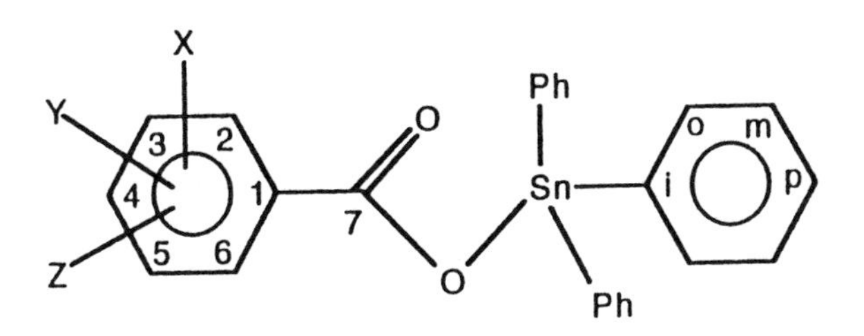

Figure 9. Antitumor activity of novel triphenyltin carboxylates.

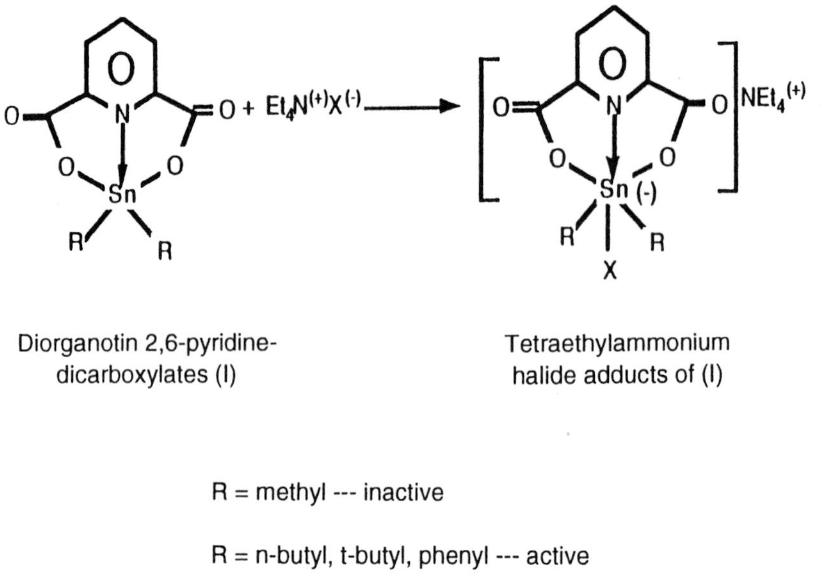

Diorganotin 2,6-pyridine-
dicarboxylates (I)

Tetraethylammonium
halide adducts of (I)

R = methyl --- inactive

R = n-butyl, t-butyl, phenyl --- active

Figure 10. Diorganotinpyridinedicarboxylates[29].

Among tumor-inhibiting non-platinum compounds, ruthenium complexes have been extensively investigated. The kinetics of water-exchange reactions shows that ruthenium in the oxidation stages +III and +II occupies a position near to platinum. This may have been an impetus to carry out further research into the development of ruthenium compounds as antitumor drugs.

The structures of two ruthenium complexes with known tumor-inhibiting activity have been shown in Figure 6. They are a *cis*-configurated tetraammine complex and a facially configurated trisammine complex[30,31]. Ruthenium DMSO complexes have also been found to exhibit promising antitumor activity[32,33]. The most important complexes are shown in Figure 11[33].

There are many indications that ruthenium interacts with DNA. An example is the adduct of a pentaammine ruthenium fragment with methyl guanine, which has recently been described in the literature[31]. It was also possible to isolate a GMP derivative of a ruthenium DMSO complex[32].

Another field of investigation is concerned with the diorganotinpyridinedicarboxylates, which can be transformed into their tetraethylammonium halide adducts[29], which are better soluble in water (Figure 10). With these compounds, an interesting activity could be observed *in vitro*. The problem is that there is virtually no tin compound which exhibits real promising antitumor activity *in vivo*. It is however a general problem in the development of

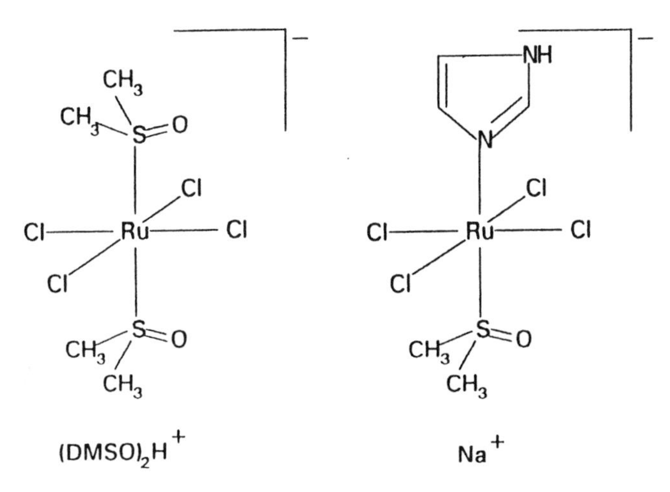

Figure 11. Structures of $(DMSO)_2H[$*trans*-$RuCl_4(DMSO)_2]$ and Na[*trans*-$RuCl_4$(DMSO)(Im)].

new tumor-inhibiting metal complexes that many such compounds are highly cytotoxic *in vitro* and show no effect or only toxicity *in vivo*.

Two ruthenium complexes which are characterized by *trans*- standing nitrogen heterocycles are highly active against a variety of experimental tumor systems[34,35] (Figure 12). The decomposition kinetics of the two compounds was investigated in HPLC experiments. The half-life of the imidazole complex in water is about 12 hours and that of the indazole complex is about 70 hours. This is slow enough to apply the compounds nearly undecomposed in the patients, and the complexes are reactive enough to interact with the biological target. A compound which is more reactive will be decomposed before it can be applied in the patient, and compounds which are significantly more stable might be excreted undecomposed without any interaction with the biological medium, as it is known of, for example, gadolinium complexes which are used as NMR imaging agents.

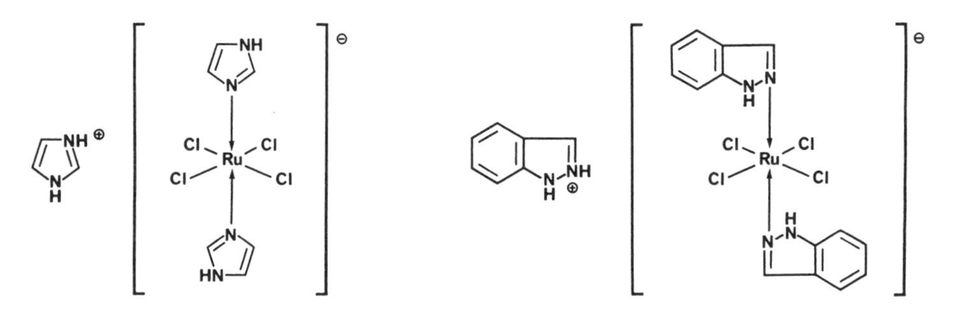

trans-Imidazolium-tetrachlorobis-(imidazole)ruthenate(III), *trans*-HIm[$RuCl_4(im)_2$], KP 418

trans-Indazolium-tetrachlorobis-(indazole)ruthenate(III), *trans*-HInd[$RuCl_4(ind)_2$], KP 1019

Figure 12. Structures of *trans*-Imidazolium-tetrachlorobis(imidazole)ruthenate(III), trans-HIm[$RuCl_4(im)_2$], and *trans*-Indazolium-tetrachlorobis(indazole)ruthenate(III), *trans*-HInd[$RuCl_4(ind)_2$].

The hydrolysis reactions have been investigated in detail[35,36,37]. In the case of the imidazole complex, a neutral aqua complex is formed first, then a diaqua complex, and after a longer time and in low concentrations, also a trisimidazole complex. In the case of the indazole complex, one of the neutral aqua complexes which are formed in the first reaction step could be crystallized and the X-ray crystal structure could be resolved.

The mode of action of these ruthenium complexes may be based on a transferrin-mediated transport system. The complex is injected into the blood, forms a transferrin complex and is then transported, more or less selectively, to the tumor cell, which exhibits a high status of transferrin receptors. After entering the cell, the ruthenium is released and may interact directly, or after reduction to Ru(II), with biomolecules like DNA.

This hypothesis was verified in different experiments[35]. The binding of the ruthenium indazole compound to plasma proteins was investigated in ultrafiltration experiments. After only three minutes, no ruthenium species could be detected in the ultrafiltrate. This means that the whole of the ruthenium complex was bound to the plasma proteins after this time.

HPLC experiments served to elucidate the binding behaviour of the ruthenium indazole compound towards transferrin, and it was found that the complex is bound to transferrin within five minutes.

In CD spectroscopic investigations, a stepwise titration of transferrin with the ruthenium indazole complex was carried out. The spectrum showed a characteristic change, up to a ratio of 2 ruthenium atoms per one transferrin.

All these findings suggest that the ruthenium compound binds to the iron-binding sites of transferrin. It was demonstrated that it fits into the N-terminal binding site of apolactoferrin, a protein which is very similar to transferrrin. It was possible to crystallize the indazole ruthenium adduct of apolactoferrin and carry out X-ray crystallographic investigations[35,38] (Figure 13).

Studies into the release of the ruthenium complex from transferrin under physiological conditions were based on the mechanism of iron transport via this protein: after entering the cell, the iron is released in endosomes at a pH of about 5. In addition, a physiological chelating agent may play a role. So the ruthenium transferrin complex was treated, at a low pH, with citrate as physiological chelating agent, and after two hours, the main part of ruthenium could be released from the transferrin molecule. This compound now is able to bind to DNA. By means of ICP-AES spectroscopy, the binding behaviour of both *trans*-HIm[$RuCl_4(im)_2$] and *trans*-HInd[$RuCl_4(ind)_2$] was investigated in comparison to cis-platin (Figure 14). An increasing affinity towards DNA with time could be demonstrated.

The most interesting antitumor activity results with the two ruthenium compounds were obtained in the acetoxymethylmethylnitrosamine (AMMN)-induced autochthonous colorectal tumors of the rat[5,35,39]. These tumors resemble the human colon tumors both macroscopically and microscopically. They show the same behaviour towards clinically established antitumor drugs such as cisplatin and 5-fluorouracil as the corresponding human tumors. It is therefore a highly predictive model for the clinical situation. Cisplatin is inactive in this tumor model, it even stimulates tumor growth. The same behaviour can

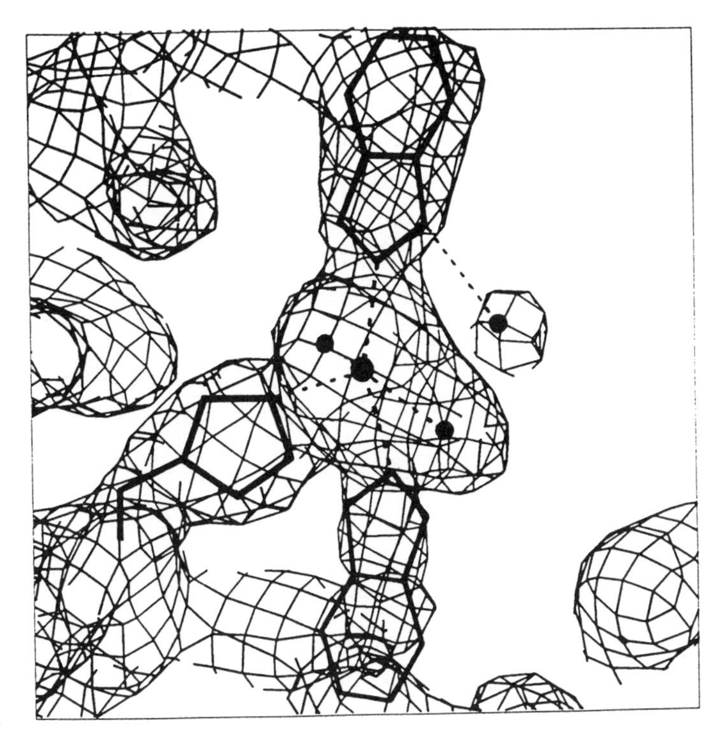

Figure 13. Difference electron density for *trans*-HInd[$RuCl_4(ind)_2$] bound in the N-terminal site of human apolactoferin, showing that the two indazole ligands are retained. The Ru atom binds to His 253.

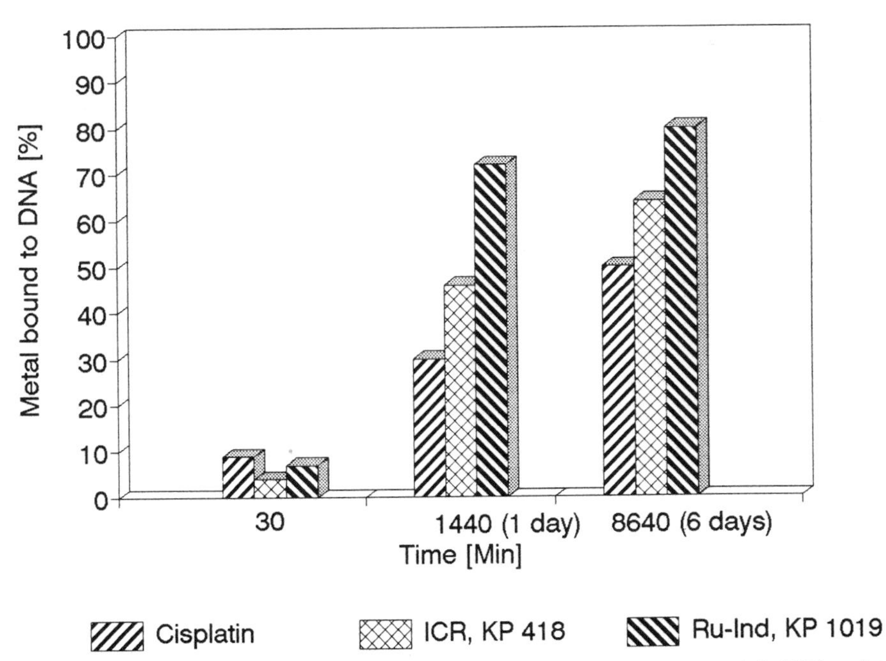

Figure 14. Binding behaviour towards salmon testes DNA of *trans*-HIm[$RuCl_4(im)_2$] (ICR) and *trans*-HInd[$RuCl_4(ind)_2$] (Ru-Ind), in 10 mM NaCl, in comparison to cisplatin.

be seen under clinical conditions. 5-Fluorouracil shows a moderate effect, with a reduction of the tumor volume to about 40 %. This moderate effect can also be seen in the clinic. The activity of *trans*-HIm[$RuCl_4(im)_2$] and *trans*-HInd[$RuCl_4(ind)_2$] in this tumor model is unexpectedly high[5,35,39,40]. The tumor volume is reduced to 5 % or 10 %. The therapy with the indazole compound could be carried out without any toxicity. In one of the groups which were treated with the indazole compound, one third of the animals was cured. This is one step towards a new and nearly untoxic treatment of colorectal tumors, which are responsible for a major share of cancer mortality today.

CONCLUSION

This overview of tumor-inhibiting non-platinum compounds has given an outline of research into metal coordination compounds as anticancer agents. A few of these compounds are under clinical investigations today. These are gallium complexes and the two titanium complexes titanocene dichloride and budotitane. Many of these compounds are still at a preclinical stage of development. Examples include gold, tin, and ruthenium complexes. It would go beyond the scope of this article to discuss all preclinically examined non-platinum complexes, but their number and the research into this field give rise to the hope that more such compounds will qualify for clinical trials and become efficient anticancer agents that will help to combat malignant tumors, which today are the second most frequent cause of death after the diseases of the circulatory system.

ACKNOWLEDGEMENTS

This work has been supported by the Deutsche Forschungsgemeinschaft, DFG, Bonn, Germany. The support of the Deutsche Krebshilfe, Dr. Mildred-Scheel-Stiftung für Krebsforschung, Bonn, Germany, is gratefully acknowledged, as well as the support of the European Community through the HCM and COST programmes.

REFERENCES

1. B. Rosenberg, Platinum complexes for the treatment of cancer, *Interdiscipl. Science Rev.* 3, 2, 134 (1978).
2. A.W. Prestayko, S.T. Crooke, and S.K. Carter, eds., "Cisplatin - Current status and new developments", Academic Press, New York (1980).
3. D.C.H. McBrien and T.F. Slater, eds., "Biochemical mechanisms of platinum antitumour drugs", IRL Press, Oxford (1986).
4. St.B. Howell, ed., "Platinum and other metal coordination compounds in cancer chemotherapy", Plenum Press, New York (1991).
5. B.K. Keppler, M.R. Berger, Th. Klenner, and M.E. Heim, Metal complexes as antitumour agents, *Adv. Drug Res.* 19:243 (1990).

6. R. Sephton and S. De Abrew, Mechanism of gallium uptake in tumours, *in:* "Metal Ions in Biology and Medicine", Ph. Collery, L.A. Poirier, M. Manfait, and J.C. Etienne, eds., John Libbey Eurotext, Paris (1990).
7. J.L. Domingo and J. Corbella, A review of the pharmacological and toxicological properties of gallium, *in:* "Metal Ions in Biology and Medicine", Ph. Collery, L.A. Poirier, M. Manfait, and J.C. Etienne, eds., John Libbey Eurotext, Paris (1990).
8. Ph. Collery and C. Pechery, Clinical experience with tumor-inhibiting gallium complexes, *in:* "Metal Complexes in Cancer Chemotherapy", B.K. Keppler, ed., VCH Weinheim (1993).
9. D. Lekim and L. Samochowiec, eds., "Germanium in biologischen Systemen", Semmelweis Verlag, Hoya (1985).
10. M. Slavik, O. Blanc, and J. Davis, Spirogermanium: A new investigational drug of novel structure and lack of bone marrow toxicity, *Invest. New Drugs* 1:225 (1983).
11. K. Miyao, T. Onishi, K. Asai, S. Tomizawa, and F. Suzuki, Toxicology and phase I studies on a novel organogermanium compound, Ge-132, *in:* "Current Chemotherapy and Infectious Disease", J.D. Nelson, C. Grassi, eds., The American Society for Microbiology, Washington DC (1980).
12. S.G. Ward and R.C. Taylor, Anti-tumor activity of the main-group metallic elements: aluminum, gallium,, indium, thallium, germanium, lead, antimony and bismuth, *in:* "Metal-Based Anti-Tumour Drugs", M.F. Gielen, ed., Freund Publishing House, London (1988).
13. B.K. Keppler, C. Friesen, H.G. Moritz, H. Vongerichten, and E. Vogel, Tumor-inhibiting bis(β-diketonato) metal complexes. Budotitane, *cis*-diethoxybis(1-phenylbutane-1,3-dionato)titanium(IV), the first transition metal complexes after platinum complexes to have qualified for clinical trials, *Structure & Bonding* 78:97 (1991).
14. B.K. Keppler, C. Friesen, H. Vongerichten, and E. Vogel, Budotitane, a new tumor-inhibiting titanium compound: preclinical and clinical development, *in:* "Metal Complexes in Cancer Chemotherapy", B.K. Keppler, ed., VCH Weinheim (1993).
15. P. Köpf-Maier, Antitumor bis(cyclopentadienyl)metal complexes, *in:* "Metal Complexes in Cancer Chemotherapy", B.K. Keppler, ed., VCH Weinheim (1993).
16. Ph. Collery, M. Morel, H. Millart, B. Desoize, C. Cossart, D. Perdu, H. Vallerand, J.C. Bouana, C. Pechery, J.C. Etienne, H. Choisy, and J.M. Dubois de Montreynaud, Oral administration of gallium in conjunction with platinum in lung cancer treatment, *in:* "Metal Ions in Biology and Medicine", Ph. Collery, L.A. Poirier, M. Manfait, and J.C. Etienne, eds., John Libbey Eurotext, Paris (1990).
17. Ph. Collery, H. Vallerand, A. Prevost, D. Milosevic, M. Morel, J.P. Dubois, B. Desoize, C. Pechery, J.M. Dubois de Montreynaud, H. Millart, and H. Choisy, Therapeutic index of gallium, orally administered, as chloride, in combination with cisplatinum and etoposide in lung cancer patients, *in:* "Metal Ions in Biology and Medicine", J. Anastassopoulou, Ph. Collery, J.C. Etienne, and Th. Theophanides, eds., John Libbey Eurotext, Paris (1992).
18. Ph. Collery, H. Millart, C. Pechery, F. Kratz, and B.K. Keppler, New gallium complexes for a cisplatin combination therapy, *in:* "Metal Ions in Biology and Medicine", J. Anastassopoulou, Ph. Collery, J.C. Etienne, and Th. Theophanides, eds., John Libbey Eurotext, Paris (1992).
19. B.K. Keppler and D. Schmähl, Preclinical Evaluation of Dichlorobis(1-phenylbutane-1,3-dionato)-titanium(IV) and budotitane, *Arzneim.-Forsch./Drug Res.* 36 (II), 12, 1822 (1986).
20. B.K. Keppler, H. Bischoff, M.R. Berger, M.E. Heim, G. Reznik, and D. Schmähl, Preclinical development and first clinical studies of budotitane, *in:* "Platinum and other metal coordination compounds in cancer chemotherapy", M. Nicolini, ed., Martinus Nijhoff Publishing, Boston (1988).
21. P. Comba, H. Jakob, B. Nuber, and B.K. Keppler, Solution structures and isomer distributions of bis(β-diketonato) complexes of titanium(IV) and cobalt(III), *Inorg. Chem.* 33:3396 (1994).

22. T. Schilling, B.K. Keppler, M.E. Heim, K. Burk, J. Rastetter, and A.-R. Hanauske, Phase I clinical and pharmacokinetic trial of the new metal complex budotitane, *Onkologie* 16:S1, Karger, Basel (1993).
23. B.M. Sutton and R.G. Franz, eds., "Bioinorganic Chemistry of Gold Coordination Compounds", Smith Kline & French Laboratories, Philadelphia (1983).
24. A.J. Lewis and D.T. Walz, Immunopharmacology of gold, *Progr. Medicinal Chem.* 19, G.P. Ellis and G.B. West, eds., Elsevier Biomedical Press, Lausanne, 1 (1982).
25. O.M. Ni Dhubhghaill and P.J. Sadler, Gold complexes in cancer chemotherapy, *in:* "Metal Complexes in Cancer Chemotherapy", B.K. Keppler, ed., VCH Weinheim (1993).
26. P.J. Sadler and R.E. Sue, The chemistry of gold drugs, *Metal-Based Drugs* 1, 2-3, 107 (1994).
27. M. Gielen, Tin-based antitumour drugs, in: "Metal Ions in Biology and Medicine", Ph. Collery, ed., John Libbey Eurotext, Montrouge (1994).
28. M. Gielen, A. El Khloufi, M. Biesemans, A. Bouhdid, D. de Vos, B. Mahieu, and R. Willem, Synthesis, characterization and high *in vitro* antitumour activity of novel triphenyltin carboxylates, *Metal-Based Drugs* 1, 4, 305 (1994).
29. M. Gielen, Tin-based antitumour drugs, *Metal-Based Drugs* 1, 2-3, 213 (1994).
30. M.J. Clarke, Oncological implications of the chemistry of ruthenium, *in:* "Metal Ions in Biological Systems", 11: Metal Complexes as Anticancer Agents, H. Sigel, ed., Marcel Dekker, New York (1980).
31. M.J. Clarke, Ruthenium complexes: Potential roles in anti-cancer pharmaceuticals, *in:* "Metal Complexes in Cancer Chemotherapy", B.K. Keppler, ed., VCH Weinheim (1993).
32. G. Mestroni, E. Alessio, G. Sava, S. Pacor, and M. Coluccia, The development of tumor-inhibiting ruthenium dimethylsulfoxide complexes, *in:* "Metal Complexes in Cancer Chemotherapy", B.K. Keppler, ed., VCH Weinheim (1993).
33. G. Mestroni, E. Alessio, G. Sava, S. Pacor, M. Coluccia, and A. Boccarelli, Water-soluble ruthenium-(III)-dimethyl sulfoxide complexes: chemical behaviour and pharmaceutical properties, *Metal-Based Drugs* 1, 1, 43 (1994).
34. B.K. Keppler, M. Henn, U.M. Juhl, M.R. Berger, R. Niebl, and F.E. Wagner, New ruthenium complexes for the treatment of cancer, *Progr. Clin. Biochem. Med.* 10:41 (1989).
35. B.K. Keppler, K.-G. Lipponer, B. Stenzel, and F. Kratz, New tumor-inhibiting ruthenium complexes, *in:* "Metal Complexes in Cancer Chemotherapy", B.K. Keppler, ed., VCH Weinheim (1993).
36. O.M. Ni Dhubhghaill, W.R. Hagen, B.K. Keppler, K.-G. Lipponer, and P.J. Sadler, Aquation of the anticancer complex *trans*-$[RuCl_4(Him)_2]^-$ (Him = imidazole), *J. Chem. Soc. Dalton Trans.*, 3305, (1994).
37. J. Chatlas, R. van Eldik, and B.K. Keppler, Spontaneous aquation reactions of a promising tumor inhibitor *trans*-imidazolium-tetrachlorobis(imidazole)ruthenium(III), *trans*-$HIm[RuCl_4(Im)_2]$, *Inorg. Chim. Acta* 233:59 (1995).
38. F. Kratz, B.K. Keppler, L. Messori, C. Smith, and E.N. Baker, Protein-binding properties of two antitumour Ru(III) complexes to human apotransferrin and apolactoferrin, *Metal-Based Drugs* 1, 2-3, 169 (1994).
39. M.R. Berger, M.H. Seelig, and A. Galeano, Metal complexes with specific activity against colorectal tumors: evaluation of a tumor model close to the clinical situation, *in:* "Metal Complexes in Cancer Chemotherapy", B.K. Keppler, ed., VCH Weinheim (1993).
40. M.H. Seelig, M.R. Berger, and B.K. Keppler, Antineoplastic activity of three ruthenium derivatives against chemically induced colorectal carcinoma in rats, *J. Cancer Res. Clin. Oncol.* 118:195 (1992).

THE CISPLATIN-INDUCED CELLULAR INJURY RESPONSE

Stephen B. Howell, Dennis P. Gately, Randolph D. Christen, and Gerrit Los

Department of Medicine and the Cancer Center, University of California San Diego, La Jolla, California 92093, USA

ABSTRACT

Treatment of cells with clinically relevant exposures to cisplatin (DDP) activates an injury response that eventually induces death via an apoptotic mechanism. The activation of apoptosis involves generation of a signal from a detector that senses the presence of DDP-induced damage, and the integration of this signal with information arriving from receptors on both the cell surface and in the interior of the cell. The ability of the cell to trigger apoptosis appears to be dependent on the integrity of these signal transduction pathways, and on the ratio of various members of the bcl-2 family of proteins and their heterodimerization partners in the cell. The molecular dissection of the signal transduction pathways involved has identified numerous opportunities where pharmacologic intervention may enhance the selectivity of the platinum-containing compounds.

INTRODUCTION

The development of resistance to the platinum-containing drugs cisplatin (DDP) and carboplatin (CBDCA) during treatment is a common problem, and constitutes a major obstacle to the cure of even sensitive tumors. It is thought to be due to the selection for and overgrowth of drug-resistant cells[1-3] that arise through either spontaneous somatic mutation[4,5] or the activation of cellular protective mechanisms during the injury response that follows exposure to effective drug levels. Over the past decade, most studies have focused on elucidation of the biochemical pharmacologic mechanisms by which resistant cells protect themselves against DDP. However, an explosion of information emerging from the disciplines of cell biology and molecular pharmacology has recently focused attention on the fact that cell death induced by DDP is an active process that involves orchestration of a large number of different signals. Investigation of the signal transduction pathways activated during the cellular injury response to DDP is yielding insight into novel strategies for preventing the emergence of resistance or reversing it once it has developed.

Platinum and Other Metal Coordination Compounds in Cancer Chemotherapy 2
Edited by H.M. Pinedo and J.H. Schornagel, Plenum Press, New York, 1996

CISPLATIN KILLS CELLS BY APOPTOSIS

Clinically relevant exposures of both murine[6] and human[7] cancer cells to DDP kills by activating the process of apoptosis. Apoptosis differs from necrosis by virtue of being an active process that requires cellular energy, transcription, and translation (reviewed in reference 8). It is characterized by margination of the chromatin in the nucleus, condensation of the cytoplasm, blebbing of the nucleus and cytoplasm to form membrane-bounded vesicles (apoptotic bodies), and eventual ingestion of these membrane-bounded vesicles by macrophages. In contrast, necrosis is characterized by early fragmentation of the chromatin, swelling of the nucleus, mitochondria, and other intracellular organelles, and early loss of plasma membrane integrity. Typically necrosis is accompanied by greater leakage of cellular contents and inflammation than apoptosis.

Despite the fact the DDP kills by apoptosis, this is not always readily demonstrable by assays that detect apoptotic death induced by other agents. Many of the agents known to trigger apoptosis result in the activation of an endogenous endonuclease that degrades DNA intranucleosomally to fragments that are multiples of 180 - 200 basepairs. However, this is not an entirely reliable hallmark of apoptosis induced by DDP. Although DDP produces this "nucleosomal ladder" pattern of DNA degradation in some cell types[6], it fails to do so in many others. In human ovarian carcinoma cell lines, it is more typical to find a wide spectrum of large fragment sizes. In one such cell line, Oremerod et al.[7] have reported a predominance of fragments in the 30 - 50 kilobase range. Other assays for apoptosis also appear to yield less impressive signal to noise ratios for DDP than for many other types of agents. In our hands, for human ovarian carcinoma and head and neck cancer cell lines, while fragmentation of the DNA can be detected by failure of [^{3}H]dThd-labelled fragments to precipitate in acid[9], this signal is only observed at very high levels of drug exposure. Quantitation of fragments by flow cytometry has not proven highly reproducible nor quantitative. This is also true for detection of apoptosis by the TUNEL assay which is sensitive to broken DNA ends and gaps that can be revealed by incorporation of a labeled nucleotide[10]. This is perhaps not surprising since at IC_{50} concentrations of DDP relatively few adducts are formed per genome (perhaps 5,000 - 10,000) such that strand breaks resulting from attempted repair may be quite sparse.

It is important to remember that apoptotic and necrotic cell death may have many elements in common, and a clear distinction between mechanisms may not be feasible in some cell systems. It is in fact quite likely that, as total drug exposure is increased to very high levels, the predominant mechanism of cell death shifts from apoptosis to necrosis at the point where there is extensive enough damage to the plasma membrane and structural components of the cell.

SIGNAL TRANSDUCTION PATHWAYS INVOLVED IN THE DDP-INDUCED CELLULAR INJURY RESPONSE

Cells injured by exposure to DDP undergo a cellular injury response (CIR) that shares characteristics with responses produced by many other injurious agents. Although very little is known about the specifics of any of the signal transduction pathways activated by DDP injury, it is possible to identify the general nature of some components that must be present. Figure 1 presents a schematic outline of some of the elements of the DDP-induced CIR.

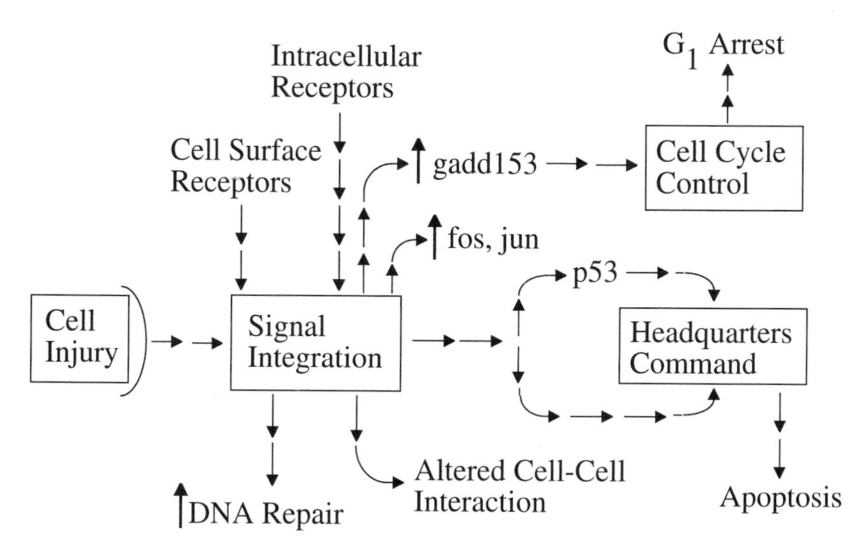

Figure 1. Schematic diagram of pathways involved in the DDP-induced cellular injury response.

Activation of the CIR requires that the cell have some mechanism for detecting the presence of damage. Since DNA appears to be the primary target of DDP, attention has focused on things that could detect the presence of either intra-strand or inter-strand crosslinks, or mono-adducts in DNA. There are currently several candidates for this role of detector. One is an HMG box-containing protein that Lippard, Whitehead and their colleagues have cloned and demonstrated to be capable of binding tightly to platinated sites in DNA[11,12]. Another candidate is the complex of two molecules (p70 and p80) that make up the Ku antigen[13]. This molecular complex binds avidly to gapped and broken DNA, and serves to attract and activate DNA protein kinase[14]. The latter molecule is constitutively present in the nucleus, and in its active form phosphorylates a variety of substrates that might be involved in the repair of DDP adducts including RPA and p53. However, the actual role of either of these candidates in the process of detecting DDP damage remains highly speculative at the present time.

A second component of the CIR signal transduction pathway that must be present is one or more centers that serve the function of signal integration. This conclusion is based on the fact that the cytotoxicity of DDP can be modulated by a variety of growth factors that bind to cell surface receptors[15,16]. This phenomenon has been best studied for EGF and its receptor[15,17]. Figure 2 shows that a one hour concurrent exposure to EGF is capable of enhancing the cytotoxicity of DDP to human head and neck carcinoma cells in log phase growth by a factor of 3.8 ± 0.2-fold. This duration of exposure is too short to produce a mitogenic response, and fails to alter the perturbation in cell cycle phase distribution induced by DDP exposure. Experiments using the 2008 human ovarian carcinoma cell system have demonstrated that the effect is not related to any alteration in the biochemical pharmacology of DDP, that it requires activation of the EGF receptor, and that its magnitude is dependent upon the number of receptors present on the cell surface[15]. Similar sensitization to DDP has been observed with use of an antibody directed at the EGF receptor in an in vivo model system[17]. This antibody presumably activates at least some component of the EGF receptor signal transduction pathway. Since EGF does not cause any alteration in the biochemical pharmacology of DDP, the presumption is that the signal generated by

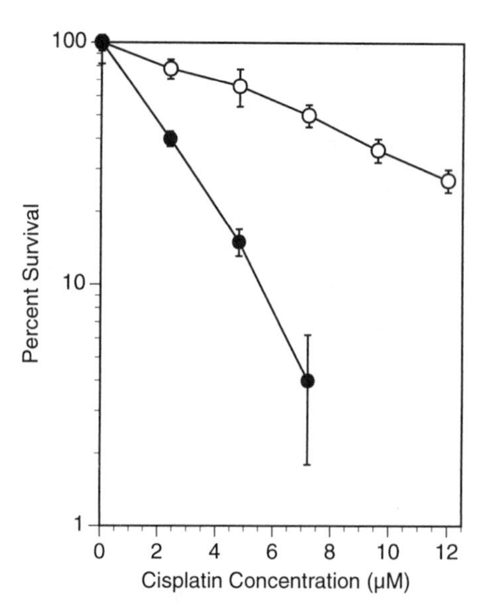

Figure 2. The effect of rhEGF on the DDP sensitivity of UMSCC-10b human head and neck carcinoma cells. Dose-response curves were determined using a colony formation assay. Cells were treated concurrently with 10 nM rhEGF and increasing concentration of DDP for 1 hour. Open circles, control cells treat with DDP alone; closed circles, cells treated with rhEGF plus DDP concurrently. Data points represent the mean ± SD of 3 experiments.

activation of the EGF receptor amplifies that generated by the damage detector. Similar sensitization to DDP has been reported for an antibody that binds to and partially activates the HER2/neu receptor in human breast and ovarian carcinoma cells[18]. At what point the receptor signals impinge on those generated by the damage detector is unknown, and is the focus of current research.

The integration center must be sensitive to signals generated by intracellular receptors as well as those located on the cell surface. Data supporting this conclusion come from experiments showing that activation of protein kinase C with phorbol ester analogs is capable of sensitizing a variety of different human cancer cell lines to DDP and several of its analogs[19,20]. Figure 3 shows the effect of incubating human ovarian carcinoma 2008 cells with 0.1 μM TPA for one hour concurrently with graded concentrations of DDP. TPA, a direct activator of protein kinase C, enhanced the sensitivity of these cells to DDP by a factor of 2.5, and to carboplatin (CBDCA) and 254-S [glycolato-*O,O'*)diammineplatinum(II)] by factors of 2.8 and 2.3, respectively[21]. Activation of another intracellular kinase that plays a central role in signal transduction, protein kinase A, is also capable of modulating sensitivity to DDP in the 2008 system[22].

A third component of the CIR signal transduction system must serve to interface signals generated by the DNA detector with the cell motor responsible for driving cell cycle progression. Exposure to cytotoxic concentrations of DDP commonly produce two easily observed perturbations in cell cycle phase distribution. The first is a concentration-dependent slowing of progression through S phase, and the second is a G_2 arrest whose magnitude and duration are related to total drug exposure. Figure 4 shows a typical progression of cell cycle phase distribution as it evolves over a period of 48 hours in human neck and neck carcinoma UMSCC-10b cells treated

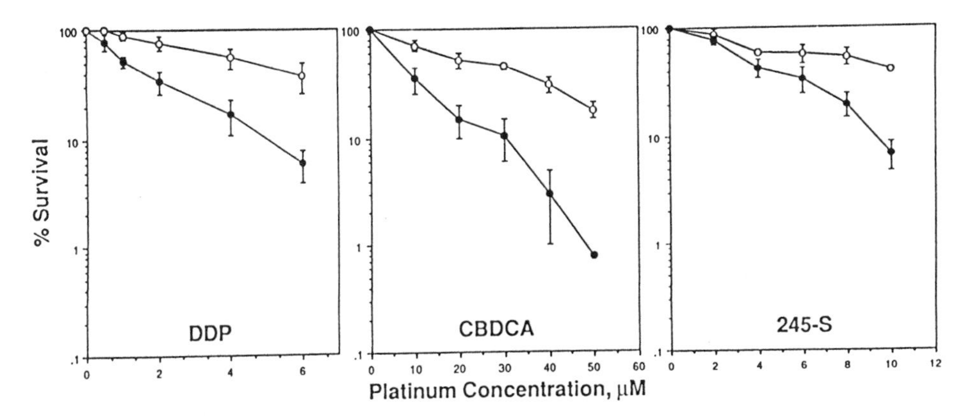

Figure 3. The effective of TPA on the DDP sensitivity of ovarian carcinoma 2008 cells. Dose-response curves were generated using clonogenic assays in which cells were exposed concurrently to TPA 0.1 μM and increasing concentrations of the platinum drug for 1 hour. Open circles, platinum drug alone; closed circles, platinum drug plus TPA. Each point represents the mean ± SD of 3 experiments performed with triplicate cultures.

with an IC_{90} concentration of DDP for 1 hour. Many other kinds of DNA damaging agents produce an arrest in G_1 as well as G_2. However, in human cancer cells such an arrest following DDP exposure is difficult to demonstrate, and in any case its magnitude is dwarfed by that of the G_2 arrest.

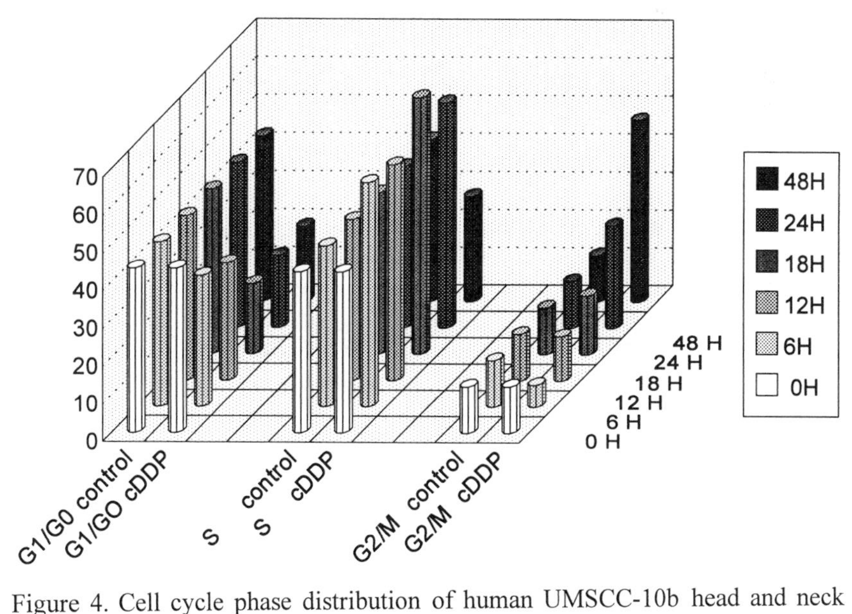

Figure 4. Cell cycle phase distribution of human UMSCC-10b head and neck carcinoma cells exposed to an IC_{90} concentration of DDP for 1 hour, and then grown in drug-free medium for the ensuing 48 hours.

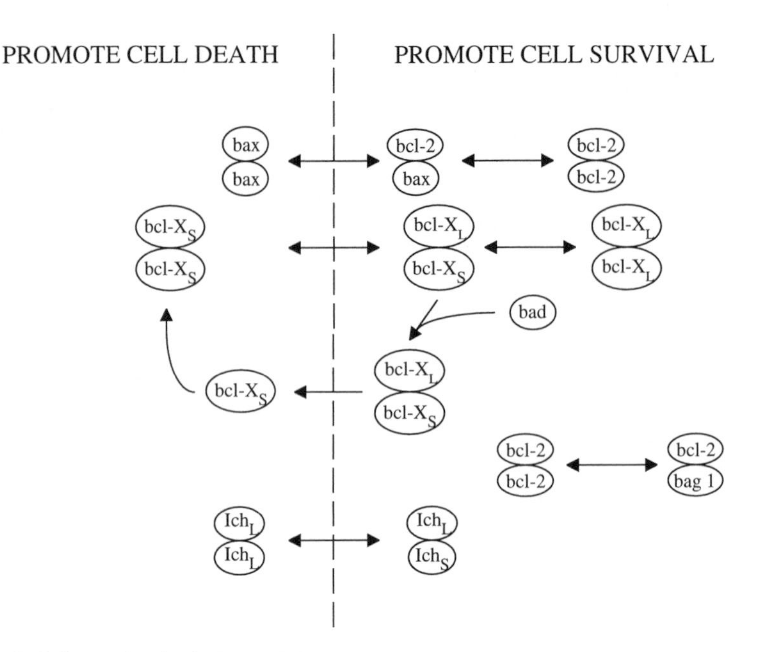

Figure 5. Schematic depiction of the interactions between some of the proteins known to participate in the regulation of apoptosis.

While the frequency with which DDP causes a G_1 arrest in human cancer cells remains poorly defined, recent molecular studies have revealed important details of the mechanisms that might underlie such an effect in cancer cells[23]. DDP is among the cell damaging agents that causes an increase in the expression and activity of p53[24]. In its role as a transcription factor, p53 activates the transcription of p21 which is a broadly active inhibitor of the cyclin-dependent kinases responsible for phosphorylation of the $p107^{Rb}$ protein which regulates exit from the G_1 phase of the cell cycle[25,26]. The working hypothesis is that if $p107^{Rb}$ cannot be phosphorylated, then it cannot release transcription factors such as E2F-DP that are responsible for initiating S phase that are bound to it[27].

Finally, the CIR mechanism must include something that activates the apoptotic response. How the apoptotic response to any agent is actually triggered is unknown, but the set-point for the trigger is highly regulated as are the individual biochemical reactions that result in nuclear degeneration[28]. The appearance of apoptotic activity in cells can now be measured using a cell-free system based on the degeneration of Xenopus nuclei, and the generation of such activity turns out to be both time and temperature dependent[29,30]. Thus, it is likely that the apoptotic response can be accurately described in terms of factors that influence its initiation and factors that regulate the full development of apoptotic activity. The initiation is clearly influenced by proteins of the bcl-2, bcl-X, and Ich-1 family[28,31] and the rapidly growing list of partners with which these proteins form dimers or multimers within the cell[32,33].

Figure 5 depicts some of the interactions between proteins that influence apoptosis, and is meant to illustrate the complexity of the situation rather than to be comprehensive. The emerging principles are that homodimers of bax, bcl-X_S, and Ich-1_L promote cell death, whereas proteins that can heterodimerize with bax, bcl-X_S, and Ich-1_L serve to reduce the concentration of the death-promoting homodimers and reduce apoptosis[28,31-33]

THE ROLE OF p53 IN TRIGGERING OF APOPTOSIS BY THE CELLULAR INJURY RESPONSE

p53 not only appears to play a central role in the mechanism by which the CIR causes cell cycle arrest, by also in the pathway by which it triggers apoptosis. In at least some experimental systems, the integrity of p53 function appears to be important for the triggering of apoptosis by chemotherapeutic agents or γ radiation[34,35]. There remains a fair amount of controversy about how loss of p53 function influences sensitivity to DDP in particular, but in one system re-introduction of wild type p53 by retrovirally-mediated gene transfer has been reported to sensitize tumors in vivo to DDP[36]. Based on data available from other types of cellular injury, it is likely that the DDP-induced CIR is capable of triggering apoptosis through several pathways, some of which are p53-dependent and some of which are not. The expectation is that the degree of p53 dependence will be found to vary among different types of tumors, or even between different cells within the same tumor.

Molecular descriptions of pathways that could explain the ability of the DDP CIR to trigger apoptosis in a p53-dependent manner are beginning to emerge. The DDP CIR induces an increase in p53 activity, and several groups have reported that p53 can suppress the expression of bcl-2[37,38] and that this may be due in part to the presence of a p53 negative-response element in the promoter of the bcl-2[39]. p53 has also recently been found to up-regulate the transcriptional activity of the bax promoter[40]. The results suggest that p53 can simultaneously down-regulate a protein that blocks apoptosis and up-regulate one that favors apoptosis, perhaps allowing the bcl-2/bax ratio to drop to a critical point. This is highly speculative at the moment, but points to the possibility that it is not the steady-state level of these proteins that is critical to induction of apoptosis but rather the extent to which the CIR modulates the ratio of these proteins.

THE USE OF GADD153 TO DETECT THE DDP-INDUCED CELLULAR INJURY RESPONSE

Dissection of the signal transduction pathways that are central to killing by DDP requires that one have assays sensitive to individual steps or a series of step within a pathway. Gadd153 is of interest for this purpose because it is a gene that is very highly transcriptionally activated by the DDP CIR as well as by other types of cellular stress[41] including UV light, serum starvation or media depletion[42], cysteine conjugates and dithiothreitol[43], hypoxia[44], and various other classes of chemotherapeutic drugs including alkylating agents[45,46]. Gadd153 is one of a family of injury response genes, but is unique in not being transcriptionally activated by phorbol esters[42]. Gadd153 was originally cloned by subtractive hybridization UV-treated CHO cells[47], and is highly conserved between mammalian species; the hamster cDNA shares 78% nucleotide sequence identity with the human cDNA [48] and >85% with the mouse cDNA homolog[49]. Ron and Habener[49] cloned CHOP-10, the mouse homolog of gadd153, and found that it was localized to the nucleus and is a dimerization partner for the C/EBP family of transcription factors. It can act as a transdominant negative inhibitor of their ability to activate promoters containing consensus sequences normally responsive to C/EBPβ and LAP[49]. The gadd153 gene is also found in at least two types of oncogenes in which translocation positions the coding region of gadd153 downstream of truncated portions of either the TLS or EWS genes, creating novel oncogenic fusion proteins[40,50,51]. Finally, gadd153 is of interest because it has been reported to regulate the progression of cells from G_1 into S phase of the cell cycle[52,53].

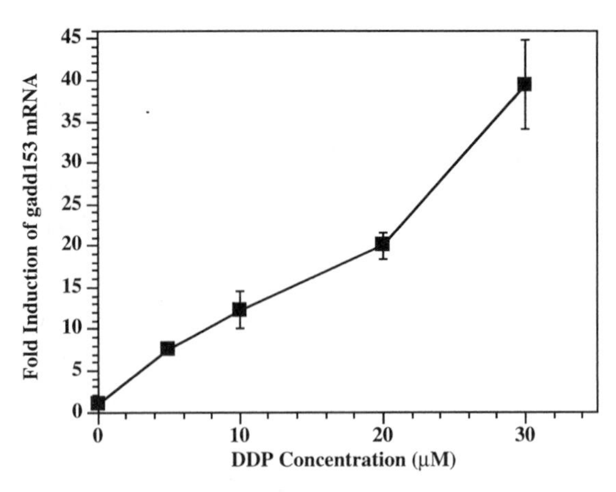

Figure 6. The effect of DDP treatment on gadd153 message levels in 2008 ovarian carcinoma cells. RNA was harvested 24 hours after a 1 hour exposure of exponentially growing cells to DDP, and analyzed by Northern blot. Error bars are ± SEM. (Reproduced by permission[54]).

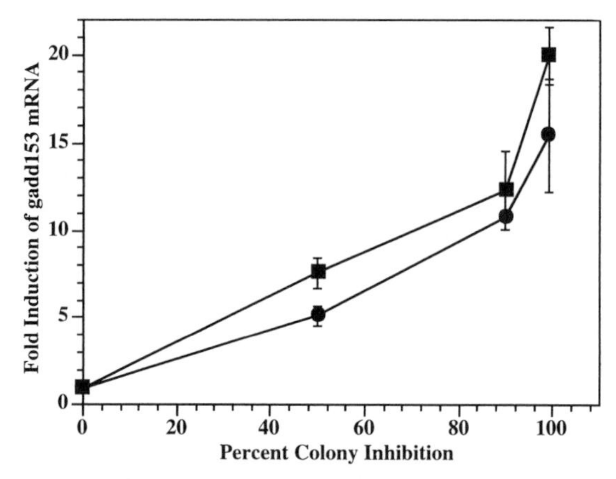

Figure 7. The effect of equitoxic DDP treatment on gadd153 message levels in 2008 ovarian carcinoma cells. RNA was harvested 24 hours after either a 1 hour (squares) or 24 hour (circles) exposure of exponentially growing cells to DDP at an IC_{50}, IC_{90}, or IC_{99} concentration. mRNA levels were determined by Northern blot. Error bars are ± SEM. (Reproduced by permission[54]).

We sought to determine whether changes in the message level of the gadd153 gene could be used to detect the presence of a DDP-induced CIR. Figure 6 shows that DDP treatment of human ovarian carcinoma 2008 cells causes a concentration-related increase in message levels measured by Northern blot analysis at 24 hours which was the time at which the peak effect was observed[54].

Figure 7 shows that there was a close relationship between the toxicity of DDP and the magnitude of gadd153 mRNA increase irrespective of the schedule of DDP exposure in vitro. For equal degrees of cytotoxicity, both a 1 hour and a 24 hour exposure to DDP produced approximately the same change in gadd153 mRNA level.

The effect of DDP treatment in vivo on gadd153 mRNA level was examined in vivo in two human tumor xenograft systems: the human ovarian carcinoma 2008 and human melanoma T289[54,55]. Groups of mice bearing tumors of ≈1 ml in volume were

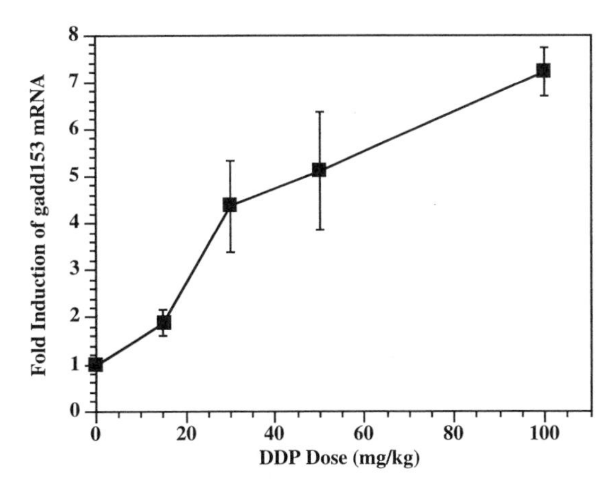

Figure 8. The effect of DDP on gadd153 mRNA levels in 2008 xenografts. RNA was harvested 24 hours after an IP injection of DDP, and analyzed by Northern blot. Values were corrected for βactin mRNA and are relative to gadd153 levels in untreated tumors. Error bars, ± SEM.

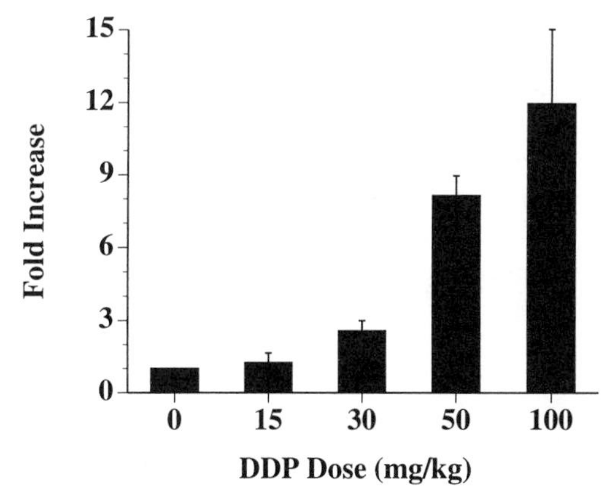

Figure 9. The effect of DDP on gadd153 mRNA levels in T289 xenografts. RNA was harvested 24 hours after an IP injection of DDP, and analyzed by Northern blot. Values were corrected for βactin mRNA and are relative to gadd153 levels in untreated tumors. Error bars, ± SEM.

injected IP with increasing doses of DDP, and RNA was harvested from biopsies obtained 24 hours later and subjected to Northern blot analysis. Figures 8 and 9 show the results from the ovarian carcinoma and melanoma, respectively. In both systems, DDP produced a dose-dependent increase in gadd153 mRNA level; the smallest dose that produced a reproducibly quantifiable change was 15 mg/kg. In both types of tumor the peak increase in gadd153 mRNA level occurred at 48 hours (data not shown).

These results indicate that gadd153 is induced in vivo by DDP treatment. This both validates the use of gadd153 as a relevant tool for dissection of the pathways involved in the DDP CIR, and also raises the question of whether the change in gadd153 mRNA level could be used as a way of quantitating the amount of injury that the tumor believes it has sustained within a short period of time following treatment. Such a use would require that the within-tumor heterogeneity be small enough that a single biopsy would be sufficient to provide a reliable indication of the response of the whole tumor nodule. The variation in the magnitude of the gadd153 mRNA response in 5 separate biopsies taken from different sections of the same tumor was analyzed for both the ovarian and melanoma xenograft system. The coefficient of variation was 11.7% in the ovarian carcinoma and 18.4% in the melanoma.

DISCUSSION

The currently available information suggests that DDP kills cells by initiating a CIR that eventually triggers apoptosis. The fact that there are multiple steps in this pathway, that the initiation of apoptosis is a highly regulated event, and that individual elements of the pathway are already known to be capable of significantly altering the sensitivity of cells to DDP all bode well for the development of strategies for increasing the sensitivity of tumor cells to the platinum-containing drugs, and protecting normal tissues from their toxicity. The gadd153 results provide but one example of how detailed information on the signal transduction pathways involved in the DDP CIR can identify tactical approaches that can be exploited in both the laboratory and the clinic. Based on the gadd153 results, two kinds of studies are being pursued at the UCSD Cancer Center. First, the gadd153 promoter is being dissected to determine which consensus sequences are essential the transcriptional activation of this gene with the goal of identifying upstream elements of the DDP CIR pathway. Second, an RT-PCR technique has been developed an applied to biopsies from patients with head and neck carcinoma receiving treatment with a DDP-containing regimen to determine the correlation between the magnitude of the increase in gadd153 mRNA at 24 hours after treatment and the probability of clinical response. There is every reason to expect that further elucidation of the DDP CIR will provide a wealth of additional opportunities for developing better platinum-based drugs and improving the therapeutic ratio for members of this class of agents.

REFERENCES

1. Prestayko AW. Cisplatin: A preclinical overview. In: Prestayko AW, Crooke ST, Carter SK, eds. Cisplatin: Current Status and New Developments. New York: Academic Press, 1980:1-7.

2. Skipper HE. A review and more quantitative analysis of the results of many internally controlled combination chemotherapy trials carried out over the past 15 years (L1210 and leukemia and P388 leukemia). Southern Res Inst 1979; monograph #2:

3. Schabel FM,Jr., Skipper HE, Trader MW. Concepts for controlling drug-resistant tumor cells. In: Mouridsen HT, Palshof T, eds. Breast Cancer: Experimental and Clinical Aspects. Oxford: Pergamon Press, 1980:199-212.

4. Goldie JH, Coldman AJ. A mathematic model for relating the drug sensitivity of tumors to their spontaneous mutation rate. Cancer Treat Rep 1979; 63:1727-1733.

5. DeMars R. Resistance of cultured human fibroblasts and other cells to purine analogs in relation to mutagenesis detection. Mutation Res 1974; 14:335.

6. Barry MA, Behnke CA, Eastman A. Activation of programmed cell death (apoptosis) by cisplatin, other anticancer drugs, toxins and hyperthermia. Biochem Pharmacol 1990; 40:2353-2361.

7. Ormerod MG, O'Neill CF, Robertson D, Harrap KR. Cisplatin induces apoptosis in a human ovarian carcinoma cell line without concomitant internucleosomal degradation of DNA. Exp Cell Res 1994; 211:231-237.

8. Kerr JFR, Winterford CM, Harmon BV. Apoptosis: its significance in cancer and cancer therapy. Cancer 1994; 73:2013-2026.

9. Matzinger P. The JAM test. A simple assay for DNA fragmentation and cell death. J Immunol Methods 1991; 145:185-192.

10. Gorczyca W, Gong J, Darzynkiewicz Z. Detection of DNA strand breaks in individual apoptotic cells by the in situ terminal deoxynucleotidyl transferase and nick translation assays. Cancer Res 1993; 53:1945-1951.

11. Toney JH, Donahue BA, Kellett PJ, Bruhn SL, Essigmann JM, Lippard SJ. Isolation of cDNAs encoding a human protein that binds selectively to DNA modified by the anticancer drug cis-diamminedichloroplatinum(II). Proc Natl Acad Sci USA 1989; 86:8328-8332.

12. Donahue BA, Augot M, Bellon SF, et al. Characterization of a DNA damage-recognition protein from mammalian cells that binds to intrastrand d(GpG) and d(ApG) DNA adducts of the anticancer drug cisplatin. Biochemistry 1990; 29:5872-5880.

13. Gottlieb Tanya M, Jackson Stephen P. The DNA-Dependent Protein Kinase: Requirement for DNA Ends and Association with Ku Antigen. Cell 1993; 72:131-142.

14. Anderson Carl W. DNA damage and the DNA-activated protein kinase. TIBS 1993; 433-437.

15. Christen RD, Hom DK, Porter DC, et al. Epidermal growth factor regulates the in vitro sensitivity of human ovarian carcinoma cells to cisplatin. J Clin Invest 1990; 86:1632-1640.

16. Isonishi S, Jekunen AP, Hom DK, et al. Modulation of cisplatin sensitivity and growth rate of an ovarian carcinoma cell line by bombesin and tumor necrosis factor alpha. J Clin Invest 1992; 90:1436-1442.

17. Fan Zhen, Baselga Jose, Masui Hideo, Mendelsohn John. Antitumor Effect of Anti-Epidermal Growth Factor Receptor Monoclonal Antibodies plus cis-Diamminedichloroplatinum on Well Established A431 Cell Xenografts. Cancer Res 1993; 53:4637-4642.

18. Pietras RJ, Fendly BM, Chazin VR, Pegram MD, Howell SB, Slamon DJ. Antibody to HER-2/neu receptor blocks DNA repair after cisplatin in human breast and ovarian cancer cells. Oncogene 1994; 9:1829-1838.

19. Isonishi S, Andrews PA, Howell SB. Increased sensitivity to cis-diamminedichloroplatinum(II) in human ovarian carcinoma cells in response to treatment with 12-0-tetradecanoylphorbol-13-acetate. J Biol Chem 1990; 265:3623-3627.

20. Basu A, Teicher BA, Lazo JS. Involvement of protein kinase C in phorbol ester-induced sensitization of HeLa cells to cis-diamminedichloroplatinum (II). J Biol Chem 1990; 265:8451-8457.

21. Isonishi S, Hom DK, Eastman A, Howell SB. Enhancement of sensitivity of platinum(II)-containing drugs by activation of protein kinase C in a human ovarianc carcinoma cell line. Br J Cancer 1994; 69:217-221.

22. Mann SC, Andrews PA, Howell SB. Modulation of cis-diamminedichloroplatinum(II) accumulation and sensitivity by forskolin and 3-isobutyl-1-methylxanthine in sensitive and resistant human ovarian carcinoma cells. Int J Cancer 1991; 48:866-872.

23. Hartwell Leland H, Kastan Michael B. Cell Cycle Control and Cancer. Science 1994; 266:1821-1828.

24. Fritsche M, Haessler C, Brandner G. Induction of nuclear accumulation of the tumor-suppressor protein p53 by DNA-damaging agents. Oncogene 1993; 8:307-318.

25. Xiong Y, Hannon GJ, Zhang H, Casso D, Kobayashi R, Beach D. P21 is a universal inhibitor of cyclin kinases. Nature 1994; 366:701-703.

26. Peter M, Herskowitz I. Joining the complex: cyclin-dependent kinase inhibitory proteins and the cell cycle. Cell 1994; 79:181-184.

27. Nevins JR. E2F: a link between the Rb tumor suppressor protein and viral oncoproteins. Science 1992; 258:424-429.

28. Reed JC. Bcl-2 and the regulation of programmed cell death. Mini-review: Cellular mechanisms of disease series. J Cell Biol 1994; 124:1-6.

29. Newmeyer DD, Farschon DM, Reed JC. Cell-free apoptosis in xenopus egg extracts: inhibition by Bcl-2 and requirement for an organelle fraction enriched in mitochondria. Cell 1994; 79:353-364.

30. Lazebnik YA, Cole S, Cooke CA, Nelson WG, Earnshaw WC. Nuclear events of apoptosis in vitro in cell-free mitotic extracts: a model system for analysis of the active phase of apoptosis. Journal of Cellular Biology 1993; 123:7-22.

31. Oltvai ZN, Korsmeyer SJ. Checkpoints of dueling dimers foil death wishes. Cell 1994; 79:189-192.

32. Yang Elizabeth, Zha Jiping, Jockel Jennifer, Boise Lawrence H, Thompson Craig B, Korsmeyer Stanley J. Bad, a Heterodimeric Partner for Bcl-XL and Bcl-2, Displaces Bax and Promotes Cell Death. Cell 1995; 80:285-291.

33. Miyashita T, Reed JC. Tumor Suppressor p53 is a Direct Transcriptional Activator of the Human bax Gene. Cell 1995; 80:293-299.

34. Lowe Scott W, Ruley HEarl, Jacks Tyler, Housman David E. p53-Dependent Apoptosis Modulates the Cytotoxicity of Anticancer Agents. Cell 1993; 74:957-967.

35. Lowe SW, Bodis S, McClatchey A, et al. p53 status and the efficacy of cancer therapy in vivo. Science 1994; 266:807-810.

36. Fujiwara Toshiyoshi, Grimm Elizabeth A, Mukhopadhyay Tapas, Zhang Wei-Wei, Owen-Schaub Laurie B, Roth Jack A. Induction of Chemosensitivity in Human Lung Cancer Cells in Vivo by Adenovirus-mediated Transfer of the Wild-Type p53 Gene. Cancer Res 1994; 54:2287-229l.

37. Miyashita T, Krajewski S, Krajewska M, et al. Tumor suppressor p53 is a regulator of bcl-2 and bax in gene expression in vitro and in vivo. Oncogene 1994; 9:1799-1805.

38. Selvakumaran M, Lin HK, Miyashita T, et al. Immediate early up-regulation of bax expression by p53 but not TGFBl: a paradigm for distinct apoptotic pathways. Oncogene 1994; 9:1791-1798.

39. Miyashita T, Harigal M, Hanada M, Reed JC. Identification of a p53-dependent negative response element in the bcl-2 gene. Cancer Res 1994; 54:3l3l-3l35.

40. Zinszner H, Albalat R, Ron D. A novel effector domain from the RNA-binding protein TLS or EWS is required for oncogenic transformation by CHOP. Genes & Dev 1995; 8:2513-2526.

41. Fornace AJ,Jr., Jackman J, Hollander MC, Hoffman-Liebermann B, Liebermann DH. Genotoxic-stress-response genes and growth-arrest-genes. Annal N Y Acad Sci 1992; 139-153.

42. Fornace AJ,Jr., Nebert DW, Hollander C, et al. Mammalian genes coordinately regulated by growth arrest signals and DNA-damaging agents. Mol Cell Biol 1989; 9:4196-4203.

43. Chen Q, Yu K, Holbrook NJ, Stevens JL. Activation of the growth arrest and DNA damage-inducible gene gadd 153 by nephrotoxic cysteine conjugates and ditiothreitol. J Biol Chem 1992; 267:8207-8212.

44. Price BD, Calderwood SK. Gadd 45 and gadd 153 messenger RNA levels are increased during hypoxia and after exposure of cells to agents which elevate the levels of the glucose-regulated proteins. Cancer Res 1992; 52:3814-3817.

45. Luethy JD, Holbrook NJ. The pathway regulating GADD153 induction in response to DNA damage is independent of protein kinase C of tyrosine kinases. Cancer Res 1994; 54:1902-1906.

46. Luethy JD, Holbrook NJ. Activation of the gadd153 promoter by genotoxic agents: a rapid and specific response to DNA damage. Cancer Res 1992; 52:5-10.

47. Fornace AJ,Jr., Alamo Jr I, Hollander MC. DNA damage-inducible transcripts in mammalian cells. Proc Natl Acad Sci USA 1988; 85:8800-8804.

48. Park JS, Luethy JD, Wang MG, et al. Isolation, characterization and chromosomal localization of the human GADD153 gene. Gene 1992; 116:259-267.

49. Ron D, Habener JF. CHOP, a novel developmentally regulated nuclear protein that dimerizes with transcription factors C/EBP and LAP and functions as a dominant-negative inhibitor of gene transcription. Genes & Dev 1992; 6:439-453.

50. Aman P, Ron D, Mandahl N. Rearrangement of the transcription factor gene CHOP in myxoid liposarcomas with t(12;16)(q13;p11). Genes, Chromosomes, and Cancer 1992; 5:278-285.

51. Panagopoulos I, Mandahl N, Ron D, et al. Characterization of the CHOP Breakpoints and Fusion Transcripts in Myxoid Lipsarcomas with the 12;16 Translocation. Cancer Res 1994; 54:6500-6503.

52. Zhan Q, Lord KA, Alamo I,Jr., et al. The gadd and MyD Genes Define a Novel Set of Mammalian Genes Encoding Acidic Proteins That Synergistically Suppress Cell Growth. Mol Cell Biol 1994; 2361-2371.

53. Ron D, Habener JF. CHOP, a novel developmentally regulated nuclear protein that dimerizes with transcription factors C/EBP and LAP and functions as a dominant-negative inhibitor of gene transcription. Genes and Development 1992; 6:439-453.

54. Gately DP, Jones JA, Christen RD, Howell SB. Induction of the growth arrest and DNA damage inducible gene gadd153 by cisplatin in vitro and in vivo. Cancer Chemother Pharmacol 1994; submitted:

55. Jones JA, Gately DP, Barton RA, et al. Induction of gadd153 in human melanoma xenografts as an indicator of genotoxic injury. Cell Pharmacol 1994; 1:233-237.

THE INTERACTION OF CISPLATIN WITH SIGNAL TRANSDUCTION PATHWAYS AND THE REGULATION OF APOPTOSIS

Alan Eastman

Department of Pharmacology and Toxicology
Dartmouth Medical School
Hanover, NH 03755

INTRODUCTION

There are many ways to kill a cell. In the absence of serum, cells usually die because they have lost a receptor-mediated survival signal. Some cells can be killed by engaging certain receptors such as during glucocorticoid-mediated killing of thymocytes. Cells can also be killed by a multitude of cytotoxic agents including cisplatin. These insults all have a common endpoint: the cells die by apoptosis. There is a network of signals, initially unique to each insult, but eventually converging on a common pathway, only the latter part of which can truly be considered apoptosis. For example, resistance to cisplatin may be mediated by altered drug accumulation, decreased DNA damage or increased DNA repair.[1,2] These mechanisms may inhibit apoptosis induced by cisplatin, but they do not rescue a cell from most drugs, and they certainly do not rescue cells from removal of serum. In contrast, the Bcl-2 protein can protect cells from each of these insults, and therefore reflects a step much further along the pathway; a step that is common to multiple insults. Bcl-2 is the prototype of an anti-apoptotic protein, defined as a means to protect cells from multiple unrelated insults. With respect to cancer chemotherapy, Bcl-2 elicits a kind of non-specific multi-drug resistance.

Apoptosis is a form of cell death defined by morphological and biochemical characteristics. The term apoptosis was coined in 1972 to emphasize the normal balance between cell replication (mitosis) and cell death (apoptosis) in the regulation of tissue homeostasis.[3] Apoptotic cells were seen in histological sections as isolated dying cells that lost contact with their neighbors and shrank.[4] Within the dying cell, chromatin could be seen condensed at the nuclear membrane, while other organelles appeared normal. Cells would often produce membrane protuberances known as blebs. Dying cells were rapidly engulfed thereby avoiding an inflammatory response. Apoptosis was also observed in cell culture which was more amenable to investigation. It was established that the chromatin condensation associated with apoptosis occurred concurrently with internucleosomal fragmentation of the chromatin-associated DNA. Subsequent experiments suggested that this

DNA digestion was a direct cause of the morphologic changes.[5] The degradation of chromatin DNA into nucleosome fragments has become synonymous with apoptosis. We first saw this pattern of DNA digestion in cells damaged with cisplatin,[6-8] and this led us to investigate the regulation of apoptosis.

A model for the regulation of apoptosis is presented in Figure 1. Although the primary target for cisplatin is in the nucleus, this alone is not sufficient to induce apoptotic changes. Many observations suggest there is a cytoplasmic component to the pathway. For example, intracellular acidification and cell shrinkage occur because of events at the cytoplasmic membrane. Hence, the primary lesion in the nucleus must transmit a signal to the cytoplasm. Presumably, it is this cytoplasmic pathway that can be suppressed by Bcl-2 thereby eliciting drug resistance. An important observation supporting this cytoplasmic pathway is that staurosporine can induce apoptotic changes in enucleated cells, at least as defined by cytoplasmic changes.[9] Furthermore, Bcl-2 can protect enucleated cells from staurosporine.[9] This model suggests that many insults can trigger the pathway at various steps; this includes the removal of any growth factor that is essential for survival. However, Bcl-2 regulates a common downstream step in this pathway, which explains its ability to suppress apoptosis by each of these insults.

Our investigations have been directed toward both ends of this pathway. At one end, we are investigating the events that occur following incubation of cells with cisplatin. These experiments have focussed on the role of cell cycle in the induction of apoptosis. At the other end, we have characterized an endonuclease potentially involved in DNA digestion, and have begun to investigate the signals regulating its activation. Recent progress on each of these subjects will be reviewed.

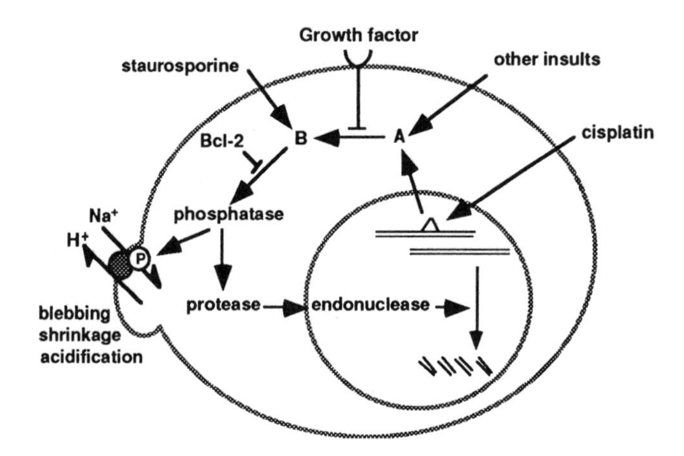

Figure 1. Model for the pathways leading from a variety of insults to intracellular acidification and DNA digestion during apoptosis (discussed in text).

THE CELL CYCLE AND INDUCTION OF APOPTOSIS BY CISPLATIN

One of the earliest observations we made following incubation of cells with cisplatin was that cells arrested at the G_2 phase of the cell cycle prior to the onset of apoptosis.[6,10] At that time, we suggested that cells underwent apoptosis from the G_2 phase, but we were unable to define the process more specifically using asynchronous cell cultures. Our recent experiments used a DNA repair-deficient cell line, CHO/UV41.[11] These cells were chosen to avoid confounding results that would arise from a reduction in DNA adducts with time. The cells were synchronized at the G_1/S border, then released and damaged with cisplatin for 1 h. The drug was then removed so that all subsequent determinations were made in cells with a constant level of DNA adducts. These cells demonstrated no slowing in S phase, but progressed to the G_2 phase where they arrested for up to 16 h.

The G_2-arrested cells contained the hyperphosphorylated (inactive) form of the mitotic kinase $p34^{cdc2}$ demonstrating that the arrest was caused by inhibition of normal dephosphorylation.[11] At later times, cells dephosphorylated $p34^{cdc2}$ thereby activating the kinase, and entered mitosis. During mitosis the chromosomes segregated abnormally leading to G_1 progeny with unequal numbers of chromosomes. This abnormal chromosome segregation arose from the formation of multipolar mitotic spindles.

Cells arrested in G_2 following cisplatin could be forced to enter mitosis within 4 hours by addition of 5 mM caffeine, or alternately, mitosis could be delayed by incubation of cells in cycloheximide. The timing of apoptosis always followed about 4 hours after mitosis, demonstrating the importance of this aberrant mitosis for the subsequent apoptosis.

Following a lethal mitosis, cells began to float off the culture dish; only the detached cells contained the DNA digestion characteristic of apoptosis. We surmised that the extracellular matrix, in this case the culture dish coated with serum proteins, acts as a survival factor. Extracellular matrix proteins such as fibronectin, which communicate through integrin-type receptors have been shown to activate a signal transduction pathway through a focal adhesion kinase.[12] Furthermore, inhibition of signalling from integrin receptors induces apoptosis.[13,14] The intracellular signalling pathway activated by integrin signalling is the same as activated by many growth factor receptors (Figure 2). Each of these receptors, as well as protein kinase C and Ras, have been shown to cause phosphorylation and activation of the Na^+/H^+-antiport leading to intracellular alkalinization.[15-17] This antiport is involved in cell volume regulation, and its inactivation may be the cause of cell shrinkage during apoptosis. The involvement of this antiport in apoptosis was initially suggested by experiments to identify the endonucleases involved in digesting the DNA (see below).

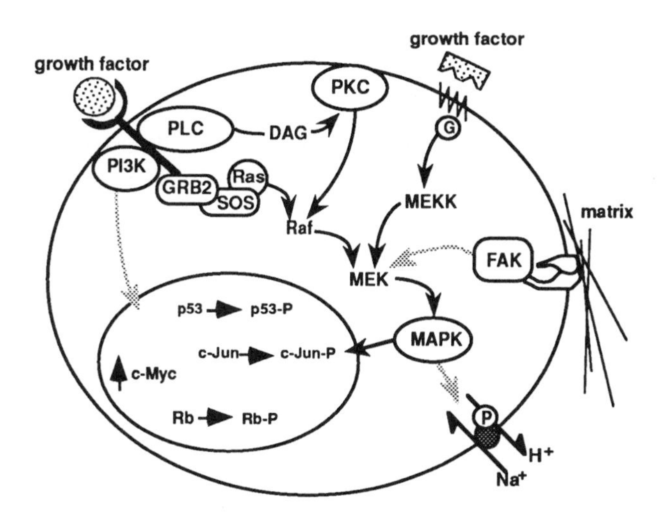

Figure 2. Schematic representation of signal transduction pathways in a cell, emphasizing those pathways activated by survival factors such as growth factors or the extracellular matrix. The majority of lines represent established fact; the shaded lines represent postulated steps with potentially additional intermediates. For a detailed discussion, see reference 34. The abbreviations used are: DAG, diacylglycerol; FAK, focal adhesion kinase; GRB2, growth factor binding protein; MAPK, mitogen-activated protein kinase; MEK, MAPK kinase; MEKK, MEK kinase; PI3K, phosphatidylinositol-3-kinase; PKC, protein kinase C; PLC, phospholipase C; SOS, mammalian son of sevenless.

The results discussed here are specific to the DNA repair-deficient cells. Very low concentrations of cisplatin were used because the cells are hypersensitive to the drug. However, in repair-competent cells, much higher concentrations of cisplatin are required to kill cells. Under these circumstances, cisplatin can prolong S phase passage before the cells reach a G_2 arrest and subsequently undergo a lethal mitosis.[10] As the concentration of cisplatin is increased further in either cell line, the cells arrest and die at all points of the cell cycle demonstrating that the G_2 arrest and lethal mitosis are not essential for cell death. However, this post-mitotic mode of cell death is probably important at pharmacologically relevant concentrations.

Considering the importance of Bcl-2 in suppressing apoptosis, we have begun to investigate how this class of proteins might modulate cell survival. Specifically, we have begun to investigate where in this pathway Bcl-2 can protect cells from cisplatin. It is interesting that Bcl-2 delays the onset of apoptosis at both high and low concentrations of cisplatin demonstrating that it is able to protect cells at all phases of the cell cycle.

THE REGULATION OF APOPTOSIS

Identification of Endonucleases

DNA digestion is probably ubiquitous during apoptosis, although the extent of DNA digestion varies with cell line. There are reports of apoptosis detected morphologically without DNA digestion, but this can be explained by limited DNA digestion that did not resolve into detectable DNA fragments.[18-20] Several recent studies have demonstrated an initial phase of DNA digestion to 300 and 50 kbp fragments without necessarily observing the digestion to 180 bp.[21,22] However, there is no evidence that apoptosis occurs in the complete absence of DNA digestion. At the initiation of these studies, we believed that identifying the endonuclease would lead to characterization of the upstream regulators of apoptosis.

Our initial experiments detected apoptotic DNA digestion in CHO cells following incubation with cisplatin.[8] Accordingly, we began to purify an endonuclease from these cells. Based on the existing dogma in apoptosis, we expected to isolate a Ca^{2+}/Mg^{2+}-dependent endonuclease. However, no evidence was obtained for such an endonuclease in CHO cells, rather an endonuclease was found that was active at acidic pH. We purified this endonuclease and identified it as DNase II.[23] Addition of this purified endonuclease to nuclei produced DNA digestion up to about pH 6.5. Furthermore, experimentally decreasing intracellular pH in CHO cells activated DNase II and induced the characteristic DNA digestion of apoptosis.

However, the detection of this or any other endonuclease does not prove its role in apoptosis. Hence, we analyzed intracellular pH and Ca^{2+} to determine which ion best correlated with the onset of DNA digestion. The conclusion from studies in a variety of cell lines following many insults was that DNA digestion was always associated with a decreased intracellular pH of 0.6 - 0.9 units.[24-26] These experiments also demonstrated that increases in intracellular Ca^{2+} only occasionally occurred, but even then could be shown to be unnecessary for DNA digestion. Furthermore, DNA digestion could be induced in the complete absence of Ca^{2+}. These results were consistent with the hypothesis that decreased pH rather than increased Ca^{2+} was associated with the activation of an endonuclease during apoptosis.

Regulation of Intracellular pH During Apoptosis

Intracellular acidification has been observed in every model of apoptosis that we have investigated.[24-26] It is very important to emphasize that the acidic shift observed during apoptosis in these experiments was not a continuum of cells progressively more acidic, rather two distinct populations were seen, one at normal pH, the other at about pH 6.4-6.7 depending upon the model. To understand the origin of this acidification, it is important to emphasize that the cells are still metabolically active. First, the acidic cells are able to take up fluorescent dye, de-esterify it, and then retain it, thereby demonstrating membrane integrity. The cells also exclude trypan blue. Second, these cells still maintain low Ca^{2+} against a Ca^{2+} gradient which requires metabolic activity. A metabolically inactive cell or necrotic cell would have intracellular pH and Ca^{2+} equal to the extracellular medium. The acidified cells therefore maintain an electrochemical gradient across the cytoplasmic membrane. Under conditions of a normal electrochemical gradient, selective inhibition of pH regulation would cause a drop of 1 pH unit.[27] Therefore, the large acidification in the apoptotic cells is due to selective inhibition of pH regulation without significant interference with the homeostasis of other ions.

Cells regulate their intracellular pH through membrane-associated ion transport proteins including Na^+/H^+-antiports, ATP-driven H^+-pumps, and bicarbonate exchangers. The Na^+/H^+-antiport is a major H^+-extruding mechanism which is driven by the inward-directed Na^+ gradient. This antiport has a high affinity for H^+ at pH 6, but is inoperative at neutral pH unless phosphorylated. As described above, a wide variety of external signals including growth factors and the extracellular matrix, as well as experimental activation of protein kinase C, activate a kinase cascade that leads to phosphorylation of the antiport and alkalinization of the cells.[15-17] These same stimuli are known to enhance cell survival.

We have now obtained direct evidence for a change in the set-point of the Na^+/H^+-antiport during apoptosis in a cytotoxic T cell line, CTLL-2, following withdrawal of IL-2.[26] We first established that the Na^+/H^+-antiport was the primary pH regulator in these cells by showing that ethylisopropylamiloride (EIPA), an inhibitor of the antiport, inhibited recovery from an acid load. Upon withdrawal of IL-2, these cells underwent intracellular acidification, and the acidified cells represented a discrete population up to 0.7 pH units below normal. The extent of acidification depended upon the extracellular pH; above pH 6.3, intracellular pH was significantly lower than extracellular pH, whereas below pH 6.3, the cells were still able to regulate their pH to slightly above the extracellular pH (Figure 3). We hypothesized that this residual pH regulation was due to the Na^+/H^+-antiport; this was confirmed by showing that EIPA prevented this pH regulation at low pH. The results demonstrate that apoptotic CTLL-2 cells retain a functional antiport but that its set-point has changed such that they can no longer raise their intracellular pH to 7.4. An important conclusion from these experiments is that the acidification is consistent with the known function of many survival factors which phosphorylate and activate the antiport; hence intracellular acidification is a likely consequence of inhibition of protein kinase cascades involved in cell survival.

Considering the frequent, if not universal correlation between DNA digestion and intracellular acidification, then we must ask whether this implicates DNase II universally in apoptosis. If low pH is the only means to activate DNase II (this still needs to be established), we should be able to prevent DNA digestion by trapping the intracellular pH. The experiment shown in Figure 3, provided an experimental strategy by which we could answer this question. When apoptosis was induced in CTLL-2 cells held at an extracellular pH of 8, the intracellular pH dropped to only 7.2, a condition under which DNase II is not expected to be active. However, DNA digestion still occurred in both CHO and CTLL-2 cells. These results suggest either that DNase II is not involved in apoptosis, or that it is activated by some other means that permits activity at neutral pH. Explanations for this latter possibility include altered catalytic activity in cells, altered activity attributable to associated proteins, or altered activity due to post-translational modification of the endonuclease. In this latter regard, we have recently observed an apparent proteolysis of the DNase II protein during apoptosis. We do not yet know whether the cleaved protein has an altered catalytic activity.

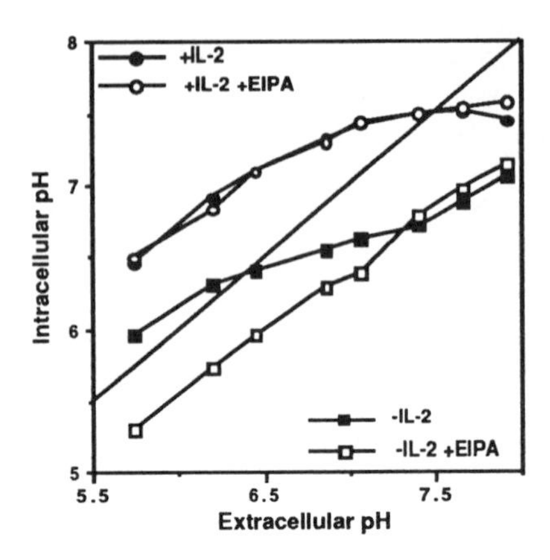

Figure 3. The dependence of intracellular pH on extracellular pH in normal and apoptotic CTLL-2 cells. Sixteen hours after withdrawal of IL-2, intracellular pH was measured in individual cells by flow cytometry. Prior to analysis, cells were equilibrated in medium at the indicated extracellular pH; in some cells, the Na^+/H^+-antiport was inhibited by inclusion of EIPA during this equilibration. Two populations of cells were identified, normal and apoptotic, and separate lines are drawn for each. For a detailed description, see reference 26.

The Relevance of Intracellular Acidification to Apoptosis

The above results raise a very important question: if low intracellular pH is not the signal for activation of an endonuclease or any other effector of an apoptotic pathway, what then is the relevance of the acidification and does it have any role in apoptosis? In the discussion above, I have suggested that survival is mediated by growth factor-activated kinase cascades; hence, if survival is mediated by kinases, then apoptosis might be caused by protein phosphatases. Potential substrates for the phosphatases include members of the kinase cascade, as well as proteins that regulate ion homeostasis such as the Na^+/H^+-antiport. It is worth noting that the antiport is involved in cell volume regulation, and it is likely that its deregulation is responsible for the characteristic shrinkage observed during apoptosis.

We have begun to investigate the possible involvement of protein phosphatases in apoptosis. We have incubated human myeloblastoid ML-1 cells with etoposide for 30 min, removed the drug, and then incubated the cells for an additional 4 hours in the presence of various inhibitors. Okadaic acid inhibited DNA digestion at 1 μM whereas calyculin A and cantharidin inhibited digestion at 10 nM and 20 μM respectively. Okadaic acid, calyculin A and cantharidin are all inhibitors of both protein phosphatases 1 and 2A; in vitro okadaic acid and cantharidin are more effective against PP2A, whereas calyculin A is more effective against PP1. The fact that only 10 nM calyculin A is required to protect cells suggested the involvement of PP1 in apoptosis. However, these concentrations can not be extrapolated directly to cells, so we selected an intracellular marker of PP1 activity to establish its potential involvement. PP1 is involved in the dephosphorylation of the retinoblastoma susceptibility protein Rb at the conclusion of mitosis; Rb is then rephosphorylated by a cyclin-dependent kinase during G_1.[28] The phosphorylated forms of Rb were measured by their retarded electrophoretic mobility followed by western blotting. Following incubation of ML-1 cells with etoposide, Rb was dephosphorylated with a time and dose dependence that correlated with the appearance of DNA digestion. This Rb dephosphorylation was inhibited at exactly the same concentrations of each inhibitor that inhibited DNA digestion. These results strongly suggest the involvement of PP1 in apoptosis. The dephosphorylation of Rb occurred primarily in the S phase cells that die most rapidly after etoposide, hence activation of PP1 was occurring aberrantly in the cell cycle.

We have also investigated the possible involvement of proteases in activating apoptosis. ML-1 cells incubated with the protease inhibitor TPCK were protected from etoposide-induced DNA digestion. We then investigated whether TPCK could also prevent dephosphorylation of Rb. In preliminary experiments, we have found that Rb is still dephosphorylated even though DNA digestion was prevented. This demonstrates several important points: first, that Rb dephosphorylation is not a consequence of DNA digestion, and second, that the protease step is down-stream of protein phosphatase PP1. As mentioned earlier, we have also observed proteolysis of DNase II, and this can be inhibited by TPCK, okadaic acid, calyculin A and cantharidin, all at the same concentrations that prevent DNA digestion. These results have led to the current hypothesis outlined in Figure 1 in which we suggest an order to these events in apoptosis: PP1 activity leads to protease activity, which in turn activates an endonuclease.

Interest in the role of proteases in apoptosis has been stimulated by results obtained on genetic regulation of cell death in the nematode *C. elegans*. The product of the ced-3 gene is required for programmed cell death in the nematode. Ced-3 encodes a protease with homology to the human protease interleukin 1β converting enzyme ICE.[29] Overexpression of ICE in human cells induces apoptosis.[30] Several proteins have been shown to be cleaved during apoptosis in human cells. These include poly(ADP-ribose) polymerase which is cleaved by a protease related to ICE (prICE).[31,32] A different protease unrelated to ICE has also been identified as increased during apoptosis and which, when added to control nuclei, induces DNA digestion.[33] These results suggest that a protease, either directly or indirectly, activates an endonuclease, and it also supports our model in which a protease may be activated in the cytoplasm leading to proteolysis of a nuclear endonuclease.

Many of the results reported here have been performed in systems in which apoptosis has been induced by a variety of stimuli. In many cases, the particular system was selected for technical reasons such as the short time from insult to apoptosis. Some of the experiments have been performed with cisplatin, but the induction of apoptosis is generally a slow process because of the requirement for cell cycle passage to convert an initial DNA adduct into a lethal lesion. However, we have no reason to suspect that any of the observations reported here are not equally true in cisplatin-damaged cells. We believe that the steps in the apoptosis pathway we are observing here are very close, if not past the point of commitment to cell death, and hence are common to all inducers of apoptosis.

ACKNOWLEDGMENTS

This paper represents a review of the work performed by many people in my laboratory over the past 4 years. They are too numerous to list as co-authors, but it is their work that is presented here, and their specific contributions are represented in the original manuscripts they have published. The list of past and present contributors include Michael Barry, Mary Kay Brown, Todd Bunch, Don Creswell, Catherine Demarcq, Ron Krieser,

Jinfang Li, Jack McBain, Salvatore Morana, Jason Reynolds, Eric Springer and Chad Wolf. Their efforts are greatly appreciated.

REFERENCES

1. V.M. Richon, N.A. Schulte, and A. Eastman, Multiple mechanisms of resistance to cis-diamminedichloroplatinum(II) in murine leukemia cells, *Cancer Res.* 47:2056 (1987).
2. A. Eastman, N. Schulte, N. Sheibani, and C.M. Sorenson, Mechanisms of resistance to platinum drugs. *in*: "Platinum and Other Metal Coordination Compounds in Cancer Chemotherapy," M. Nicolini, ed., Martinus Nijhoff Publishing, Boston. (1988)
3. J.F.R. Kerr, A.H. Wyllie, and A.R. Currie, Apoptosis: a basic biological phenomenon with wide-ranging implications in tissue kinetics, *Br. J. Cancer* 26:239 (1972).
4. A.H. Wyllie, J.F.R. Kerr, and A.R. Currie, Cell death: the significance of apoptosis, *Int. Rev. Cytol.* 68:251 (1980).
5. M.J. Arrends, R.G. Morris, and A.H. Wyllie, Apoptosis: the role of the endonuclease, *Am. J. Path.* 136:593 (1990).
6. C.M. Sorenson and A. Eastman, Mechanism of cis-diamminedichloroplatinum(II)-induced cytotoxicity: role of G_2 arrest and DNA double-strand breaks, *Cancer Res.* 48:4484 (1988).
7. C.M. Sorenson, M.A. Barry, and A. Eastman, Analysis of events associated with cell cycle arrest at G_2 and cell death induced by cisplatin, *J. Nat. Cancer Inst.* 82:749 (1990).
8. M.A. Barry, C.A. Behnke, and A. Eastman, Activation of programmed cell death (apoptosis) by cisplatin, other anticancer drugs, toxins and hyperthermia, *Biochem. Pharmacol.* 40:2353 (1990).
9. M.D. Jacobson, J.F. Burne, and M.C. Raff, Programmed cell death and Bcl-2 protection in the absence of a nucleus, *EMBO J.* 13:1899 (1994).
10. C.M. Sorenson and A. Eastman, Influence of cis-diamminedichloroplatinum(II) on DNA synthesis and cell cycle progression in excision repair proficient and deficient Chinese hamster ovary cells, *Cancer Res.* 48:6703 (1988).
11. C. Demarcq, R.T. Bunch, D. Creswell, and A. Eastman, The role of cell cycle progression in cisplatin-induced apoptosis in Chinese hamster ovary cells, *Cell Growth Different.* 5:983 (1994).
12. I. Zachary and E. Roxengurt, Focal adhesion kinase ($p125^{FAK}$): a point of convergence in the action of neuropeptides, integrins and oncogenes, *Cell* 71:891 (1992).
13. J.E. Meredith, B. Fazeli, and M.A. Schwartz, The extracellular matrix as a cell survival factor, *Molec. Biol. Cell* 4:953 (1993).
14. A.M.P. Montgomery, R.A. Reisfeld, and D.A. Cheresh, Integrin $\alpha_v\beta_3$ rescues cells from apoptosis in three-dimensional dermal collagen, *Proc. Natl. Acad. Sci. USA.* 91:8856 (1994).
15. M.A. Schwartz, C. Lechene, and D.E. Ingber, Insoluble fibronectin activates the Na/H antiporter by clustering and immobilizing integrin $\alpha_5\beta_1$, independent of cell shape, *Proc. Natl. Acad. Sci. USA.* 88:7849 (1991).
16. K. Swann and M. Whitaker, Stimulation of the Na/H exchanger of sea urchin eggs by phorbol ester, *Nature* 314:274 (1985).
17. C. Sardet, L. Counillon, A. Franchi, and J. Pouyssegur, Growth factors induce phosphorylation of the Na^+/H^+ antiporter, a glycoprotein of 110 kD, *Science* 247:723 (1990).

18. D.S. Ucker, P.S. Obermiller, W. Eckhart, J.R. Apgar, N.A. Berger, and J. Meyers, Genome digestion is a dispensable consequence of physiological cell death mediated by cytotoxic T lymphocytes, *Mol. Cell. Biol.* 12:3060 (1992).
19. F. Oberhammer, W. Bursch, W. Parzefall, P. Breit, M. Stadler, and R. Schulte-Hermann, Effects of transforming growth factor β on cell death of cultured rat hepatocytes, *Cancer Res.* 51:2478 (1991).
20. Z.F. Zakeri, D. Quaglino, T. Latham, and R.A. Lockshin, Delayed internucleosomal DNA fragmentation in programmed cell death, *FASEB J.* 7:470 (1993).
21. F. Oberhammer, J.W. Wilson, C. Dive, I.D. Morris, J.A. Hickman, A.E. Wakeling, P.R. Walker, and M. Sikorska, Apoptotic death in epithelial cells: cleavage of DNA to 300 and/or 50 kb fragments prior to or in the absence of internucleosomal fragmentation, *EMBO J.* 12:3679 (1993).
22. D.G. Brown, X.-M. Sun, and G.M. Cohen, Dexamethasone-induced apoptosis involves cleavage of DNA to large fragments prior to internucleosomal fragmentation, *J. Biol. Chem.* 268:3037 (1993).
23. M.A. Barry and A. Eastman, Identification of deoxyribonuclease II as an endonuclease involved in apoptosis, *Arch. Biochem. Biophys.* 300:440 (1993).
24. M.A. Barry and A. Eastman, Endonuclease activation during apoptosis: the role of cytosolic Ca^{2+} and pH, *Biochem. Biophys. Res. Commun.* 186:782 (1992).
25. M.A. Barry, J.E. Reynolds, and A. Eastman, Etoposide-induced apoptosis in human HL-60 cells is associated with intracellular acidification, *Cancer Res.* 53:2349 (1993).
26. J. Li and A. Eastman, Apoptosis in an IL-2-dependent cytotoxic T lymphocyte cell line is associated with intracellular acidification: role of the Na^+/H^+-antiport, *J. Biol. Chem.* 270:3203 (1995).
27. I.H. Madshus, Regulation of intracellular pH in eukaryotic cells, *Biochem. J.* 250:1 (1988).
28. M. Dohadwala, E.F. da Cruz e Silva, F.L. Hall, R.T. Williams, D.A. Carbonara-Hall, A.C. Nairn, P. Greengard, and N. Berndt, Phosphorylation and inactivation of protein phosphatase 1 by cyclin-dependent kinases, *Proc. Natl. Acad. Sci. USA.* 91:6408 (1994).
29. J. Yuan, S. Shaham, S. Ledoux, H.M. Ellis, and H.R. Horvitz, The C. elegans cell death gene ced-3 encodes a protein similar to mammalian interleukin-1 β-converting enzyme, *Cell* 75:641 (1993).
30. M. Miura, H. Zhu, R. Rotello, E.A. Hartwieg, and J. Yuan, Induction of apoptosis in fibroblasts by IL-1 β-converting enzyme, a mammalian homolog of the C. elegans cell death gene ced-3, *Cell* 75:653 (1993).
31. S.H. Kaufmann, Induction of endonucleolytic DNA cleavage in human acute myelogenous leukemia cells by etoposide, camptothecin, and other cytotoxic anticancer drugs: a cautionary note, *Cancer Res.* 49:5870 (1989).
32. Y.A. Lazebnik, S.H. Kaufmann, S. Desnoyers, G.G. Poirier, and W.C. Earnshaw, Cleavage of poly(ADP-ribose) polymerase by a proteinase with properties like ICE, *Nature* 371:346 (1994).
33. S.C. Wright, Q.S. Wei, J. Zhong, H. Zheng, D.H. Kinder, and J.W. Larrick, Purification of a 24-kD protease from apoptotic tumor cells that activates DNA fragmentation, *J. Exp. Med.* 180:2113 (1994).
34. A. Eastman, Survival factors, intracellular signal transduction, and the activation of endonucleases in apoptosis, *Sem. Cancer Biol.* 6:in press (1995).

CELL CYCLE CHECKPOINTS AND CANCER CHEMOTHERAPY

Patrick M. O'Connor and Saijun Fan

Laboratory of Molecular Pharmacology
Developmental Therapeutics Program
Division of Cancer Treatment
National Cancer Institute
Bethesda, Maryland 20892

INTRODUCTION

The majority of anticancer agents in current clinical practice arrest cell cycle progression at one or more definable points (Table 1). The DNA damaging agents including, bleomycin, adriamycin, etoposide, nitrogen mustards and cisplatin, arrest cell cycle progression in G1 and/or G2 phases. These agents can also prolong S phase progression. The antimetabolites including, methotrexate, 5-fluorouracil and 6-mercaptopurine, arrest cell cycle progression at the G1/S phase border and in S phase. The microtubule inhibitors including, vincristine, vinblastine and taxol, arrest cells primarily in mitosis and more specifically in a pseudometaphase state. Arrest at these stages in the cell cycle is dependent on the integrity of a series of negative feedback control systems that have become commonly termed checkpoints (1, 2). These checkpoints could protect cells from cytotoxicity by extending the time for drug-induced perturbations to be corrected before cell division. This possibility is supported by findings from yeast genetics, which have shown that inactivating mutations in checkpoint control genes sensitized cells to DNA damaging agents, antimetabolites or antimitotic agents (1, 2). Furthermore, the findings by Pardee and colleagues, that chemical agents like pentoxifylline, can abrogate G2 checkpoint control and synergise with DNA damaging agents (3), also supports the protective role of cell cycle checkpoints. Based on such observations, we have suggested that the integrity of checkpoint control systems in cancer cells may in large part determine chemosensitivity (1). Our thinking here is that uncontrolled progression through one or more of these checkpoints in the presence of damage will predispose cells to killing. We discuss below recent observations we and other workers have made in investigating the role of the G1 and G2 cell cycle checkpoints in chemosensitivity. We focus on attempts to determine whether the G1 and G2 cell cycle checkpoints are commonly defective in cancer cells, whether such checkpoint alterations affect chemosensitivity and whether checkpoint alterations in cancer cells could provide new opportunities for drug discovery.

DO CELL CYCLE CHECKPOINT ALTERATIONS EXIST IN CANCER CELLS?

THE P53 TUMOR SUPPRESSOR AND THE G1 CHECKPOINT

The p53 tumor suppressor is required to arrest cell cycle progression in late G1 phase following DNA damage (2). However, since the p53 gene is also the most commonly

Table 1. Antitumor agents, targets and cell cycle perturbations.

Drug	Targets	Mode of action	Cell cycle arrest
Methotrexate 5-Fluorouracil 6-Mercaptopurine	Enzymes	Inhibition of Nucleic acid metabolism	S phase delay
Cisplatin Nitrogen mustards Mitomycin	DNA	Crosslinking	G1, G2 (S phase delay)
Irradiation Bleomycin	DNA	Free radical attack	G1, G2 (S phase delay)
Etoposide Adriamycin Camptothecin	Topoisomerases	Stabilization of DNA Cleavable complexes	G1, G2 (S phase delay)
Vinblastine Vincristine Taxol	Tubulin	Microtubule Inhibitors	Mitosis

mutated gene in human cancer (4), many cancer cells lack the ability to G1 arrest in the presence of DNA damage. This lack of arrest has been implicated in the increased genomic instability commonly observed in p53 disrupted cells (presumably a consequence of reduced DNA repair before S phase progression). Lack of arrest could aid in the evolution of cancer as well as impart altered sensitivity to commonly used anticancer agents. Exemplifying the tumor suppressor properties of p53, transgenic mice in which p53 function was disrupted by homologous recombination suffer a markedly increased frequency of tumors, particularly lymphoma's, compared to their wild-type littermates (5).

Our work in a series of Burkitt's lymphoma and lymphoblastoid cell lines, which were characterized in regard to p53 gene status, testifies to the importance of a functional p53 pathway in G1 arrest following exposure to ionizing radiation (6). Cell lines with wild-type p53 arrest strongly in G1 of the cell cycle following γ-irradiation, while cells with mutant p53 fail to G1 arrest (Figure 1). Importantly, cancer cells with heterozygous p53 status, in which cells have one mutant p53 allele and one wild-type p53 allele, also fail to G1 arrest (Figure 1). This finding illustrates the dominant-negative influence of the mutant p53 protein upon the function of the wild-type protein. Indeed, patients with Li-Fraumeni syndrome carry germ line mutations in one of their p53 alleles (7), and a potential dominant-negative effect of the mutated p53 protein might explain the increased cancer susceptibility of these individuals. The integrity of the p53 pathway is important for G1 arrest following treatment with agents that induce DNA strand breaks, such as ionizing radiation and etoposide (8). Agents that induce DNA base damage such as the nitrogen mustards and cisplatin appear to be less active in inducing a p53-dependent G1 arrest (8). These findings and those of Nelson and Kastan (9), have lead to the suggestion that the p53 pathway may preferentially recognize DNA strand breaks induced either directly or indirectly by DNA damage. Indirect mechanisms to the activation of p53 could arise as a consequence of DNA strand breaks formed during DNA repair or DNA strand breaks resulting from collisions between DNA replication or transcription complexes and DNA lesions. This latter "collision-induced DNA break" hypothesis may be important in the activation of the p53 tumor suppressor in cells treated with the topoisomerase I inhibitors, such as camptothecin (9).

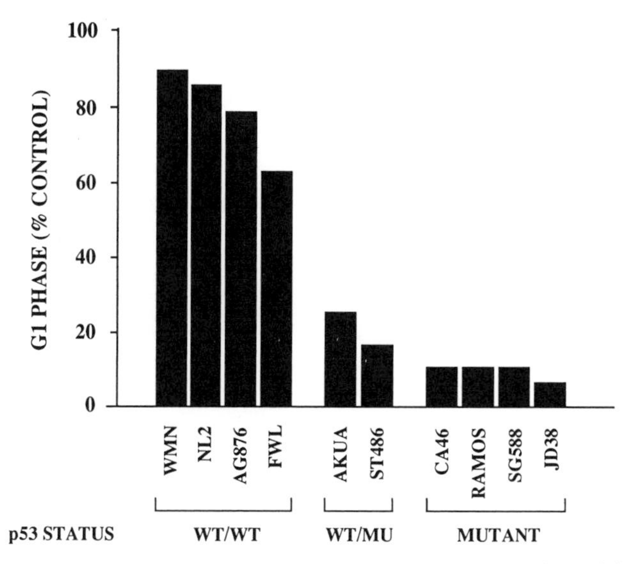

Figure 1. Relationships between G1 arrest and p53 gene status in a panel of Burkitt's lymphoma and lymphoblastoid cell lines. Shown is the percentage of the original G1 population that arrested in G1 at 16 hours following 6.3 Gy of gamma-rays

A further syndrome in which p53 function is incapacitated is in ataxia telangiectasia (AT, 10). The AT defect has been mapped to a locus on chromosome 11 (11q21-23), and a disruption in this region blocks the relay of a signal from DNA damage to the p53 tumor suppressor (10). This presumably renders cells from patients with ataxia telangiectasia devoid of p53 function and may again help explain the predisposition of these patients to tumorigenesis. A candidate gene for AT has recently been cloned by Yossi Shiloh and colleagues (11). Excitingly, the AT gene contains a region of similarity with the fission yeast gene RAD3 and the budding yeast MEC1 gene, both of which encode PI-3 kinases implicated in DNA damage checkpoint control.

The p53 tumor suppressor pathway can also be disrupted in a variety of other ways which leaves the p53 gene sequence normal. These include infection of cells with viruses such as the human papillomavirus type-16/18 virus which contain the E6 gene, whose product binds to and stimulates the destruction of p53 (12). Furthermore, amplification of the cellular MDM2 gene, commonly seen in soft tissue sarcomas, leads to high level expression of a gene product which binds to the amino terminus of p53 and inactivates p53 function (13). Taken together, the findings described above point to the common inactivation of p53 function in cancer cells.

THE P53 GENE PRODUCT IS A TRANSCRIPTIONAL REGULATOR

The p53 gene product exerts its activity through transcriptional regulation, having the ability to induce transcription of genes which contain a cognate p53 binding sequence, as well as suppress gene transcription through binding of proteins associated with genes that contain TATA elements in their promoter (14). Genes that p53 induces include the DNA damage-inducible gene Gadd45, whose product binds to PCNA, a protein which is involved in DNA replication and DNA repair (15). P53 also induces transcription of the WAF1/CIP1 gene, whose product is a potent inhibitor of the cyclin-dependent kinases which regulate the G1/S phase transition (16, 17). Indeed, the discovery of WAF1/CIP1 allowed the p53 tumor suppressor pathway to be linked to the cell cycle machinery in a way that might explain G1 arrest. Definitive proof of this hypothesis awaits publication of studies in transgenic mice in which the WAF1/CIP1 gene has been "knocked out". Cells from such mice should lack G1 arrest if Waf1/Cip1 is the sole mediator of p53-induced G1 arrest. P53 also induces transcription of the MDM2 gene which encodes a negative regulator of p53 function believed to feedback on p53 to inhibit transcriptional activity. This feedback loop would then limit the longevity of p53's actions to the stability of the newly synthesized gene products which p53 has induced (18). P53 has also recently been found to induce transcription of the BAX gene, whose product is a potent inducer of apoptosis (19, 20). These finding point to a possible mechanism by which p53 could induce apoptosis in cells with DNA damage. Indeed, we have found that BAX is only induced in those wild-type p53 cells which are committed to apoptosis following DNA damage (19). Cells in which p53 activation induces a G1 arrest but not apoptosis fail to show up-regulation of BAX. These finding suggest a factor may be present in some cell types that protects against p53-induced apoptosis by preventing BAX induction. Studies are underway in our laboratory to define this factor, with the aim of manipulating its activity to promote apoptosis in cell types not inherently prone to DNA damage-induced apoptosis.

THE G2 CHECKPOINT

Arrest of lymphoma cells in G2 phase following DNA damage appears to occur independent of p53 function (6). This has lead many observers to conclude that the p53 tumor suppressor is not involved in the G2 checkpoint. Perhaps it is more correct to say that p53 function is not essential for G2 arrest because "redundant" mechanisms operate at the G2 checkpoint. This would still leave open the possibility that p53 could add an additional layer of protection to the G2 checkpoint which might explain our recent observations, discussed below, that p53-disrupted cells are more sensitive to G2 checkpoint abrogators than cells with intact p53 (21). The molecular dissection of elements involved in G2 checkpoint regulation have focused on the Cdc2 kinase, which when activated promotes entry into mitosis. In cells treated with DNA damaging agents we have

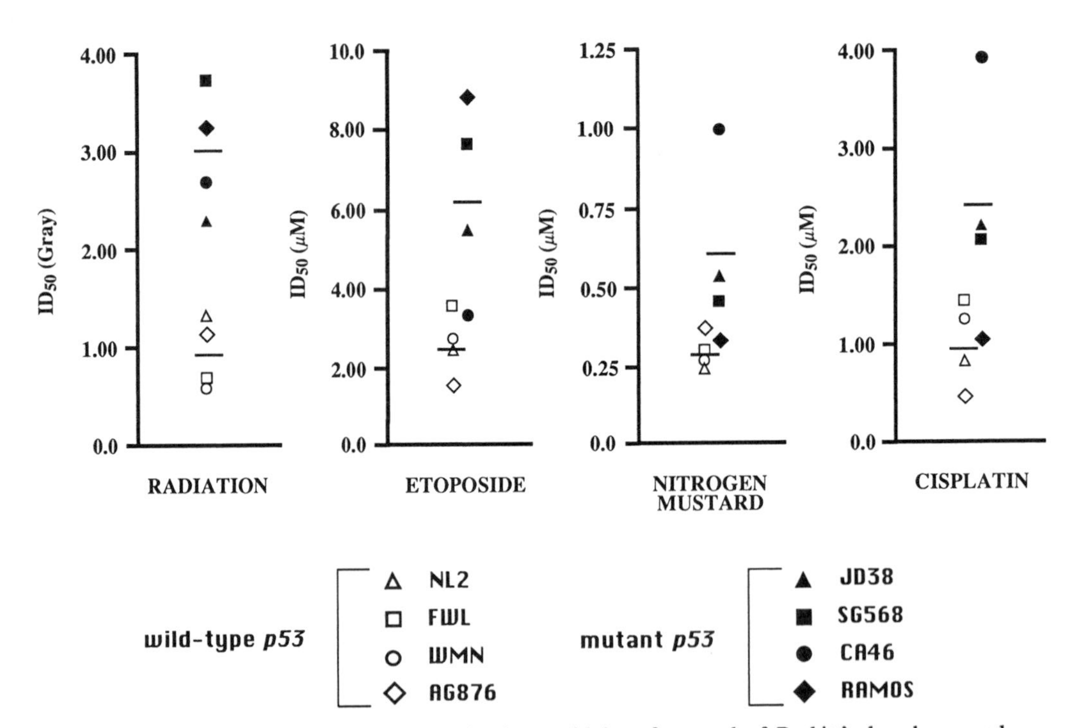

Figure 2. The role of p53 gene status in the sensitivity of a panel of Burkitt's lymphoma and lymphoblastoid cell lines to several types of DNA damaging agents. Shown is the dose of each agent that reduced cell survival by 50% of the control untreated population. The bars shown are the mean ID50 value for each set of either wild-type p53 or mutant-p53 cell lines.

found that Cdc2 activation is suppressed due to the inability to remove inhibitory phosphorylations from the ATP binding domain of the kinase (22, 23, 24). The Cdc25C phosphatase is responsible for the removal of these Cdc2-inhibitory phosphorylations and the activation of Cdc25C is suppressed in DNA damaged cells (24).

Reports of defects in the G2 checkpoint of cancer cells have been sparse. Some reports point to alterations in G2 arrest following DNA damage in cells from patients with ataxia telangiectasia (25) or Burkitt's lymphoma (22). One possible explanation for differences between the number of cancer cell lines that show G1 versus G2 checkpoint disruption, is that unlike the G1 checkpoint, which is entirely dependent upon p53 function, the G2 checkpoint may be composed of a multilayered control system. If several entities contribute to regulation of the G2 checkpoint then loss of one of these entities would not expose defective G2 arrest unless cells were challenged through that particular defective entity. A recent survey of the 60 cell lines in the NCI anticancer drug screen, using flow cytometric approaches to measure the degree of G2 arrest following 6.3 Gy of γ-rays did, however, reveal marked differences in G2 arrest capacity (O'Connor et al., in preparation). These differences, if confirmed by additional methodologies, could provide the first detailed assessment of G2 arrest capacity in cancer cell lines. Arrest at this checkpoint can be overcome by G2 checkpoint abrogators such as pentoxifylline (3) and cells in which p53 has been disrupted appear to be more susceptible to G2 checkpoint abrogation (21, 26, 27). These findings have suggested that a fundamental difference may exist in the regulation of the G2 checkpoint in cells with intact versus disrupted p53 function. A detailed study is presently on-going in our laboratory to determine if the formation and/or activation of the cyclin B/Cdc2 kinase, which regulates the G2/M transition, is altered in cells with mutant-p53.

DO CHECKPOINT ALTERATIONS IN CANCER CELLS AFFECT CHEMOSENSITIVITY?

ALTERATIONS IN P53 FUNCTION AND CHEMOSENSITIVITY

The findings described above suggest that p53 acts a checkpoint control protein that induces G1 arrest in the presence of DNA damage, This would presumably allow more time for cells to complete DNA repair before S phase entry. The role of p53 in this regard is analogous to the role of the RAD9 gene in yeast, which halts cell cycle progression following DNA damage (2). The participation of p53 and RAD9 in cell cycle arrest following DNA damage would help ensure the fidelity with which genetic material is transmitted from one generation to the next. This is supported by findings from cells which lack RAD9 or p53 function, since such cells suffer greater genomic instability than their normal counterparts (2). This being the case, one might expect that p53 disruption would sensitize cancer cells to DNA damaging agents. However, since p53 can also induce apoptosis in some cell types the outcome of p53 activation is highly dependent on cellular context. In hematopoietic and lymphoid cells, activation of p53 following DNA damage with such agents as ionizing radiation, etoposide, nitrogen mustard and cisplatin leads to apoptosis (8). Cells which lack p53 function are less able to relay the DNA damage signal to the apoptotic machinery and thereby stand a greater chance of evading apoptosis. This offers a selective advantage to the survival of mutant p53 cells following treatment with DNA damaging agents. Indeed, this means that to achieve equivalent cell killing of wild-type and mutant p53 lymphoid cell lines, much higher doses of DNA damaging agents are required in the mutant p53 cells (Figure 2).

In conjunction with Steve Friend, John Weinstein, Al Fornace, Kurt Kohn and colleagues we found that those cells in the NCI anticancer drug screen with disrupted p53 function also tended to be less sensitive to DNA damaging agents than cells with wild-type p53 function (O'Connor et al., in preparation). We also found that the activity of the microtubule inhibitors including taxol, vincristine and vinblastine was not dependent on p53 gene status. The independence of p53 gene status in the activity of taxol and vincristine has now been confirmed in our panel of Burkitt's lymphoma and lymphoblastoid cells as well as isogenic cell lines in which p53 function has been disrupted by transfection with the HPV-E6 or mutant p53 genes (Fan et al., in preparation). These results are encouraging because they show that at least some agents exist in current clinical practice that would not be limited in their activity against mutant p53 tumor cells.

Cell types, not inherently prone to apoptosis, are a more difficult case in which to formulate a model on the role of p53 in sensitivity to DNA damaging agents. Kastan and colleagues (28) were the first to show that the wild-type p53 colon carcinoma RKO cell line, which is not inherently prone to DNA damage induced apoptosis (19), did not show altered sensitivity to ionizing radiation when transfected with the HPV-E6 or mutant p53 transgenes. Similar conclusions were made in regard to p53 status by Weichselbaum and colleagues using a series of head and neck cancer cell lines (29). Using the MCF-7 and RKO cell line systems as an investigative model we disrupted p53 function and looked for changes in sensitivity to a variety of DNA damaging agents (21). These cell lines despite having a robust p53 response to DNA damage are not inherently prone to apoptosis (19). We found similar results for ionizing radiation to that described by Kastan, Weichselbaum and colleagues (28, 29). We also found that disruption of p53 function in MCF-7 and RKO cells did not alter the sensitivity of the cells to adriamycin, etoposide or methylmethanesulphonate. However, we found that p53 disruption in MCF-7 and RKO cells sensitized cells to cisplatin (Figure 3). Our findings with cisplatin agreed with earlier studies by Brown et al., who found that mutant p53 transfection into a cisplatin-resistant derivative of the A2780 cell line, which contained an endogenous wild-type p53 gene, also sensitized these cells to cisplatin (30). Unlike the other agents we tested in MCF-7 cells, cisplatin-induced DNA adducts are repaired primarily through nucleotide excision, a repair pathway which p53 and p53 regulated gene products have been implicated (15). The enhanced ultraviolet light sensitivity of RKO cells lacking p53 function and host cell reactivation assays which showed reduced reactivation of ultraviolet light- and cisplatin-treated CAT reporter plasmids in p53 disrupted cells (21, 31), added additional support for differences in nucleotide excision repair in p53 disrupted cells. These findings may be of importance since they suggest that certain mutant-p53 tumor cells may be vulnerable to DNA damaging agents that induce DNA lesions repaired exclusively through the nucleotide

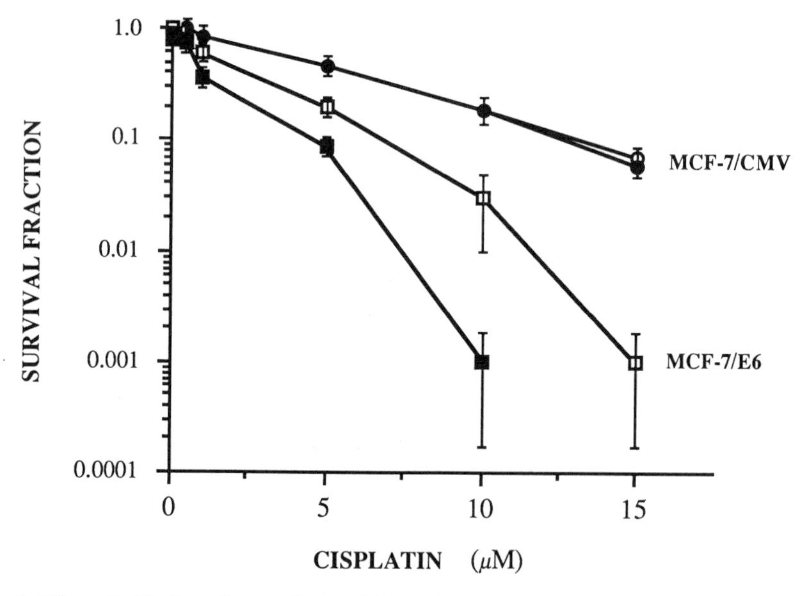

Figure 3. Effect of p53 disruption on the sensitivity of breast carcinoma MCF-7 cells to cisplatin and pentoxifylline. Shown are the results of clonogenic survival assays of MCF-7 cells transfected with either control CMV-plasmid (circles) or a CMV-plasmid that contains the human papillomarvirus E6 gene (squares) whose product stimulates p53 degradation. Cells were treated with cisplatin for one hour (open symbols) or a combination of cisplatin and a 24 hour exposure to 2 mM pentoxifylline (closed symbols). Values shown are the mean and standard deviation.

excision repair pathway (21). These results also encourage the continued investigation of p53 disrupted cells for additional abnormalities which could be pharmacologically exploited in stratagems designed to preferentially kill p53 disrupted cells.

ALTERATIONS AND MANIPULATION OF THE G2 CHECKPOINT IN CANCER CELLS AND CONSEQUENCES FOR CHEMOSENSITIVITY

During the course of our investigations into the chemosensitivity of MCF-7 cells with intact versus disrupted p53, we also tested the effects of pentoxifylline (21). Pentoxifylline is related to caffeine and both agents abrogate G2 arrest induced by DNA damage and synergize with DNA damaging agents (3). We used pentoxifylline to "knock-out" the G2 checkpoint in MCF-7 cells with the view that lack of G2 arrest should be more deleterious to the p53 disrupted MCF-7 cells than control transfectants still capable of G1 arrest. In agreement with this hypothesis, we found that a combination of pentoxifylline and cisplatin was preferentially toxic to MCF-7 cells with disrupted p53 (Figure 3). We initially felt that the lack of an effect of pentoxifylline on cisplatin toxicity in control transfected MCF-7 cells was due to G1 arrest and DNA repair. However, flow cytometry studies revealed that pentoxifylline preferentially abrogated the G2 checkpoint in MCF-7 cells which lacked p53 function (21). This was a striking finding, which was supported by studies by Russell et al., (26) and Powell et al., (27) who found that caffeine preferentially abrogated the G2 checkpoint in p53 disrupted cells. These findings pointed to possible differences in G2 checkpoint regulation in p53 intact versus p53 disrupted cells and offered a pharmacologically attractive stratagem to preferentially kill cancer cells with mutant p53. A limitation to the use of pentoxifylline in the clinical application of this stratagem is the relatively high plasma levels required to mimic these *in vitro* observations. We are presently exploring several compounds in our laboratory for preferentially abrogation of the G2 checkpoint in cells with disrupted p53. Our aim is to define a more potent and selective G2 checkpoint abrogator which could be employed in combination chemotherapy for the preferential killing of mutant p53 tumors.

SUMMARY AND FUTURE DIRECTIONS

Our investigations of the role of the G1 and G2 cell cycle checkpoints in the chemosensitivity of cancer cells have revealed that not only are these checkpoints altered in human cancer cell lines but these changes impart altered sensitivity to many currently used anticancer agents. Highlighted, within the common genetic changes in human cancer is disruption of the p53 checkpoint gene which globally impacts upon chemosensitivity. Our current research efforts are focused on the exploration of relationships between p53 gene status and the activity of the 45,000 compounds tested in the NCI anticancer cell screen. We are searching for agents that might preferentially kill cancer cells with mutant p53. Candidate agents will undergo confirmatory analysis using transfected cell lines in which the endogenous p53 gene product has been disrupted and chemosensitivity comparisons made with control transfectants. We are also pursuing our recent observations that G2 checkpoint abrogators, like pentoxifylline, synergize with cisplatin to preferentially kill cancer cells with disrupted p53. Our aims here are to generate new lead compounds to replace pentoxifylline in stratagems designed to exploit p53 deficiencies in cancer cells. Continued exploration of cell cycle checkpoints and their role in chemosensitivity and the exploitation of transfected cell line systems differing in the function of individual gene products might continue to yield important new discoveries for drug development.

REFERENCES

1. P. M. O'Connor and K. W. Kohn. A fundamental role for cell cycle regulation in the chemosensitivity of cancer cells? Semin. Cancer Biol., 3:409 (1992).

2. L. H., Hartwell and M. B. Kastan. Cell cycle control and cancer. Science, 266:1821 (1994)

3. H. J. Fingert, A. T. Pu, Z. Chen, P. B. Googe, M. C. Alley and A. B. Pardee. In vivo and in vitro enhanced antitumor effects by pentoxifylline in human cancer cells treated with thiotepa. Cancer Res., 48:4375 (1988)

4. M. Hollstein, D. Sidransky, B. Vogelstein and C. C. Harris. p53 mutations in human cancers. Science, 253:49 (1991)

5. L. A. Donehower, M. Harvey, B. L. Slagle, M. J. McArthur, C. A. Montgomery, Jr., J. S. Butel and A. Bradley. Mice deficient for p53 are developmentally normal but susceptible to spontaneous tumors. Nature, 356:215 (1992)

6. P. M. O'Connor, J. Jackman, D. Jondle, K. Bhatia, I. Magrath and K. W. Kohn. Role of the p53 tumor suppressor gene in cell cycle arrest and radiosensitivity of Burkitt's lymphoma cell lines. Cancer Res., 53:4776 (1993)

7. D. Malkin, F. P. Li, L. C. Strong, J. F. Fraumeni, Jr., C. E. Nelson, D. H. Kim, J. Kassel, M. A. Gryka, F. Z. Bischoff, M. A. Tainsky and S. H. Friend. Germ line p53 mutations in a familial syndrome of breast cancer, sarcoma's and other neoplasms. Science, 250:1233 (1990)

8. S. Fan, W. S. El-Deiry, I. Bae, J. Freeman, D. Jondle, K. Bhatia, A. J. Fornace, Jr., I. Magrath, K. W. Kohn and P. M. O'Connor. p53 gene mutations are associated with decreased sensitivity of human lymphoma cells to DNA damaging agents. Cancer Res., 54:5824 (1994)

9. W. G. Nelson and M. B. Kastan. DNA strand breaks: the DNA template alterations that trigger p53-dependent DNA damage response pathways. Mol. Cell. Biol., 14:1815 (1994)

10. M. B. Kastan, Q. Zhan, W. S. El-Deiry, F. Carrier, T. Jacks, W. V. Walsh, B. Plunkett, B. Vogelstein and A. J. Fornace, Jr. A mammalian cell cycle checkpoint pathway utilizing p53 and GADD45 is defective in ataxia telangiectasia. Cell, 75:817 (1993)

11. K. Savitsky, A. Bar-Shira, S. Gilad, G. Rotman, Y. Ziv, L. Vanagaite, D.A. Tagle, S. Smith, T. Uziel, S. Sfez, M. Ashkenazi, I. Pecker, M. Frydman, R. Harnik, S.R. Patanjali, A. Simmons, G.A. Clines, A. Sartiel, R.A. Gatti, L. Chessa, O. Sanal, M.F. Lavin, N.G.J. Jaspers, A.M.R. Taylor, C.F. Arlett, T. Miki, S.M. Weissman, M. Lovett, F.S. Collins, Y. Shiloh. A single Ataxia Telangiectasia gene with a product similar to PI-3 kinase. Science, 268:1169 (1995)

12. T. D. Kessis, R. J. Slebos, W. G. Nelson, M. B. Kastan, B. S. Plunkett, S. M. Han, A. T. Lorincz, L. Hedrick and K. R. Cho. human papillomavirus 16 E6 expression disrupts the p53 mediated cellular response to DNA damage. Proc. Natl. Acad. Sci. (USA), 90:3988 (1993)

13. X. Wu, J. H. Bayle, D. Olson and A. J. Levine. the p53-mdm-2 autoregulatory feedback loop. Genes and Dev., 7:1126 (1993)

14. G. P. Zambetti and A. J. Levine. A comparison of the biological activities of wild-type and mutant p53. FASEB J., 7:855 (1993)

15. M. L. Smith, I. Chen, Q. Zhan, I. Bae, T. Gilmer, M. B. Kastan, P. M. O'Connor and A. J. Fornace, Jr. Interaction of the p53 regulated protein Gadd45 with proliferating cell nuclear antigen. Science, 266:1376 (1994)

16. W. S. El-Deiry, T. Tokino, V. E. Veculescu, D. B. Levy, R. Parson, J. M. Trent, D. Lin, W. E. Mercer, K. W. Kinzler and B. Vogelstein. WAF1, a potential mediator of p53 tumor suppression. Cell, 75:817 (1993)

17. J. W. Harper, G. R. Adami, N. Wei, K. Keyomarsi and S. J. Elledge. The p21 Cdk-interacting protein Cip1 is a potent inhibitor of G1 cyclin dependent kinases. Cell, 75:805 (1993)

18. I. Bae, S. Fan, K. Bhatia, K.W. Kohn, A.J. Fornace, Jr., and P.M. O'Connor. Relationships between G1 Arrest and Stability of the p53 and $p21^{Cip1/Waf1}$ Proteins following γ-Irradiation of Human Lymphoma Cells. Cancer Res., 55:2387 (1995)

19. Q. Zhan, S. Fan, I. Bae, C. Guillouf, D. A. Liebermann, P. M. O'Connor and A. J. Fornace, Jr. Induction of bax by genotoxic stress in human cells correlates with normal p53 status and apoptosis. Oncogene, 9:3743 (1994)

20. M. Selvakumaran, H-K. Lin, T. Miyashita, H. G. Wang, S. Krajewski, J. C. Reed, B. Hoffman and D. Liebermann. Immediate early up-regulation of bax expression by p53 but not TGF beta 1:a paradigm for distinct apoptotic pathways. Oncogene, 9:1791 (1994)

21. S. Fan, M. L. Smith, D. J. Rivet, D. Duba, Q. Zhan, K. W. Kohn, A. J. Fornace, Jr. and P. M. O'Connor. Disruption of p53 function sensitizes breast cancer MCF-7 cells to cisplatin and pentoxifylline. Cancer Res., 55:1649 (1995)

22. P. M. O'Connor, D. K. Ferris, G. A. White, J. Pines, T. Hunter, D. L. Longo and K. W. Kohn. Relationships between the cdc2 kinase, DNA crosslinking and cell cycle perturbations induced by nitrogen mustard. Cell Growth & Differen. 3:43 (1992)

23. P. M. O'Connor, D. K. Ferris, M. Pagano, G. Draetta, J. Pines, T. Hunter, D. L. Longo and K. W. Kohn. G2 delay induced by nitrogen mustard in human cells affects cyclin A/Cdk2 and cyclin B1/cdc2 kinases differently. J. Biol. Chem. 268:8298 (1993)

24. P. M. O'Connor, D. K. Ferris, I. Hoffmann, J. Jackman, G. Draetta and K. W. Kohn. Role of the cdc25C phosphatase in G2 arrest induced by nitrogen mustard. Proc. Natl. Acad. Sci. (USA), 91:9480 (1994)

25. F. Zampetti-Bosseler and D. Scott. Cell death, chromosome damage and mitotic delay in normal human, ataxia telangiectasia and retinoblastoma fibroblasts after x-irradiation. Int J. Radiat Biol. Relat. Stud. Phys. Chem. Med. 39:547 (1981)

26. K. J. Russell, L. W. Weins, D. A. Galloway and M. Groudine. Abrogation of the G2 checkpoint results in differential radiosensitization of G1 checkpoint deficient and competent cells. Cancer Res., 55:1639 (1995)

27. S. N. Powell, J. S. DeFrank, P. Connell, M. Preffer, D. Dombkowski, W. Tang and S. Friend. Differential sensitivity of p53+ and p53- cells to caffeine-induced radiosensitization and override of G2 delay. Cancer Res., 55:1643 (1995)

28. W. J. Slichenmyer, W. G. Nelson, R. J. Slebos and M. B. Kastan. Loss of a p53 associated G1 checkpoint does not decrease cell survival following DNA damage. Cancer Res., 53:4164 (1993)

29. D. G. Brachman, M. Beckett, D. Graves, D. Haraf, E. Vokes and R. Weichselbaum. p53 mutation does not correlate with radiosensitivity in head and neck cancer cell lines. Cancer Res., 53:3666 (1993)

30. R. Brown, C. Clugston, P. Burns, A. Edlin, P. Vasey, B. Vojtesek and S. Kaye. Increased accumulation of p53 protein in cisplatin-resistant ovarian cell lines. Int. J. Cancer, 55:678 (1993)

31. M. L. Smith, I. T. Chen, Q. Zhan, P. M. O'Connor and A. J. Fornace, Jr. Involvement of the p53 tumor suppressor in repair of UV-type DNA damage. Oncogene, 10:1053 (1995)

CARBOPLATIN VERSUS CISPLATIN IN THE CHEMOTHERAPY OF SOLID TUMORS - PRO CARBOPLATIN

David S. Alberts, M.D.

Professor of Medicine and Pharmacology
Deputy Director, Arizona Cancer Center
Tucson, Arizona
Chairman, Gynecologic Cancer Committee
Southwest Oncology Group
San Antonio, Texas.

BACKGROUND

The fact that we are still debating, in 1995, the merits of carboplatin versus cisplatin in the chemotherapy of solid tumors, exemplifies the extreme slowness of planning, implementing and completing important clinical trials in oncology. Cisplatin was approved for use in the treatment of ovarian cancer in the late 1970s, and carboplatin was approved for salvage therapy of ovarian cancer, in the late 1980s. Six years later, we continue to struggle with decisions concerning the appropriate role for these two platinum analogs in the management of ovarian, cervix, endometrial, head and neck, lung, and testicular cancers. Although there have been abundant phase III clinical trials comparing these agents in a number of target disease sites, the sample size and trial designs in most cases have proven inadequate to allow final conclusions to be drawn. Nevertheless, I have had no difficulty composing a "Pro Carboplatin" position paper. I even can say that this was a pleasurable experience!

Listed in Table 1 are my choices concerning the relative merits of carboplatin versus cisplatin therapy of solid tumors with respect to intensity of pre-medications, ease of drug administration, acute and chronic toxicities, anti-tumor activities and pharmacoeconomics.

The check marks in Table 1 are placed under the carboplatin or cisplatin column headings according to the weight of published data favoring one analog over the other with respect to each of the important evaluation variables. The following discussion concerns the rationale behind these check marked choices.

EASE OF ADMINISTRATION

There is little room for argument concerning the fact that carboplatin is easily administered in the outpatient setting. This drug can be administered safely in as little as 50 ml sterile water over as short a period as 15 minutes. No pre- or post- intravenous hydration or osmotic diuresis is required.[1] In contrast, cisplatin must be administered slowly, usually at a rate of no more than 1 mg/minute, or over approximately 2-4 hours for doses of 75-100 mg/m^2 BSA. Pre-hydrating with 1 to 2 L of intravenous saline and at least 37.5 gm of intravenous mannitol are usually prescribed, depending on the cisplatin dose, age of patient and status of renal function.[2]

Platinum and Other Metal Coordination Compounds in Cancer Chemotherapy 2
Edited by H.M. Pinedo and J.H. Schornagel, Plenum Press, New York, 1996

Table 1. Relative Merits of Carboplatin and Cisplatin in Clinical Oncology*

	Pro Carboplatin	*Pro Cisplatin*	*Equivalent*
Ease of Administration	√		
Premedication Intensity Required	√		
Toxicities			
Emetogenesis	√		
Renal Toxicity	√		
Ototoxicity			
Peripheral Neuropathy	√		
Neuromuscular Toxicity	√		
Myelosuppression		√	
Antitumor Activity			
Ovarian Cancer			√
Lung Cancer			
Non-Small Cell			√
Small Cell			√
Head and Neck Cancer		√	
Testicular Cancer		√	
Endometrial Cancer	√		
Cervix Cancer		√	
Pharmacoeconomics	√		

*Check marks are placed in the columns under "Pro Carboplatin" or "Pro Cisplatin" on the basis of the "weight" of the medical literature supporting the relative merits of one analog over the other.

TOXICITIES

More than two dozen large, randomized clinical trials have been performed comparing carboplatin to cisplatin in the management of a wide variety of solid tumors. Virtually every one of these published studies has documented carboplatin's favorable safety profile at standard doses of 300 - 450 mg/m^2 administered every 4 weeks. Clearly, carboplatin is less emetogenic than cisplatin and is associated with minimal renal, oto- or neuromuscular toxicities at these effective dose levels. These studies also have documented carboplatin's more pronounced myelotoxicity which tends to be cumulative, but generally well tolerated through six courses of standard type therapy.

ANTI-TUMOR ACTIVITY

Obviously, anti-tumor activity is the controversial aspect of the ongoing debate concerning the relative merits of carboplatin and cisplatin. Listed below are the individual tumor categories with accompanying discussion supporting the placement of the check marks in Table 1.

Ovarian Cancer

At least 11 randomized studies have been performed with the primary objective being the comparison of carboplatin to cisplatin anticancer activity in patients with advanced ovarian cancer.[3-13] Additionally, a large metanalysis of data from 2,060 patients entered worldwide into phase III studies comparing carboplatin to cisplatin in ovarian cancer has been published as a part of a consensus conference held in Cambridge, England in 1990.[14]

The three largest randomized, phase III trials [6-8] and the metanalysis performed by the Ovarian Cancer Trialists group have individually reported equivalent patient survival durations following carboplatin or cisplatin treatment programs.[14,15]

In great contrast, carboplatin treatment in these multiple clinical trials was associated with significantly less peripheral neuropathy and renal, oto- and neuromuscular toxicities.[6-8] On the basis of these data, panelists and speakers participating in the 1994 NIH Consensus Conference on Ovarian Cancer concluded that carboplatin is the platinum analog of choice in the first-line treatment of patients with advanced ovarian cancer.[16] In fact, because of its major myelosuppressive effects, carboplatin has become one of three commonly used drugs in high-dose therapy administered prior to autologous bone marrow transplantation in various ovarian cancer treatment protocols. Although early on, concerns had been raised about carboplatin's long-term impact on patient survival, as compared to the effect of relatively equivalent doses of cisplatin, the mature data from various metanalyses fail to show any disadvantage for carboplatin.

Non-Small Cell Lung Cancer

It is extremely difficult to interpret the results of randomized clinical trials in patients with advanced non-small cell lung cancer. Complete responses are uncommon and partial responses do not translate into prolonged survival durations. This partially is related to the relatively low response rates achieved (i.e., 30-40% at best) and their short durations (i.e., generally 2-4 months). Additionally, survival data must be interpreted with caution in most of these trials since there is ongoing controversy concerning the impact on survival of chemotherapy versus the "best supportive care". Thus, when evaluating results of phase III trials in patients with non-small cell lung cancer, quality of survival and overall survival durations must be considered as the ultimate study endpoints.

The Eastern Cooperative Oncology Group (ECOG) reported the results of a 743 patient, non-small cell lung cancer study in 1989.[17] Patients were randomized to five different treatment regimens, including MVP (mitomycin/vinblastine/cisplatin), VP (vinblastine/cisplatin), CAMP (cyclophosphamide/doxorubin/methotrexate/procarbazine)/ MVP, carboplatin (400mg/m^2) and iproplatin. While single-agent carboplatin was associated with only an overall response rate of 9%, it also produced the longest median survival (i.e., 31 weeks) ever observed in advanced non-small cell lung cancer with drugs evaluated by the ECOG (i.e., up to that point). As observed in other studies, carboplatin was extremely well-tolerated and produced significantly lower percentages of severe and life-threatening toxicities as compared to MVP, VP, and MVP/CAMP. This study clearly establishes carboplatin as an active and useful agent in the treatment of non-small cell lung cancer and suggests that cisplatin is at best equivalent to carboplatin in the treatment of this disease.

In a European Organization for Research and Treatment of Cancer (EORTC) lung cancer working party study, 239 patients with non-small cell lung cancer were randomized to receive etoposide (100 mg/m^2/day for 3 days) plus either cisplatin (120 mg/m^2) or carboplatin (325 mg/m^2).[18] Patients randomized to the cisplatin/etoposide arm had an objective response rate of 27% and a median survival of 30 weeks. These figures were not significantly different from the objective response rate of 16% and median survival of 27 weeks experienced by the patients who received the carboplatin/etoposide therapy; however, the carboplatin-based therapy again was associated with significantly less toxicity. There was a higher rate of toxic deaths and more frequent and severe leukopenia, ototoxicity, and renal function impairment in the cisplatin/etoposide treated patients. From these studies, it is possible to conclude that carboplatin, used as a single agent or in combination therapy, appears to be equivalent in activity to cisplatin and significantly less toxic in the treatment of patients with advanced non-small cell lung cancer.

Small Cell Lung Cancer

It is well-documented that chemotherapy is associated with up to a 5-fold increase in the median length of survival in patients with limited disease, small cell lung cancer. Combination therapy appears to produce higher objective response rates and survival durations as compared to the results of single-agent therapy. Cisplatin plus etoposide combination therapy has proven superior to virtually any other combination treatment.[19] Carboplatin has proven one of the most active single agents in the treatment of this disease with objective response rates in the range of 25-35%.[20]

In a recently reported phase III study, 147 patients with limited and extensive disease small cell lung cancer were randomized to receive etoposide, 100 mg/m^2 (days 1, 2 & 3) plus either cisplatin, 50 mg/m^2 on days 1 & 2, or carboplatin, 300 mg/m^2 on day

1. Treatment cycles were administered every 3 weeks for up to a total of 6 cycles. Patients with limited and extensive disease were equally randomized to the two treatment arms. Patients with limited disease at the end of the third therapy cycle were eligible for thoracic radiation therapy. In patients with limited disease, there was a complete response rate of 44% for those treated with etoposide/cisplatin, and 37% for those treated with etoposide/carboplatin. Median survival duration for the two treatment arms were equivalent (i.e., 12.5 months for etoposide/cisplatin and 11.8 months for etoposide/carboplatin). There were significantly less leukopenia, clinical infection, nausea/vomiting, neurological and ototoxicity in patients treated with etoposide/carboplatin.

On the basis of both the single agent and the combination agent studies, carboplatin appears at least as active as cisplatin in the treatment of small cell lung cancer and once again is definitely less toxic.

Testicular Cancer

Because the cure rate is in the range of 70-80% in all male patients with metastatic germ cell tumors, the major focus of research has been on patients with good-risk disease (i.e., those with a high probability of cure). The major purposes of these trials has been to determine ways to decrease chemotherapy-induced toxicities while maintaining a high cure rate. A large randomized trial was published by Bajorin et al.[21] comparing cisplatin at 20 mg/m^2/day on days 1-5 plus etoposide at 100 mg/m^2/ day on days 1-5 with cycles repeated every 3 weeks versus carboplatin at 500 mg/m^2 on day 1 plus etoposide 100 mg/m^2/day on days 1-5 with cycles repeated every 4 weeks. Of 265 patients assessable, 134 were treated with cisplatin/etoposide, and 131 were treated with carboplatin/etoposide. Complete response was observed in 90% of the cisplatin/etoposide treated patients compared to 88% of those treated with carboplatin/etoposide. Only 3% of the cisplatin/etoposide treated patients relapsed following complete response compared to 12% of the carboplatin/etoposide treated patients. At a median follow-up of 22.4 months event-free and relapse-free survival were inferior for patients treated with carboplatin/etoposide. There was no difference in overall survival durations between the two study arms.

Investigators concluded that there were several reasons for the increased relapse rate in patients treated with carboplatin/etoposide, including the lower dose intensity for carboplatin (i.e., every 4 weeks dosing) versus cisplatin (i.e., every 3 week dosing), and the use of a fixed dosing schedule despite the well-known effects of renal function on carboplatin clearance. At this point, it appears that cisplatin is the platinum analogue of choice in the treatment of males with metastatic germ cell tumors.

OTHER TUMOR TYPES

Cisplatin continues to be the treatment of choice in patients with advanced head and neck cancer [22] and advanced and/or recurrent cervix cancer.[23] Although carboplatin appears to have important clinical activity in both of these tumor types, cisplatin alone or in combination therapy has been associated with significantly higher response rates in some clinical studies; nevertheless, cisplatin treatment does not appear significantly better than carboplatin treatment with respect to survival durations.

Although phase III studies are unlikely to be performed comparing carboplatin with cisplatin in the treatment of endometrial cancers, carboplatin appears to be a more useful agent with objective response rates in the range of 30-35% in patients with chemotherapy naive, metastatic and/or recurrent disease.[24] On the basis of limited clinical trials data, it would appear that carboplatin is one of the two most active agents in the treatment of advanced and/or recurrent endometrial cancer. Because of its favorable toxicity profile, it would appear a superior agent as compared to cisplatin in the treatment of these older aged patients.

PHARMACOECONOMICS

Although limited pharmacoeconomic studies have been performed comparing carboplatin or cisplatin, a recent publication in Pharmacy and Therapeutics has emphasized the pharmacoeconomics advantages of carboplatin over cisplatin in the treatment of advanced ovarian cancer.[25] In this prospective pharmacoeconomic analysis of a phase III trial of carboplatin/cyclophosphamide versus cisplatin/cyclophosphamide in patients with advanced, previously untreated ovarian cancer, carboplatin was found to be significantly less expensive than cisplatin treatment with respect to toxicity-related costs for 6 cycles of chemotherapy. Overall, the toxicity-related cost savings from the substitution of carboplatin for cisplatin compared favorably with the price differential between these two anti-cancer agents. Similar conclusions have been drawn concerning the pharmacoeconomic analysis of clinical trials in other solid tumor types.

CONCLUSIONS

On the basis of this review, the following conclusions can be drawn concerning the relative merits of carboplatin versus cisplatin in the treatment of solid tumors:

1. Carboplatin is equivalent to cisplatin in anti-cancer activity in the treatment of advanced ovarian, small cell and non-small cell lung and endometrial cancers.
2. Carboplatin has a remarkably more favorable safety profile than cisplatin, demonstrated in virtually all phase III head-to-head comparisons, revealing significantly less nausea/vomiting, ototoxicity and renal impairment as compared to cisplatin.
3. On the basis of its equivalent anti-tumor activity (i.e. especially with respet to survival prolongation) and its significantly more favorable safety profile, carboplatin has proven the drug of choice in the management of patients with advanced ovarian cancer. A similar argument can be made in favor of the use of carboplatin as a single agent and/or in combination therapies for patients with advanced non-small cell and small cell lung cancers as well as those with metastatic and/or recurrent endometrial cancer.
4. Primary pharmacoeconomic analysis reveals significant advantages for carboplatin over cisplatin in the outpatient management of patients with advanced ovarian cancer as well as those with other solid tumors known to be sensitive to platinum based chemotherapy programs.

REFERENCES

1. U.S.A. FDA approved package insert for Paraplatin (carboplatin for injection).

2. U.S.A. FDA approved package insert for Platinol (cisplatin for injection USP).

3. E. Witshaw. Ovarian trials at the Royal Marsden. *Cancer Treat Rev*, 12: 67-71 (suppl A). (1985).

4. M. Adams , I.J. Kerby and I. Rocker, et al. A comparison of the toxicity and efficacy of cisplatin and carboplatin in advanced ovarian cancer. The Swons Gynaecological Cancer Group. *Acta Oncol,* 28 (1): 57-60 (1989).

5. C. Mangioni, G. Bolis and S. Pecorelli, et al. Randomized trial in advanced ovarian cancer comparing cisplatin and carboplatin. *J Natl Cancer Inst*, 81: 1464-72 (1989).

6. K. Swenerton, J. Jeffrey and G. Stuart, et al. Cisplatin-cyclophosphamide versus carboplatin-cyclophosphamide in advanced ovarian cancer: A randomized phase III study of the National Cancer Institute of Canada Clinical Trials Group. *J Clin Oncol*, 10: 718-26 (1992).

7. W.W. Ten Bokkel Huinink, M.E.L. Van der Burg and A.T. Van Oosterom. et al. Carboplatin combination therapy for ovarian cancer. *Cancer Treat Rep*, 15 (8): 9-15 (1988).

8. D.S. Alberts, S. Green and E.V. Hannigan, et al. Improved therapeutic index of carboplatin plus cyclophosphamide versus cisplatin plus cyclophosphamide: Final report by the Southwest Oncology Group of a phase III randomized trial in stages III and IV ovarian cancer. *J Clin Oncol,* 10: 706-17 (1992).

9. P.F. Conte, M. Bruzzone and F. Carnino, et al. Carboplatin, doxorubicin and cyclophosphamide versus cisplatin, doxorubicin, and cyclophosphamide. A randomized trial in stage III-IV epithelial ovarian carcinoma. *J Clin Oncol*, 9: 658-63 (1991).

10. D. Belpomme, R. Bugat and M. Rives, et al. Carboplatin versus cisplatin in association with cyclophosphamide and doxorubicin as first line therapy in stage III-IV ovarian carcinoma: Results of an ARTAC phase III trial. *Proc ASCO*, 11:227 (1992).

11. J.H. Edmonson, G.M. McCormack and H.S. Wieand, et al. Cyclophosphamide-cisplatin versus cyclophosphamide-carboplatin in stage III-IV ovarian carcinoma. A comparison of equally myelosuppressive regimens. *J Natl Cancer Inst*, 81-1500-4 (1989).

12. H. Gurney, D. Crowther and H. Anderson, et al. Five year follow-up and dose delivery analysis of cisplatin, iproplatin or carboplatin in combination with cyclophosphamide in advanced ovarian carcinoma. *Ann Oncol*, 1: 427-33 (1990).

13. T. Kato, H. Nishimura and T. Yamabe, et al. Phase III study of carboplatin for ovarian cancer. *Japanese J of Cancer Chemother*, 15: 2297-304 (1988).

14. K. Aabo, P. Adnitt, M, Adams, D.S. Alberts, et al. Chemotherapy in advanced ovarian cancer: An overview of randomized clinical trials. *BMJ* 303: 884-893 (1991).

15. C.J. Williams, L. Steward and M. Parmar, et al. Meta-analysis of the role of platinum compounds in advanced ovarian carcinoma. *Semin Oncol*, 19 (2): 120-8 (1992).

16. NIH Consensus Conference on Ovarian Cancer: Screening, Treatment and Follow-Up NIH Consensus Development Panel on Ovarian Cancer. *JAMA*, 273: 491-491, (1995).

17. P.D. Bonomi, D.M. Finkelstein, J.C. Ruckdeschel, et al. Combination Chemotherapy versus Single Agents Followed by Combination Chemotherapy in Stage IV Non-Small Cell Lung Cancer: A Study of the Eastern Cooperative Oncology Group. *J Clin Onc,* 7:1602-1613, (1989).

18. J. Klastersky, J.P. Sculier, H. Lacroix, et al., Randomized Study Comparing Cisplatin or Carboplatin with Etoposide in Patients with Advanced Non-Small Cell Lung Cancer: European Organization for Research and Treatment of Cancer, Protocol 07861. *J Clin Onc,* 8:1556-1562, (1990).

19. P.A. Bunn, M. Cullen, M. Fukuoka, et al., Chemotherapy in Small Cell Lung Cancer: A Consensus Report. *Lung Cancer,* 5:127-134, (1989).

20. P.A. Bunn, Clinical Experience with Carboplatin in Lung Cancers, *Semin Oncol* 19:1-11, (1992).

21. D.S. Bajorin, M.F. Sarosty, D.G. Pfeister, et al., Randomized of Etoposide and Cisplatin versus Etoposide and Carboplatin in Patients with Good-Risk Germ Cell Tumors: A Multi-institutional Study, *J Clin Onc*, 11:598-606, (1993).

22. Al-Sarafam, *Semin Oncol*, 21(5):28-34, (1994).

23. D.S. Alberts, D.J. Garcia. Salvage Chemotherapy in Recurrent or Refractory Squamous Cell Cancer of the Uterine Cervix. *Semin Oncol,* 21(4):37-46, (1994).

24. J.B. Green, S. Green, D.S. Alberts, et al. Carboplatin in Advanced Endometrial Cancer: A Southwest Oncology Group Phase II Study. In: Carboplatin JM-8: Current Prospective and Future Directions, P.A. Bunn, R. Canetta, R.F. Ozols, and M. Rozencweig (Eds). W.B. Saunders Company, pp. 113-121, (1990).

25. D. Alberts, E. Hannigan, R. Canetta, et al. Cisplatin versus carboplatin in advanced ovarian cancer? An economic analysis. *Pharmacy & Therapeutics*, 19: 692-706, (1994).

CISPLATIN AND CARBOPLATIN IN TESTIS AND OVARIAN CANCER: PRO CISPLATIN

Stephen D. Williams, M.D.

Indiana University Cancer Center
550 N. University Blvd., Rm. 1640
Indianapolis, IN 46202

INTRODUCTION

The availability of the analogs cisplatin and carboplatin has sparked considerable controversy as to the relative merits of the two agents. A number of randomized clinical trials have been done that address this issue. As expected, most of these trials have been done in the tumor types that are the most responsive to these agents, ovarian and testis cancer.

The clinical pharmacology and pharmacodynamics of these agents have been well studied and will not be reviewed here. The toxicity profiles of these drugs differ, with cisplatin causing more nausea and vomiting, nephrotoxicity, and neurotoxicity. Carboplatin, on the other hand, is more dependent on renal excretion and its elimination and toxicity are more effected by renal function. Its major toxicity is myelosuppression. Because of its route of excretion, it is now usually dosed using a formula based upon renal function and using calculated or measured creatinine clearance or glomerular filtration rate.

This manuscript will review the status of clinical trials in testis and ovarian cancer that address the relative merits of these platinum analogs.

GERM CELL CANCERS

Primarily because of the presumed lessened toxicity of carboplatin, this drug was substituted for cisplatin in combination chemotherapy regimens. An early trial was one of the Royal Marsden Hospital that evaluated carboplatin, etoposide, and bleomycin (Childs, Nicholls et al. 1992). The population chosen for study was one deemed to have favorable prognosis metastatic disease. The goal of this and other subsequent studies was to lessen toxicity in a population of patients deemed to have an excellent prognosis for cure. This trial was not a randomized one. The initial portion of the study was in reality a phase I trial using escalating doses of carboplatin for subsequent groups of patients. All patients received etoposide 120 mg/M^2 on days 1 to 3 of each three week course plus bleomycin 30 units on days 2, 9, and 16. Four courses were given. There were a total of 121 patients at the time of the most recent publication. Of these, the first 12 received carboplatin 300 mg/M^2, the next four 350 mg/M^2, the next six 400 mg/M^2, and the next four 450 mg/M^2. For the remaining patients, the Calvert formula (using EDTA clearance) was used to determine carboplatin dose. Twenty patients received a dose calculated to an AUC of 4.6 mg.min/ml and 75 an AUC dose of 5.0 mg.min/ml. Patients had Royal Marsden stage IM through IVL3. Patients were excluded in they had a nodal mass greater than 10 cm, more than 20 lung metastases, liver, bone, or brain metastases, or an alphafetoprotein greater than 1000 or HCG more than 10,000.

Platinum and Other Metal Coordination Compounds in Cancer Chemotherapy 2
Edited by H.M. Pinedo and J.H. Schornagel, Plenum Press, New York, 1996

Toxicity of this regimen was quite acceptable. Interestingly, carboplatin dose did not correlate well with myelosuppression. With a median follow-up of 40 months, 118 patients are alive and no patient died of germ cell cancer. There were a total of nine treatment failures, an important parameter for regimens designed to lessen toxicity. The authors correlated risk of failure and carboplatin dose. For a given body surface area, the range of calculated EDTA clearances was high. Seven of the nine failures occurred in patients who received less than 400 mg/M^2 of carboplatin. Further, seven of nine failures also occurred in patients treated to a predicted AUC of less than 5.0. Median AUC of failing patients was 4.02 as opposed to 4.85 in patients successfully treated. The authors concluded that carboplatin could be substituted for cisplatin with reduced toxicity and that a dose to give an AUC of 5.0 was likely as active as the cisplatin containing regimen.

This issue has been studied in two large randomized trials. A trial of Memorial Sloan-Kettering Cancer Center and the Southwest Oncology Group compared four cycles of etoposide 100 mg/M^2 days 1-5 plus cisplatin 20 mg/M^2 days 1-5 (EP) or carboplatin (EC) (Bajorin, Sarosdy et al. 1993). Carboplatin dose was 350 mg/M^2 in the first 17 patients, 400 mg/M^2 in five patients, and 500 mg/M^2 in the remaining 108 patients. Patients received four courses of therapy, with EP given every three weeks and EC every four weeks. Eligible patients had good risk disease as defined by MSKCC criteria. Complete response rates of EC and EP patients were similar (88% and 90%). However, 16 patients (12%) treated with EC relapsed versus four patients (3%) with EP. 24% of EC patients experienced an adverse event (incomplete response or relapse) versus 13% of EP patients (p=.02). Event free and relapse free survival were inferior for EC (p=.02 and .005). There were no survival differences. AUC values (using creatinine clearance) were retrospectively available for 117 carboplatin patients. There was no correlation between outcome and AUC levels greater than or less than 5.0. 96 of these 117 patients had AUC's greater than 5.0. The authors concluded than EC given in this dose and schedule was inferior, but that it is conceivable that administration of EC every three weeks might be more efficacious.

The largest trial to date has been published only in abstract. This trial is one of the Medical Research Council and EORTC (Horwich and Sleijfer 1994). 589 favorable prognosis patients received etoposide 120 mg/M^2 days 1-3 plus bleomycin 30 units on day 1 plus either cisplatin (100 mg/M^2 total dose over 2 or 5 days) or carboplatin (AUC 5.0 using GFR). Courses were given every three weeks. Complete response rate on BEP and CEB (93% versus 88%) were similar (p=.12). However, failure free rate at one year was 80% for CEB versus 90% for BEP (p=0.01). There are no survival differences currently observed. The carboplatin regimen is thought to be inferior.

There are other pieces of evidence that suggest that standard dose carboplatin is inferior to cisplatin. In three xenographs of platinum sensitive human embryonal carcinoma, carboplatin was less active than cisplatin in equitoxic doses (Bokemeyer, Harstrick et al. 1994). One other small random trial, though badly flawed, suggested that EP is superior to EC (Tjulandin, Garin et al. 1993). Finally, a French group noted adverse events in 9/24 "good prognosis" patients (37.5%) treated with etoposide 120 mg/M2 days 1-3 plus carboplatin 450 mg/M^2 day 2 every four weeks for four courses (Kattan, Mahjoubi et al. 1993).

The issue of carboplatin versus cisplatin is confounded by the varying designs used in the major trials, particularly the different etoposide doses, schedule (three versus four weeks) and inclusion or omission of bleomycin. However, there is emerging evidence of the inferiority of carboplatin. Further presentation of the EORTC trial is awaited, but certainly for the time being carboplatin should not be used as first-line therapy in patients with germ cell tumors. Also, with modern emetics and supportive care, toxicity issues are less important, particularly considering the infrequent incidence of significant neuro- and ototoxicity in these young patients.

OVARIAN CANCER

Substitution of carboplatin for cisplatin has particular appeal in the treatment of advanced ovarian cancer. There is unfortunately substantially less possibility for cure and the patients are of an age group in which neurotoxicity is potentially much more troublesome.

Three randomized trials compared single agent cisplatin to carboplatin in previously untreated patients (Adams, Kerby et al. 1989; Mangioni, Bolis et al. 1989; Taylor, Wiltshaw et al. 1994). Carboplatin dose was 400 mg/M^2 and cisplatin 100 mg/M^2, both given every four weeks. Most patients had bulky residual stage III tumor. All three trials plus a subsequent meta-analysis (Advanced Ovarian Cancer Trialists Group 1991) showed no therapeutic differences and that carboplatin had more hematologic and less non-hematologic toxicity.

Several randomized trials that compared combination chemotherapy regimens containing cisplatin versus a similar regimen containing carboplatin have been completed. A study of the Mayo Clinic and the North Central Cancer Treatment Group compared cyclophosphamide plus either cisplatin or carboplatin (Edmonson, McCormack et al. 1989). The study was terminated prematurely because of poorer results of patients on the carboplatin arm. However, the very low dose of carboplatin used (150 mg/M2) make the results of this study meaningless.

Alberts et. al. reported the results of a trial of the Southwest Oncology Group that compared cyclophosphamide 600 mg/M^2 plus either cisplatin 100 mg/M^2 or carboplatin 300 mg/M^2 (Alberts, Green et al. 1992). A total of six courses were to be given at four week intervals. All patients had sub-optimal stage III (>2cm residual tumor) or stage IV disease. 342 patients were entered and 291 eligible. The regimens were therapeutically equivalent but the carboplatin arm produced less non-hematologic toxicity, particularly neuromuscular toxicity and hearing loss.

A trial of the National Cancer Institute of Canada employed a nearly identical study design except that the dose of cisplatin was 75 mg/M^2 instead of 100 mg/M^2 (Swenerton, Jeffrey et al. 1992). Of the 419 eligible patients on this study, 59% had greater than 2 cm residual stage III or stage IV tumor. There were no differences seen in any therapeutic parameter. Cisplatin treated patients were more likely to discontinue treatment because of toxicity and experienced more non-hematologic toxicity, while carboplatin patients had more myelosuppression.

An Italian study (Conti, Bruzzone et al. 1991) compared doxorubicin and cyclophosphamide plus either cisplatin 50 mg/M^2 or carboplatin 200 mg/M^2. There were 164 randomized patients, 63 % of whom had greater than 2 cm residual tumor. The arms were generally equivalent both in terms of therapeutic efficacy and toxicity, although the cisplatin arm caused more nausea and vomiting.

An interesting trial of the EORTC compared doxorubicin, cyclophosphamide and hexamethylmelamine plus either cisplatin 20 mg/m^2 days 1-5 (CHAP-5) or carboplatin 350 mg/M^2 day 1 (CHAC-1). A total of 342 patients were entered on this study, 63% of whom had more than 1 cm residual tumor. A preliminary analysis that included the first 264 patients showed no differences in initial response, but was too early to make an assessment of remission duration or survival (ten Bokkel Huinink, van der Berg et al. 1988). Cisplatin produced more neurotoxicity.

A meta-analysis of these and other smaller randomized trials showed no important differences in the cisplatin and carboplatin regimens(Group 1991).

The EORTC study was partially updated recently in a review article that had as a co-author the protocol chairman (Vermorken, ten Bokkel Huinink et al. 1993). There were no overall or within strata differences (stage, residual tumor volume, performance status, and grade) in therapeutic outcome. However, in patients with no or less than 1 cm residual tumor, CHAP-5 patients had overall and progression free survivals of 59.2 and 39.5 months versus 32.5 and 17.9 months for CHAC-1 patients. These differences were not significant and the number of total patients (127) relatively small.

It is clear from the available data that cisplatin or carboplatin are of similar activity for patients with suboptimal residual tumor, both as single agents and when used in combinations that do not contain paclitaxel. Patients treated with carboplatin have less neurotoxicity, an important issue in this patient population that has a median age of 60 and relatively short survival. However, in such patients, it has now been shown that cisplatin plus paclitaxel when compared with cisplatin plus cyclophosphamide is associated with a higher clinical response rate and a longer progression free interval (McGuire, Hoskins et al. 1993) and, recently, improved survival (Gynecologic Oncology Group; unpublished data). Further, there is a fairly small comparative experience with carboplatin in patients with small volume residual tumor, the patients with the longest survival, a chance for cure, and the most to gain from treatment. The data from the EORTC study, while certainly not definitive, are provocative. These two considerations should temper the wholesale adoption of carboplatin regimens for the treatment of ovarian cancer. Considering the results in germ cell tumors, significant differences in platinum analogs may be seen only in that subset of patients with the highest likelihood for cure or long survival.

CONCLUSIONS

1. While there are confounding issues, cisplatin is almost certainly superior to carboplatin in the treatment of germ cell tumors.

2. Toxicity is a moot issue in these young patients.

3. In ovarian cancer, available evidence suggests that cisplatin and carboplatin are of similar activity as single agents.

4. Randomized trials of combination chemotherapy regimens comparing cisplatin and carboplatin have shown no clear meaningful therapeutic differences.

5. In patients with suboptimal stage III and stage IV tumors, carboplatin is appealing because of lessened neurotoxicity.

6. The relative efficacy in patients with optimal stage III disease is less well studied.

7. It is conceivable that cisplatin is superior in these patients, particularly considering the data in germ cell tumors.

8. The substitution of paclitaxel for cyclophosphamide has clearly improved results; the relative merits of cisplatin and carboplatin is this situation are unknown.

REFERENCES

Adams, M., Kerby, I. J., et al. (1989). A comparison of the toxicity and efficacy of cisplatin and carboplatin in advanced ovarian cancer. *Acta Oncol* 28: 57.

Alberts, D. S., Green, S., et al. (1992). Improved therapeutic index of carboplatin plus cyclophosphamide versus cisplatin plus cyclophosphamide: final report by the Southwest Oncology Group of a phase III randomized trial in stages III and IV ovarian cancer. *J Clin Oncol* 10: 706.

Bajorin, D. F., Sarosdy, M.F., et al. (1993). Randomized trial of etoposide and cisplatin versus etoposide and carboplatin in patients with good-risk germ cell tumors: a multiinstitutional study. *J Clin Oncol* 11: 598.

Bokemeyer, Harstrick, C. A., et al. (1994). The use of carboplatin in malignant germ cell tumours. *Eur J Cancer* 30A: 721.

Childs, W. J., Nicholls. E. J., et al. (1992). The optimisation of carboplatin dose in carboplatin, etoposide and bleomycin combination chemotherapy for good prognosis metastatic nonseminomatous germ cell tumours of the testis. *Ann Oncol* 3: 29.

Conti, P. F., Bruzzone, M., et al. (1991). Carboplatin, doxorubicin, and cyclophosphamide versus cisplatin, doxorubicin, and cyclophosphamide: a randomized trial in stage III-IV ovarian carcinoma. *J Clin Oncol* 9: 658.

Edmonson, J. H., McCormack, G.M., et al. (1989). Cyclophosphamide-cisplatin versus cyclophosphamide-carboplatin in stage III-IV ovarian carcinoma: a comparison of equally myelosuppressive regimens. *J Natl Cancer Inst* 81: 1500.

Advanced Ovarian Cancer Trialists Group. (1991). Chemotherapy in advanced ovarian cancer: an overview of randomised clinical trials. *Brit Med J* 303: 884.

Horwich, A. and Sleijfer, D. (1994). Carboplatin-based chemotherapy in good prognosis metastatic non-seminoma of the testis: an interim report of an MRC/EORTC randomised trial. *Proc. Leeds Germ Cell Tumour Conference* (abstract).

Kattan, J., Mahjoubi, M., et al. (1993). High failure rate of carboplatin-etoposide combination in good risk non-seminomatous germ cell tumours. *Eur J Cancer* 29A: 1504.
Mangioni, C., Bolis, G., et al. (1989). Randomized trial in advanced ovarian cancer comparing cisplatin and carboplatin. *J Natl Cancer Inst* 81: 1464.
McGuire, W. P., Hoskins, W. J., et al. (1993). A phase III trial comparing cisplatin/Cytoxan and cisplatin/Taxol in advanced ovarian cancer. *Proc Amer Soc Clin Oncol* 12: 255 (abstract).
Swenerton, K., Jeffrey, J., et al. (1992). Cisplatin-cyclophosphamide versus carboplatin-cyclophosphamide in advanced ovarian cancer: a randomized phase III study of the National Cancer Institute of Canada Clinical Trials Group. *J Clin Oncol* 10: 718.
Taylor, A. E., Wiltshaw, E., et al. (1994). Long-term follow-up of the first randomized study of cisplatin versus carboplatin for advanced epithelial ovarian cancer. *J Clin Oncol* 12: 2066-2070.
ten Bokkel Huinink, W. W., van der Berg M. E. L., et al. (1988). Carboplatin in combination therapy for ovarian cancer. *Cancer Treat Rev* 15: 9.
Tjulandin, S. A., Garin, A. M., et al. (1993). Cisplatin-etoposide and carboplatin-etoposide induction chemotherapy for good-risk patients with germ cell tumors. *Ann Oncol 4*: 663-667.
Vermorken, J. B., ten Bokkel Huinink, W. W., et al. (1993). Carboplatin versus cisplatin. *Ann Oncol* 4: S41.

HYPERSENSITIVITY TO CISPLATIN IN MOUSE LEUKEMIA L1210/0 CELLS: AN XPG DNA REPAIR DEFECT

Richard D. Wood, Juhani A. Vilpo[1], Leena M. Vilpo[1],
David E. Szymkowski[2], Anne O'Donovan[3] and Jonathan G. Moggs

Imperial Cancer Research Fund, Clare Hall Laboratories,
South Mimms, Herts, EN6 3LD, United Kingdom
[1]present address: Department of Clinical Chemistry,
Tampere University Hospital, Tampere, Finland
[2]present address: Roche Research Centre
Welwyn Garden City, AL7 3AY, United Kingdom
[3]present address: Amersham International,
Cardiff, CF4 7YT, United Kingdom

INTRODUCTION

A major limitation to the clinical efficacy of cisplatin is the intrinsic or acquired resistance of many neoplasms to the drug. As a result, many studies to investigate the mechanisms of cisplatin resistance have been carried out with human and rodent cells in culture. Acquired resistance has been ascribed in different cases to changes in drug accumulation, intracellular drug inactivation by enhanced levels of glutathione or metallothionein, and enhanced DNA repair.

Removal of cisplatin adducts from DNA is accomplished by only one characterized mechanism, nucleotide excision repair. In mammalian cells this process involves many proteins including the xeroderma pigmentosum (*XP*) and excision repair cross-complementing (*ERCC*) gene products[1]. If cancer cells are able to acquire increased resistance to drugs or radiation by increasing their capacity for nucleotide excision repair, it is important to understand the molecular mechanism of the increase. In order to investigate the possible role of enhanced DNA repair as a general mechanism for cisplatin resistance, we have been studying the difference in DNA repair capacity amongst several well-known mouse L1210 cell lines. The ability of cell extracts to repair cisplatin-DNA adducts was analyzed using techniques that measure nucleotide excision repair of DNA *in vitro*[2,3].

RESULTS AND DISCUSSION

A widely used and cited model system that has provided support for enhanced DNA repair as a means of cellular resistance to platinum compounds has employed cisplatin-resistant and sensitive derivatives of mouse leukemia L1210 cells. L1210 cells were originally developed in 1949 as a carcinogen-induced acute lymphoid leukemia in the DBA mouse strain[4], and the cells were maintained by serial transplantation until an *in vitro* culture system was described in 1966[5]. The cell subline L1210/0 was first described in 1977 by Burchenal and co-workers at the Memorial Sloan-Kettering laboratory, and used as representative of the "wild-type" form of L1210 cells[6,7], although it now appears that L1210/0 was an unusually cisplatin-sensitive isolate. As part of a large-scale study of acquired platinum resistance, Burchenal's laboratory then grew L1210/0 in gradually increasing cisplatin concentrations and selected a resistant subline denoted L1210/DDP (or L1210/PDD) [7].

A series of further sublines isolated by Eastman and colleagues gave even more dramatic examples of acquired resistance. For instance, the cell line designated $L1210/DDP_{10}$ was developed by growth of L1210/DDP in increasing drug concentrations over several years, until the cells could eventually be maintained in medium containing 10 μg/ml *cis*-DDP [8]. The $L1210/DDP_{10}$ line was thus ~100-fold more resistant to cisplatin than the reference L1210/0 line (on the basis of the IC_{50}, the drug concentration at which cell growth is inhibited by 50%). The reported sensitivity of these cell lines to cisplatin is shown in Table 1.

Table 1. Cisplatin sensitivity and DNA repair capacity of cell lines (from Sheibani et al.[9])

cell line	IC_{50} (μg/ml)	adducts/plasmid inhibiting CAT activity by 63%
normal CHO	4.7	19.3
ERCC2-defective CHO	2.3	1.6
ERCC1-defective CHO	0.1	2.2
ERCC4-defective CHO	0.05	3.0
L1210/0	0.12	3.4
L1210/DDP	1.2	10
$L1210/DDP_{10}$	13.2	12

Subsequently, Eastman and co-workers demonstrated that $L1210/DDP_{10}$ had a much greater rate of removal of cisplatin adducts from cellular DNA than L1210/0 [10] and that $L1210/DDP_{10}$ was better able to reactivate a *cis*-DDP-damaged plasmid vector than L1210/0 [9]. Some of the plasmid reactivation data from the latter study are summarized in Table I. Also included are plasmid reactivation data for three known nucleotide-excision repair defective mutants of rodents cells, having mutations in the ERCC2, ERCC1, and ERCC4 genes. It is striking that the ability of L1210/0 cells to reactive cisplatin-treated DNA is low, and indeed Sheibani et al.[9] noted that "The sensitive L1210/0 cells exhibited activity similar to that in the repair-deficient CHO cell lines; that is, they appeared to repair no damage...".

Cisplatin and UV-sensitivity of L1210 Cell Lines

To further investigate the difference in DNA repair between the L1210 cell lines, we used several in vivo and in vitro methods. The results are given in a recent publication[11], and briefly summarized here. A growth inhibition assay was used to confirm the large difference in cisplatin sensitivity between L1210/0 and L1210/DDP_{10} cells. L1210/DDP_{10} cells had an IC_{50} about 100-fold greater than that of L1210/0 cells (Fig. 1). A sample of original L1210 cells was also obtained from the American Type Culture Collection (where they had been deposited in 1979 at about passage 300) via the European Collection of Animal Cell Cultures (ECACC, Salisbury, U.K.). These normal L1210 cells, referred to here as L1210/ECACC, were found to be significantly more cisplatin resistant (~6-7–fold greater IC_{50}) than the L1210/0 subline. All three cell lines were tested for sensitivity to UV irradiation. No difference between the L1210/ECACC and L1210/DDP_{10} lines was noted (Fig. 1), but these lines tolerated approximately 10 times more UV radiation than the L1210/0 line. The UV sensitivity of L1210/0 (IC_{50} about 1 J/m^2) is similar to that of known nucleotide excision repair-deficient XP or ERCC cell lines[12], which typically have an IC_{50} between 0.8 and 2 J/m^2. This mutagen-sensitive phenotype of L1210/0 appears to have existed since its isolation in 1977. The results in Fig. 1, and other data [11] show that the high resistance of L1210/DDP_{10} to cisplatin is specific to that drug, and is not associated with a general increase in resistance to drugs or UV radiation in comparison to the type culture collection cell line L1210/ECACC.

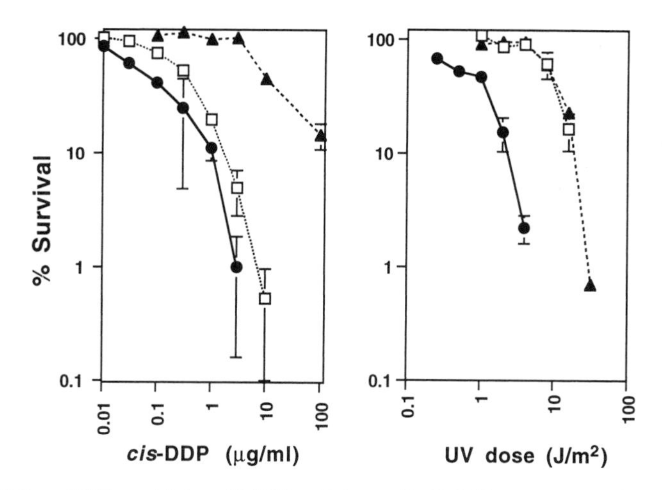

Figure 1. *cis*-DDP and UV sensitivity of L1210 cell lines. (Left) Cells were grown in suspension in the presence of various *cis*-DDP concentrations for three days, and living cells were then counted. (Right) Exponentially growing cells were UV irradiated with various single doses and counted after three days in culture. L1210/0 (●), L1210/ECACC (❑), and L1210/DDP_{10} (▲). Data represent averages of 4 replicate cultures (±SD). Adapted from Vilpo et al.[11].

DNA Repair Synthesis *in vitro*

The DNA repair capacity of L1210/0 cell extracts in vitro was investigated in several ways. In one approach, a duplex M13 DNA circle was constructed to contain a single acetylaminofluorene (AAF) modification per molecule on a defined guanine residue (Fig. 2).

This single-lesion substrate was incubated with whole cell extract in buffer that included [α-^{32}P]dATP to allow repair synthesis to occur. Plasmid DNA recovered from the reaction mixture was linearized with a restriction enzyme and subjected to gel electrophoresis and fluorography for analysis of the incorporation of nucleotides into repair patches as previously described [3,13]. Figure 2 (right) shows that L1210/0 extracts could not carry out detectable repair of the AAF adduct. L1210/DDP$_{10}$ extracts performed adduct-specific repair synthesis in the 31 bp fragment containing the lesion, and in the 5' flanking 68 bp fragment, a distribution consistent with a repair patch size of about 30 nucleotides. In addition, L1210/DDP10 extracts are able to repair a single site-specifically located 1,3-intrastrand d(GpTpG) cisplatin adduct, while L1210/0 extracts are totally defective in this repair. Both extracts show around 10-fold less repair of the more common 1,2-intrastrand d(GpG) cisplatin adduct (D.E.S. and R.D.W., unpublished data).

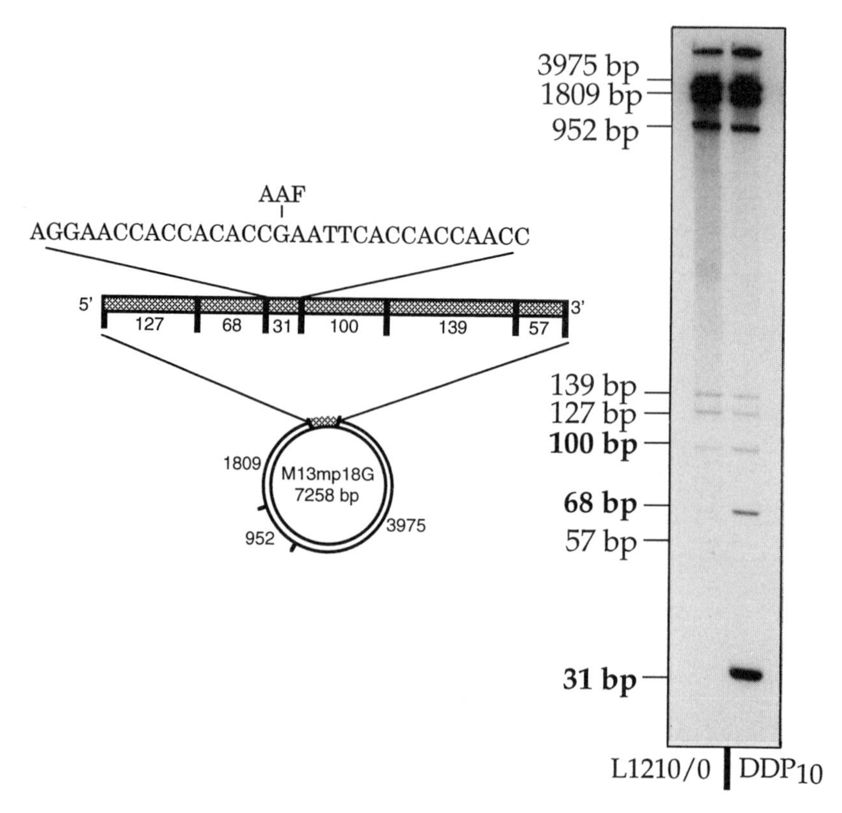

Figure 2. Repair of a site-specifically placed AAF adduct. (Left) Construct containing on the (-) strand a single AAF-guanine adduct in closed circular duplex M13mp18G DNA. The nucleotide sequence is shown for the (-) strand flanking the adduct. The fragment sizes shown (in bp) are for *Bst*NI digestion. (Right) The DNA was incubated with 200 µg L1210/0 or L1210/DDP$_{10}$ cell extract protein and digested with *Bst*NI. Repair synthesis was detected after gel electrophoresis and autoradiography. Data from Vilpo et al.[11].

Protein extracts from the sensitive and resistant cells were also examined for their ability to carry out DNA repair synthesis in plasmid DNA treated "globally" with cisplatin or UV radiation. L1210/0 cell extracts were considerably less efficient in repairing UV-irradiated DNA or cisplatin-treated DNA than either L1210/ECACC or L1210/DDP$_{10}$ cell extracts (Fig. 3). The repair synthesis performed by extracts from the latter two mouse cell lines was in the same range as that seen with similar amounts of extract protein from repair-proficient normal human cell lines under identical conditions [14]. The low level of repair of damaged DNA by L1210/0 cell extracts was comparable to that seen with severely nucleotide excision repair-deficient human XP or rodent ERCC cell lines [14].

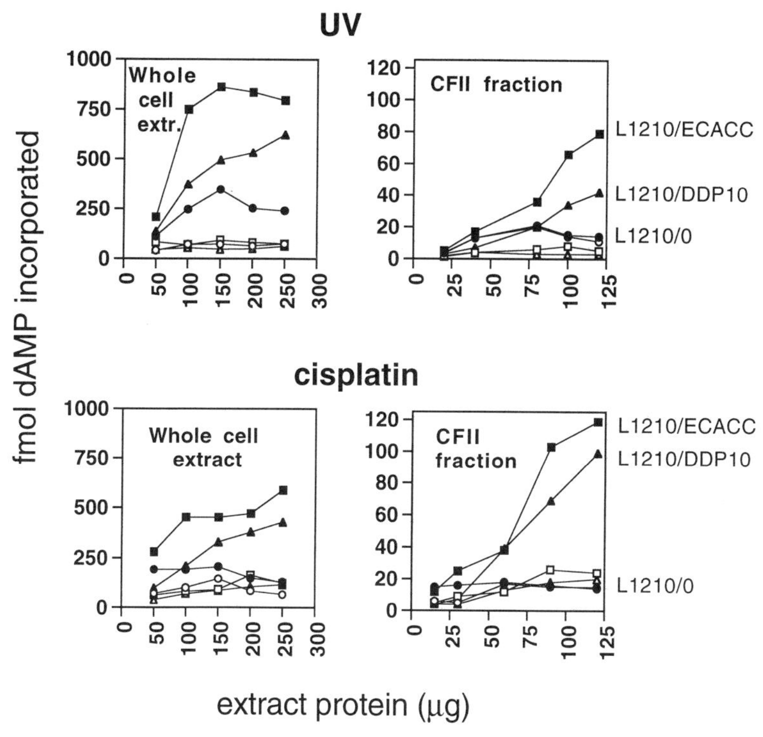

Figure 3. DNA repair synthesis by extracts from L1210/0, L1210/ECACC, and L1210/DDP_{10} cells. For the UV experiments, each reaction included 250 ng each of UV-irradiated pBluescript KS^+ plasmid and undamaged pHM14, and the indicated amounts of whole cell extracts or CFII protein fractions. For the cisplatin experiments, DNA repair synthesis was meaured in platinated closed circular M13mp18GG duplex DNA (250 ng/reaction) by the indicated amounts of whole cell extracts or CFII protein fractions. The graphs show femtomoles of dAMP incorporated into UV- or *cis*-DDP-damaged plasmid (closed symbols) and undamaged plasmid (open symbols). The cell extracts analyzed were L1210/0 (circles), L1210/ECACC (squares), and L1210/DDP_{10} (triangles). Data adapted from Vilpo et al.[11].

Nature of the repair defect in L1210/0 cells

The nucleotide excision repair defect in the L1210/0 subline was then assigned to a known repair-defective complementation group. This analysis was performed by mixing extract from L1210/0 cells with extracts of cells representing defined repair-defective complementation groups[14-16]. L1210/0 extract protein could correct the *in vitro* repair deficiency of all tested XP and ERCC cell extracts, with the significant exceptions of ERCC5 rodent cell extracts and human XP-G cell extracts. Since the *XPG* and *ERCC5* genes are known to be equivalent[14,17-19] these results strongly suggested that L1210/0 had a defect in XPG.

To test this hypothesis, XPG protein was added to reaction mixtures with L1210/0 cell extracts (or lower background "CFII fractions"), either as a partially purified protein [14] from human HeLa cells (compare Fig. 4, lane 1 with lane 4) or as a purified polypeptide[20] produced from a recombinant baculovirus (compare Fig. 4, lane 10 with lane 11). In both cases, the XPG protein could correct the DNA repair defect of L1210/0 cell extracts, but could not correct extracts from repair-deficient cells representing other genetic complementation groups.

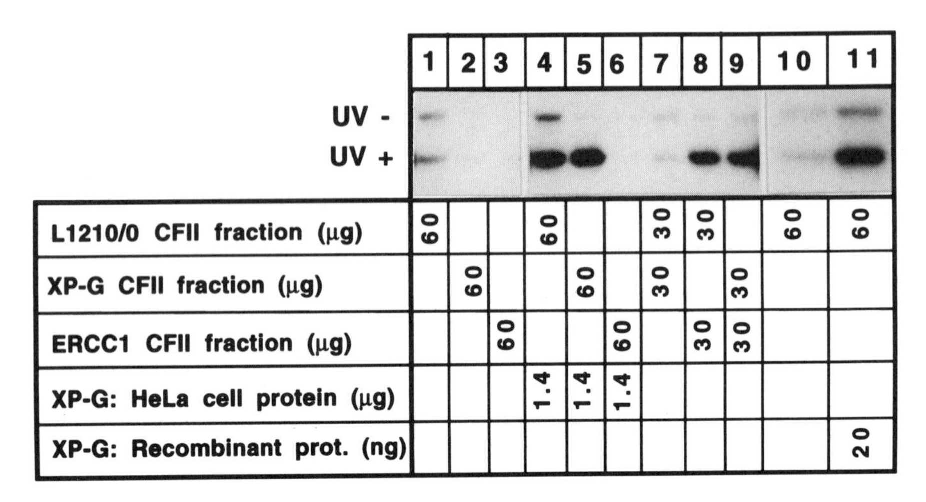

	1	2	3	4	5	6	7	8	9	10	11
L1210/0 CFII fraction (μg)	60			60			30	30		60	60
XP-G CFII fraction (μg)		60			60		30		30		
ERCC1 CFII fraction (μg)			60			60		30	30		
XP-G: HeLa cell protein (μg)				1.4	1.4	1.4					
XP-G: Recombinant prot. (ng)											20

Figure 4. Complementation of DNA repair synthesis defect in L1210/0 with fractionated extracts and with purified XPG complementing proteins. Autoradiographs of the DNA repair reactions show the incorporation of dAMP into UV-damaged plasmid (UV+) and undamaged control plasmid (UV-). Reactions contained the indicated amounts of CFII fractions from extracts of L1210/0, XPG83 (XP-G) cells, or 43-3B (ERCC1 mutant) cells, with XPG protein where indicated. The reactions were supplemented with RPA and PCNA proteins as described[16]. Data from Vilpo et al.[11].

To determine whether the XPG defect in L1210/0 was a result of grossly altered expression of the *XPG* gene, blots of poly(A)+ RNA from the three L1210 cell lines were hybridized with *XPG* cDNA probes. The level of expression of *XPG* mRNA was similar in the three mouse cell lines. The increased UV and *cis*-DDP sensitivity of the L1210/0 line may thus be due to a relatively small change in the coding sequence of the *XPG* gene[11]. In this respect, L1210/0 resembles all *XPG* mutants examined to date in human or mouse cells in that all retain expression of *XPG* mRNA[17-19,21].

The origin of the L1210/0 line is unique because the XPG DNA repair defect apparently developed spontaneously. During the selection of the L1210/DDP or L1210/DDP_{10} cell lines the *XPG* gene has become functional again, perhaps by mutagen-induced reversion of a point mutation as has been described for human XPA cells [22]. The repair capacity of L1210/DDP_{10} does not appear to be quite up to the level of normal L1210/ECACC cells (Fig. 3), and so the XPG in L1210/DDP_{10} may not have reverted to the precise wild-type sequence.

Function of the XPG DNA repair protein in nucleotide excision repair

The XPG protein is a key component in the nucleotide excision repair process. The protein of 1186 amino acids is encoded by a gene that can fully correct the repair defect in cell lines established from xeroderma pigmentosum group G patients [17]. The recombinant XPG polypeptide has been purified in our laboratory using a baculovirus expression system [20]. XPG is a DNA endonuclease with remarkable "structure-specific" properties. For example, an endonucleolytic incision is made in a synthetic DNA substrate containing a duplex region and single-stranded arms. One strand of the duplex is cleaved at the border with single-stranded DNA. This cut has a specific polarity, where the single strand is oriented with the 3' to 5' direction pointing away from the duplex region. XPG will also cut a "bubble" structure with a consistent polarity (Fig. 4A). This and other experiments provide strong evidence that XPG is involved in making the incision 3' to DNA lesions during removal of a damaged oligonucleotide[23]; see Fig. 4B. XPG was originally described as a nuclease that can cleave M13 bacteriophage DNA[20,24], and presumably it does so by cleaving at single-strand/duplex junctions at hairpin loops in such DNA. XPG is a member of a family of structure-specific DNA endonucleases that act in a similar manner[25,26].

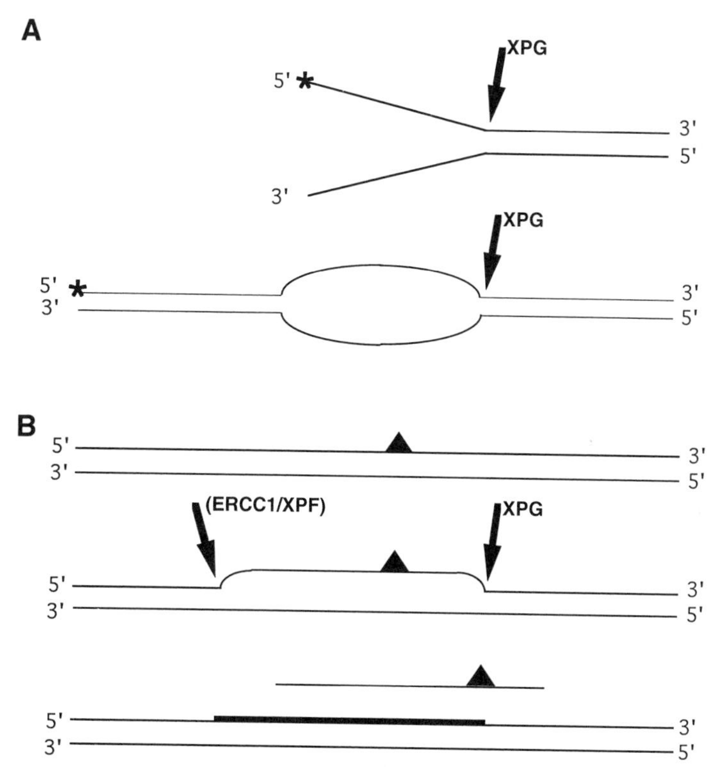

Figure 5. A. Diagram showing the position of endonucleolytic incisions by purified XPG protein on two synthetic DNA substrates: a splayed-arm DNA structure with a duplex region of 30 bp and two unpaired arms of 30 nucleotides, and a "bubble" DNA substrate with two duplex regions of 30 base pairs flanking an unpaired region of 30 nucleotides. When the top strand is labeled at the position shown by the asterisk, cleavage of that strand is observed at the position shown by the arrow. **B.** Model showing XPG making the incision 3' to DNA lesions during removal of a damaged oligonucleotide and subsequent repair synthesis to restore the DNA duplex. The ERCC1/XPF protein complex is predicted to make the 5' incision; see O'Donovan et al.[23].

Implications for Studies of Acquired Cisplatin Resistance

The results show that a specific defect in the DNA nucleotide excision repair pathway exists in the L1210/0 line, even though these cells have been used as reference "normal" controls in many studies[6-10,27-34] where mechanisms of *cis*-DDP resistance have been investigated. Because of the DNA excision repair deficiency of the L1210/0 line, these data may need to be reinterpreted. Development of new platinum complexes to circumvent drug resistance has relied on data obtained with the mouse L1210 model. For example, experiments that included L1210/0 were the main impetus for the development of tetraplatin for clinical trials [6,35]. Many platinum complexes have been identified as promising because, in contrast to cisplatin, they have similar toxicities towards L1210/0 and L1210/DDP cells[7,32].

Nucleotide excision repair of DNA adducts is clearly important in determining the intrinsic resistance of cells to cisplatin, and can account for the 6.5-fold different in IC_{50} for cisplatin between the XPG-defective L1210/0 and L1210/ECACC. There is no evidence that the L1210/DDP_{10} line has a higher than normal capacity to repair DNA adducts: extracts of L1210/ECACC cells repaired cisplatin or UV-damaged DNA as well as, or even better than the cisplatin-resistant L1210/DDP_{10} cells. This suggests that enhancement of DNA repair (above the level of normal, repair-proficient cells) is not a mechanism of cisplatin resistance in the L1210 system. Nevertheless, the L1210/DDP_{10} line can survive in the presence of approximately 30-fold higher *cis*-DDP concentrations than can the L1210/ECACC line. Other known features of L1210/DDP_{10} must account for this resistance, including decreased accumulation of *cis*-DDP [8], an increased cellular level of glutathione[8], and increased tolerance of cells to *cis*-DDP-DNA adducts, by a mechanism not fully defined[10,30].

Some of the main arguments that have been put forward to suggest a role for increased DNA repair in acquired cisplatin resistance include the increased repair in resistant cell lines, increased levels of DNA polymerases α and β in resistant cells, and a potentiation of the cytotoxic effect of cisplatin by the DNA polymerase inhibitor aphidicolin. In the mouse cell system examined here, enhancement of DNA repair above the normal level does not in fact appear to be a mechanism of acquired resistance to cisplatin. Increased DNA repair capacity has been suggested to occur in some other *cis*-DDP resistant cell lines in addition to the L1210 example[36-38], and it could be of interest to use the approach described here in order to further investigate those cases. The other two arguments are not compelling, because neither DNA polymerase α nor β appears to play a role in nucleotide excision repair[39], and because aphidicolin inhibits DNA replication as well as repair, which could account for its additional toxic effect. The present results emphasize that DNA repair is important as part of the intrinsic cellular defense against cisplatin damage to the genome, but that any role of increased nucleotide excision repair in acquired resistance should be re-evaluated.

ACKNOWLEDGMENTS

We are grateful to Alan Eastman (Dartmouth Medical School) for providing cell lines and for useful discussions. We thank Daniel Scherly, Stuart Clarkson, Adelina A. Davies, and Stephen C. West for collaboration on the XPG protein work, and many colleages at ICRF Clare Hall Laboratory for donating materials and advice. This work was supported by the Imperial Cancer Research Fund, with assistance to J.V. and L.V. from the Maj and Tor Nessling Foundation, the Academy of Sciences of Finland, and the Finnish Foundation for

Cancer Research. J. V. was a recipient of a U.K. Royal Society/Academy of Sciences of Finland Exchange Visitor Fellowship.

REFERENCES

1. A. Aboussekhra and R.D. Wood, Repair of ultraviolet light-damaged DNA by mammalian cells and *Saccharomyces cerevisiae*. *Curr. Opin. Genet. Devel.*, 4:212 (1994).
2. J. Hansson, S.M. Keyse, T. Lindahl and R.D. Wood, DNA excision repair in cell extracts from human cell lines exhibiting hypersensitivity to DNA damaging agents. *Cancer Res.*, 51:3384 (1991).
3. D.E. Szymkowski, K. Yarema, J.E. Essigmann, S.J. Lippard and R.D. Wood, An intrastrand d(GpG) platinum crosslink in duplex M13 DNA is refractory to repair by human cell extracts. *Proc. Natl. Acad. Sci. USA*, 89:10772 (1992).
4. L. Law, T. Dunn, P. Boyle and J. Miller, Observation on the effect of a folic-acid antagonist on transplantable lymphoid leukemias in mice. *J. Natl. Cancer Inst.*, 10:179 (1949).
5. G. Moore, A. Sandberg and K. Ulrich, Suspension cell culture and in vivo and in vitro chromosome constitution of mouse leukemia L1210. *J. Natl. Cancer. Inst.*, 36:405 (1966).
6. J.H. Burchenal, K. Kalaher, T. O'Toole and J. Chisholm, Lack of cross-resistance between certain platinum coordination compounds in mouse leukemia. *Cancer Res.*, 37:3455 (1977).
7. J. Burchenal, K. Kalaher, L. Lokys and G. Gale, Studies of cross-resistance, synergistic combinations and blocking of activity of platinum derivatives. *Biochimie*, 60:961 (1978).
8. V. Richon, N. Schulte and A. Eastman, Multiple mechanisms of resistance to cis-diamminedichloroplatinum(II) in murine leukemia L1210 cells. *Cancer Res.*, 47:2056 (1987).
9. N. Sheibani, M.M. Jennerwein and A. Eastman, DNA repair in cells sensitive and resistant to *cis*-diamminedichloroplatinum(II): host cell reactivation of damaged plasmid DNA. *Biochemistry*, 28:3120 (1989).
10. A. Eastman and N. Schulte, Enhanced DNA repair as a mechanism of resistance to *cis*-diamminedichloroplatinum(II). *Biochemistry*, 27:4730 (1988).
11. J.A. Vilpo, L.M. Vilpo, D.E. Szymkowski, A. O'Donovan and R.D. Wood, An XPG DNA repair defect causing mutagen hypersensitivity in mouse leukemia L1210 cells. *Mol. Cell Biol.*, 15:290 (1995).
12. G. Weeda and J.H.J. Hoeijmakers, Genetic analysis of nucleotide excision repair in mammalian cells. *Semin. Cancer Biol.*, 4:105 (1993).
13. J. Hansson, M. Munn, W.D. Rupp, R. Kahn and R.D. Wood, Localization of DNA repair synthesis by human cell extracts to a short region at the site of a lesion. *J. Biol. Chem.*, 264:21788 (1989).
14. A. O' Donovan and R.D. Wood, Identical defects in DNA repair in xeroderma pigmentosum group G and rodent ERCC group 5. *Nature*, 363:185 (1993).
15. R.D. Wood, P. Robins and T. Lindahl, Complementation of the xeroderma pigmentosum DNA repair defect in cell-free extracts. *Cell*, 53:97 (1988).
16. M. Biggerstaff, D.E. Szymkowski and R.D. Wood, Co-correction of the ERCC1, ERCC4 and xeroderma pigmentosum group F DNA repair defects *in vitro*. *EMBO J*, 12:3685 (1993).
17. D. Scherly, T. Nouspikel, J. Corlet, C. Ucla, A. Bairoch and S.G. Clarkson, Complementation of the DNA repair defect in xeroderma pigmentosum group G cells by a human cDNA related to yeast *RAD2*. *Nature*, 363:182 (1993).
18. T. Shiomi, Y.-n. Harada, T. Saito, N. Shiomi, Y. Okuno and M. Yamaizumi, An ERCC5 gene with homology to yeast RAD2 is involved in group G xeroderma pigmentosum. *Mutat. Res.*, 314:167 (1994).
19. M.A. MacInnes, J.A. Dickson, R.R. Hernandez, D. Learmonth, G.Y. Lin, J.S. Mudgett, M.S. Park, S. Schauer, R.J. Reynolds, G.F. Strniste and J.Y. Yu, Human ERCC5 cDNA-cosmid complementation for excision repair and bipartite amino acid domains conserved with RAD proteins of *Saccharomyces cerevisiae* and *Schizosaccharomyces pombe*. *Mol. Cell Biol.*, 13:6393 (1993).
20. A. O' Donovan, D. Scherly, S.G. Clarkson and R.D. Wood, Isolation of active recombinant XPG protein, a human DNA repair endonuclease. *J. Biol. Chem.*, 269:15965 (1994).
21. T. Nouspikel and S.G. Clarkson, Mutations that disable the DNA repair gene *XPG* in a xeroderma pigmentosum group G patient. *Hum. Molec. Genet.*, 3:963 (1994).
22. J.E. Cleaver, F. Cortes, L.H. Lutze, W.F. Morgan, A.N. Player and D.L. Mitchell, Unique DNA repair properties of a xeroderma pigmentosum revertant. *Mol. Cell. Biol.*, 7:3353 (1987).
23. A. O' Donovan, A.A. Davies, J.G. Moggs, S.C. West and R.D. Wood, XPG endonuclease makes the 3' incision in human DNA nucleotide excision repair. *Nature*, 371:432 (1994).
24. Y. Habraken, P. Sung, L. Prakash and S. Prakash, Human xeroderma-pigmentosum group-G gene encodes a DNA endonuclease. *Nucleic Acids Res*, 22:3312 (1994).
25. J.J. Harrington and M.R. Lieber, Functional domains within FEN-1 and Rad2 define a family of structure-specific endonucleases - implications for nucleotide excision-repair. *Genes Dev*, 8:1344 (1994).

26. P. Robins, D.J.C. Pappin, R.D. Wood and T. Lindahl, Structural and functional homology between mammalian DNase IV and the 5' nuclease domain of *Escherichia coli* DNA polymerase I. *J. Biol. Chem.*, 269:28535 (1994).
27. M.M. Jennerwein, A. Eastman and A.R. Khokhar, The role of DNA-repair in resistance of L1210 cells to isomeric 1,2-diaminocyclohexaneplatinum complexes and ultraviolet-irradiation. *Mutat. Res.*, 254:89 (1991).
28. A.F. Nichols, W.J. Schmidt, S.G. Chaney and A. Sancar, Limitations of the in vitro repair synthesis assay for probing the role of DNA repair in platinum resistance. *Chem. Biol. Interact.*, 81:223 (1992).
29. P. Calsou, J.-M. Barret, S. Cros and B. Salles, DNA excision-repair synthesis is enhanced in a murine leukemia L1210 cell line resistant to cisplatin. *Eur. J. Biochem.*, 211:403 (1993).
30. G.R. Gibbons, J.D. Page, S.K. Mauldin, I. Husain and S.G. Chaney, Role of carrier ligand in platinum resistance in L1210 cells. *Cancer Research*, 50:6497 (1990).
31. G. Gibbons, W. Kaufman and S. Chaney, Role of DNA replication in carrier-ligand-specific resistance to platinum compounds in L1210 cells. *Carcinogenesis*, 12:2253 (1991).
32. N. Farrell, Y. Qu, L. Feng and B. Van Houten, Comparison of chemical reactivity, cytotoxicity, interstrand cross-linking and DNA sequence specificity of bis(platinum) complexes containing monodentate or bidentate coordination spheres with their monomeric analogues. *Biochemistry*, 29:9522 (1990).
33. N. Farrell, Y. Qu and M.P. Hacker, Cytotoxicity and antitumor activity of bis(platinum) complexes. A novel class of platinum complexes active in cell lines resistant to both cisplatin and 1,2-diaminocyclohexane complexes. *J Med Chem*, 33:2179 (1990).
34. N. Farrell, L.R. Kelland, J.D. Roberts and M.V. Beusichem, Activation of the *trans* geometry in platinum antitumor complexes: a survey of the cytotoxicity of *trans* complexes containing planar ligands in murine L1210 and human tumor panels and studies on their mechanism of action. *Cancer Res.*, 52:5065 (1992).
35. L.R. Kelland, P. Mistry, G. Abel, S.Y. Loh, C.F. O' Neill, B.A. Murrer and K.R. Harrap, Mechanism-related circumvention of acquired cis-diamminedichloroplatinum(II) resistance using 2 pairs of human ovarian-carcinoma cell-lines by ammine amine platinum(IV) dicarboxylates. *Cancer. Res.*, 52:3857 (1992).
36. P.A. Andrews and S.B. Howell, Cellular pharmacology of cisplatin - perspectives on mechanisms of acquired-resistance. *Canc. Cells*, 2:35 (1990).
37. H. Timmer-Bosscha, N. Mulder and E. de Vries, Modulation of cis-diamminedichloroplatinum(II) resistance: a review. *Brit J Cancer*, 66:227 (1992).
38. G. Chu, Cellular-responses to cisplatin - the roles of DNA-binding proteins and DNA-repair. *J Biol Chem*, 269:787 (1994).
39. D. Coverley, M.K. Kenny, D.P. Lane and R.D. Wood, A role for the human single-stranded DNA binding protein HSSB/RPA in an early stage of nucleotide excision repair. *Nucleic Acids Res.*, 20:3873 (1992).

DNA REPAIR AND THE CARRIER LIGAND SPECIFICITY OF PLATINUM RESISTANCE

Stephen G. Chaney and Edward L. Mamenta

Department of Biochemistry and Biophysics, Lineberger Comprehensive Cancer Center and Curriculum in Toxicology
University of North Carolina
Chapel Hill, NC 27599-7260

INTRODUCTION

The major mechanisms of platinum resistance at the DNA level appear to be enhanced DNA repair and increased tolerance of Pt-DNA adducts. It is clear that enhanced repair activity, as measured by removal of Pt-DNA adducts from genomic DNA, is observed in a great many platinum-resistant cell lines (1-3). This enhanced removal of Pt-DNA adducts most likely represents an increase in nucleotide excision repair capacity (4), since that is the only repair process which has been shown to remove Pt-DNA adducts. Furthermore, those cell lines with defects in nucleotide excision repair show enhanced sensitivity to platinum complexes (5,6). Recent research has defined the enzymes needed for nucleotide excision repair (7). However, it is not yet clear which of these enzymes are limiting for nucleotide excision repair *in vivo* and are, therefore, increased in amount and/or activity in cisplatin-resistant cell lines. For example, Dabholkar *et al.* (8,9) have reported increased expression of ERCC1 and XPAC in tissue from patients who were resistant to cisplatin chemotherapy. However, both Katz *et al.* (10) and Sheibani and Eastman (11) have found no increased expression of ERCC1 in cisplatin-resistant cell lines. Similarly, we have found no increased expression or cisplatin-inducibility of XPAC in cisplatin-resistant cell lines (Vaisman *et al.*., manuscript in preparation).

Pt-DNA adducts, like other bulky adducts, appear to be preferentially repaired in actively transcribed genes (12-15). However the role of gene-specific repair in platinum resistance is uncertain at this point. Zhen *et al.* (12) have reported a better correlation between platinum resistance and repair of interstrand cross-links in actively transcribed genes than for platinum resistance and repair of intrastrand adducts in genomic DNA, interstrand cross-links in genomic DNA or intrastrand adducts in actively transcribed genes. In contrast, Johnson *et al.* (16,17) recently reported that interstrand cross-links were repaired at the same rates in actively transcribed genes, inactive genes and non-

Platinum and Other Metal Coordination Compounds in Cancer Chemotherapy 2
Edited by H.M. Pinedo and J.H. Schornagel, Plenum Press, New York, 1996

coding regions of the DNA. However, the question of whether or not interstrand cross-links are preferentially repaired in actively transcribed genes may be irrelevant since both groups are in agreement that repair of interstrand cross-links is enhanced in platinum-resistant cell lines (12,16,17).

None of these observations provides quantitative information on how important this enhanced repair activity is to the overall platinum resistance of the cell. This is best gauged by comparing the cell's ability to tolerate Pt-DNA adducts (measured as the number of Pt-DNA adducts required to give 50% inhibition of cell growth) with its ability to repair Pt-DNA adducts. A number of recent studies, in which both tolerance of and repair of Pt-DNA adducts have been carefully measured, have suggested that the increased nucleotide excision repair activity is insufficient by itself to explain the enhanced tolerance of Pt-DNA adducts in platinum-resistant cell lines (18-23). This is most readily apparent in the human ovarian carcinoma C13* (12,24,25) and SK-OV-3 (21) and human small cell lung carcinoma SW1271 (23) cell lines which display enhanced tolerance of Pt-DNA adducts, but no significant increase in overall repair activity.

RESULTS AND DISCUSSION

The discrepancy between repair and tolerance of Pt-DNA adducts in cisplatin-resistant cell lines can also be seen by comparing tolerance and repair in panels of mouse leukemia (26) and human ovarian carcinoma (16) cell lines with increasing 1levels of platinum resistance (Figure 1). In the mouse leukemia panel of cell lines, (Figure 1A) the least resistant cell line was 22.5-fold resistant, had a 18.4-fold increase in the ability to tolerate Pt-DNA adducts and only a 2.1-fold increase in the repair of Pt-DNA adducts. In the human ovarian carcinoma panel of cell lines (Figure 1B), the least resistant cell line was 16-fold resistant, had a 6.8-fold increase in tolerance and only a 2.2-fold increase in repair. In the more resistant cell lines tolerance of Pt-DNA adducts increased linearly with resistance up to a 70-fold-increase in tolerance for the mouse leukemia cell lines and 64-fold for the human ovarian cell lines, while repair activity remained relatively constant.

The difference between tolerance and repair of Pt-DNA adducts can be seen even more dramatically when one examines the carrier ligand specificity of resistance. It had been known for some time that platinum compounds with the 1,2-diaminocyclohexane(dach) carrier ligand are able to circumvent platinum resistance in some (27-29), but not all (30,31), cisplatin-resistant cell lines. Previous data had shown that this carrier ligand specificity of resistance could not be explained by carrier ligand effects on platinum accumulation, efflux or incorporation into DNA (18,19). In every platinum-resistant cell line studied there was a clear carrier ligand specificity with respect to the tolerance of Pt-DNA adducts that matched the carrier ligand specificity of resistance (18,19). However, nucleotide excision repair showed no carrier ligand specificity, either *in vitro* (32) or in cell culture (18,19).

This is best seen in the L1210 cell lines where we have measured removal of intrastrand adducts from genomic DNA (18), interstrand adducts from genomic DNA and interstrand adducts from the actively transcribed DHFR gene. The later experiment was performed in collaboration with Dr. Lone Peterson and Vilhelm Bohr. The L1210 cell lines used in these experiments are particularly appropriate for studying the carrier ligand specificity of resistance since the L1210/DDP cell line is much more resistant to ethylenediamine (en)-Pt and cis-diammine-Pt compounds than to dach-Pt compounds, while the L1210/DACH cell line has the opposite specificity. The initial formation of interstrand cross-links following a 5 hour exposure to Pt(en)Cl_2 or 1,2-diaminocylohexanetetrachloroplatinum(IV) (ormaplatin) is shown in Figure 2.

The formation of interstrand cross-links as a percentage of total adducts was decreased in both the L1210/DDP and L1210/DACH cell lines. However, no selectivity was observed formation of en-Pt versus dach-Pt interstrand cross-links. Decreased formation of interstrand cross-links has been reported previously for other platinum-resistant cell lines (34-36). It is usually attributed to "quenching" of Pt-DNA

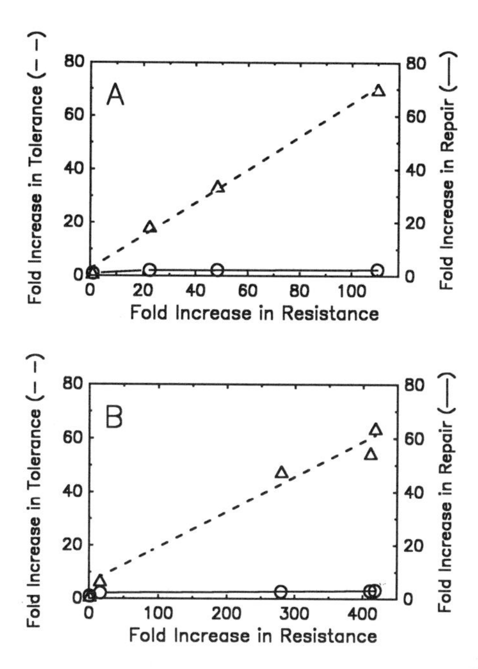

Figure 1. Tolerance and Repair of Pt-DNA Adducts in Cisplatin-Resistant Cell Lines. The figure depicts the fold increase in tolerance of Pt-DNA adducts (– –), measured as the number of Pt-DNA adducts/base pair of DNA required to inhibit cell growth by 50%, and the fold increase in repair of Pt-DNA adducts (—), measured as the % Pt-DNA adducts removed over the first 12 hr(B) or 24 hr(A), in a series of cell lines with increased resistance to cisplatin. A: mouse leukemia cell lines derived from L1210/0 (from Eastman and Schulte (26)). B: human ovarian carcinoma cell lines derived from A2780 (from Johnson *et al.* (16)).

monoadducts by reaction with glutathione or other non-protein sulfhydryls, which are often elevated in the resistant cell lines (34-36). However, no elevation of glutathione or other non-protein sulfhydryl is seen in the L1210/DACH cell line (18,37). Since previous studies have shown that the nucleotide excision repair system repairs Pt-G monoadducts more rapidly than Pt-GG or Pt-AG diadducts (32), we consider preferential repair of Pt-DNA monoadducts as a more likely explanation for the decrease in interstrand cross-link formation seen in these platinum-resistant cell lines. Thus, the increase in nucleotide excision repair seen in the L1210/DDP and L1210/DACH cell lines probably contributes to resistance both through increased removal of Pt-DNA intrastrand cross-links and through preferential removal of Pt-DNA monoadducts before they can form interstrand cross-links. However, neither of these processes appears to contribute to the carrier ligand specificity of resistance.

The net repair of total interstrand cross-links at 24 hours is shown in Table 1.

Interstrand cross-links continued to form in all cell lines following removal of drug, reaching a maximum at 3-8 hours after removal of drug (data not shown) and were subsequently repaired only slowly. In fact, in the L1210/DDP cell line interstrand cross-links had not returned to their initial levels even by 24 hours. No significant enhancement in the repair of interstrand cross-links was observed in either the L1210/DDP or L1210/DACH cell lines. Nor was any carrier ligand specificity in the repair of interstrand cross-links observed in any of these cell lines. These data are fully consistent with the early studies of Strandberg *et al.* (38,39) in these cell lines. Those studies showed that complete repair of the interstrand cross-links often required 48-72 hours.

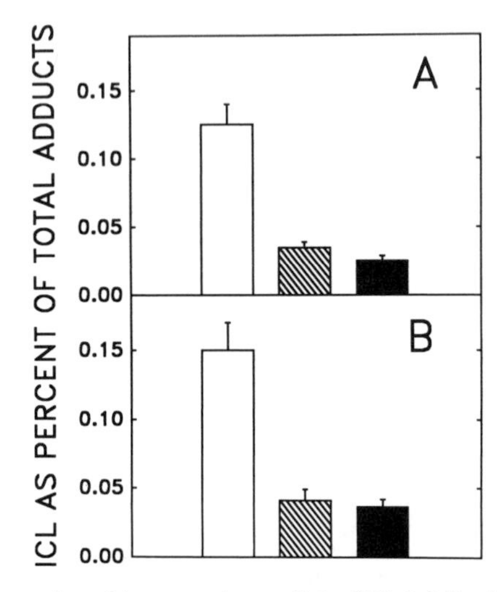

Figure 2. Net formation of interstrand cross-links (ICLs) following a 5 hr exposure to $Pt(en)Cl_2$ or ormaplatin. ICLs were determined by alkaline elution as described by Kohn (33). Total Pt-DNA adducts were determined as described previously (18). ICLs and total Pt-DNA adducts were measured immediately following a 5 hr exposure of cells to either $Pt(en)Cl_2$ (A) or ormaplatin (B). L1210/0, blank bars; L1210/DDP, hatched bars; L1210/DACH, filled bars.

Finally, repair of interstrand cross-links in the actively transcribed DHFR gene is shown in Figure 3. All three cell lines appear to show preferential repair of interstrand cross-links in the actively transcribed DHFR gene. Furthermore, both platinum-resistant cell lines appear to show enhanced repair of interstrand cross-links in the DHFR gene, while showing no detectable enhancement of interstrand cross-link repair in genomic DNA. Both observations are similar to those made previously by Zhen *et al.* (12) in cisplatin-resistant human ovarian carcinoma cell lines. However, these interpretations should be made with caution because it is not clear that the alkaline elution assay and renaturing agarose gel electrophoresis assay are measuring the same type of adducts (Peterson *et al.*, in preparation). In any case, the most significant observation from these experiments is the lack of carrier ligand specificity. While there may have been some preferential repair of dach-Pt interstrand cross-links in the L1210/0 cell line (Fig 3A), there was no detectable difference in the repair of en-Pt and dach-Pt interstrand cross-links in the two resistant cell lines (Fig 3 B, C). Thus, in the L1210 cell lines there is no carrier ligand specificity for the repair of intrastrand adducts, preferential; repair of monoadducts

Table 1. Repair of total interstrand cross-links[a]

Cell Line	Drug[b]	% Repair at 24 hr[c]
L1210/0	$Pt(en)Cl_2$	73.2 ± 7.5% (n=5)
L1210/0	$Pt(dach)Cl_4$	63.4 ± 4.2% (n=5)
L1210/DDP	$Pt(en)Cl_2$	-117 ± 31% (n=4)
L1210/DDP	$Pt(dach)Cl_4$	-119 ± 25% (n=4)
L1210/DACH	$Pt(en)Cl_2$	77.5 ± 5.1% (n=4)
L1210/DACH	$Pt(dach)Cl_4$	60.7 ± 4.5% (n=3)

[a]Cells were incubated with 100 μg/ml of $Pt(en)Cl_2$ or ormaplatin for 5 hours at 37° C (except that L1210/DACH was incubated with 200 μg/ml ormaplatin). The cells were then collected by centrifugation, washed once with PBS and incubated an additional 24 hours in fresh RPMI-1640 medium. ISC were determined as described in Figure 2.
[b]$Pt(dach)Cl_4$ = ormaplatin
[c]Expressed as average ± standard derivation; n = number of biological repeats.

(resulting in decreased interstrand cross-link formation), repair of total interstrand cross-links, or the repair of interstrand cross-links in an actively transcribed gene. We have also observed no significant differences in the repair of dach-Pt and cis-diammine-Pt adducts in sensitive and cisplatin-resistant human ovarian carcinoma cell lines (19). This lack of specificity for the repair of Pt-DNA adducts is fully consistent with *in vitro* and *in vivo* data on the specificity of the nucleotide excision repair system (4,32). These data clearly show that while nucleotide excision repair probably makes some contribution to the overall extent of platinum resistance, it does not contribute to carrier-ligand specificity of that resistance.

Since nucleotide excision repair does not appear to account for the carrier ligand specificity of resistance, or even to be sufficient to account for the full degree of platinum resistance at the DNA level, other possibilities must be considered. Increased tolerance of Pt-DNA adducts could be due to cell cycle arrest, decreased apoptosis or enhanced replicative bypass of Pt-DNA adducts. Since cell cycle arrest and apoptosis will be covered elsewhere in the symposium, I will focus on replicative bypass. Replicative bypass is the rate-limiting step in post-replication repair, which is a well-established mechanism for dealing with severe DNA damage (40). The ability of the cells to replicate past Pt-DNA adducts has been studied in some detail. *In vitro* experiments have shown that intrastrand diadducts (either GG or AG) are very effective at blocking the progression of DNA polymerases (41,42). However, most cells have at least a limited ability to replicate past bulky DNA lesions *in vivo* (40). Since any complete block to replication is likely to prove lethal to the cell, we postulated that this replicative bypass might play a significant role in platinum resistance. In experiments with L1210 cells, Gibbons *et al.*(43) found a selective 3.7-fold increase in the ability of L1210/DDP cells to replicate past en-Pt adducts, with essentially no change in their ability to replicate past dach-Pt adducts.

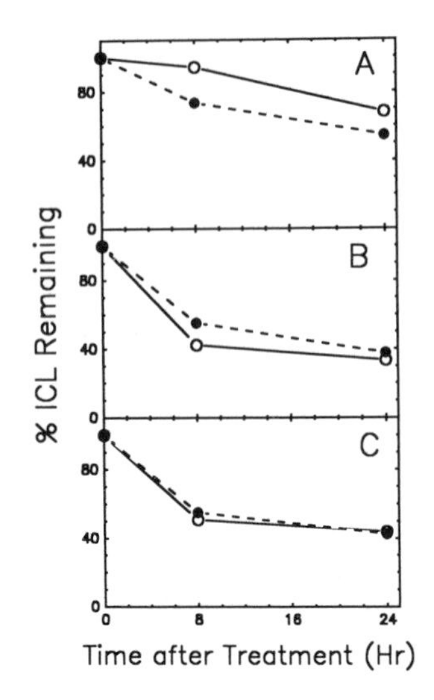

Figure 3. Repair of interstrand cross-links (ICLs) in the actively-transcribed DHFR gene. Cells were incubated with $Pt(en)Cl_2$ or ormaplatin as described in Figure 2. Aliquots of cells were collected immediately following incubation with the Pt drug and at 8 and 24 hour post-incubation. Interstrand cross-links in the 16 kb Kpn1 fragment of the DHFR gene were determined by renaturating agarose gel electrophoresis as described by Zhen *et al.* (12). A: L1210/0; B: L1210/DDP; C: L1210/DACH; (● - - - ●), dach-Pt ICLs; (0 —— 0), en-Pt ICLs.

Conversely, they observed a 3.6-fold increase in the ability of L1210/DACH cells to replicate past dach-Pt adducts, but only a 20% increase in their ability to replicate past en-Pt adducts. These observations have more recently been extended to human ovarian carcinoma cell lines where Mamenta *et al.* (25) have also observed a strong correlation between the carrier ligand specificity of resistance and the carrier ligand specificity of replicative bypass. Thus, in all cell lines tested to date, the ability to replicate past Pt-DNA adducts appears to play an important role in Pt resistance. It appears to account for much of the enhanced ability of Pt-resistant cell lines to tolerate Pt-DNA adducts, and is the only mechanism of resistance which reliably predicts the carrier ligand specificity of replicative bypass in mouse leukemia cell lines (L1210/0) and human ovarian carcinoma cell lines (A2780 and 2008) and their Pt-resistant derivations (Table 2).

In evaluating the significance of these observations, there are several points worth noting: 1) These data rely on the assumption that there are no major differences between the types of Pt-DNA adducts formed in resistant and sensitive cell lines or for dach-Pt versus en-Pt or cis-diammine-Pt adducts. The elegant experiments of Eastman (26,44), Fichtinger-Schepman (45) and their coworkers appear to support this assumption. 2) Several laboratories have suggested that the inhibition of DNA synthesis cannot be the critical step in Pt-induced cytoxicity because the levels of Pt adducts required to inhibit DNA synthesis are at least an order of magnitude greater than those required to cause cell cycle arrest (46-48). We have argued, however, that inhibition of DNA synthesis in only a minority of replicons should be sufficient to cause cell lethality and this level of inhibition of DNA synthesis would be undetectable by most standard assays of DNA synthesis (43). More recently Demarcq *et al.* (6) have reported that not only are repair deficient cells

Table 2. The relative contribution of accumulation, repair and replicative bypass to resistance and the carrier ligand specificity of resistance.

	L1210/DDP vs L1210/0	L1210/DACH vs L1210/0	A2780/DDP vs A2780	C13* vs 2008
Fold:[a]				
↑Resistance	15.1[c]	10[c]	7.4[e]	12.0[f]
↓Accumulation	2.2[c]	2.2[c]	2.4[e]	1.8[f]
↑Tolerance	7.5[c]	10.1[c]	4.8[e]	6.9[f]
↑Repair	1.8[c]	1.9[c]	1.2[e]	1.1[f]
↑Replicative Bypass	3.8[d]	3.6[d]	2.3[f]	4.5[f]
Relative:[b]				
↑Resistance	3.4[c]	4.4[c]	0.9[e]	3.8[f]
↓Accumulation	1.8[c]	2.1[c]	1.2[e]	1.4[f]
↑Tolerance	1.8[c]	2.1[c]	1.4[e]	2.2
↑Repair	0.96[c]	1.0[c]	1.2[e]	1.0[f]
↑Replicative Bypass	3.6[d]	3.0[d]	N/A	2.1[f]

[a]For $Pt(en)Cl_2$ in L1210/DDP compared to L1210/0, $Pt(dach)Cl_2$ in L1210/DACH compared to L1210/0 and cisplatin in A2780/DDP compared to A2780 and C13* compared to 2008.
[b]For $\Delta PtCl_2(en)/\Delta PtCl_2(dach)$ in L1210/DDP compared to L1210/0, $\Delta PtCl_2(dach)/\Delta PtCl_2(en)$ in L1210/DACH compared to L1210/0 and Δcisplatin/Δormaplatin in A2780/DDP compared to A2780 and C13* compared to 2008.
[c]From Gibbons *et al.* (18)
[d]From Gibbons *et al.* (43)
[e]From Schmidt and Chaney (19)
[f]From Mamenta *et al.* (25)

more sensitive to platinum complexes, but that no significant inhibition of DNA synthesis was observed at doses which caused greater than 99% cell kill. It is important to remember, however, that post-replication repair requires both replicative bypass and subsequent repair of the damaged DNA. Thus, replicative bypass can only reduce cytotoxicity in a repair-proficient cell line. In a repair deficient cell line, replicative bypass is likely to result in catastrophic problems during mitosis or the subsequent cell cycle as observed by Demarcq *et al.* (6). 4) In all of the cell lines we have studied, replicative bypass appears to make a stronger contribution to resistance than differences in nucleotide excision repair process (Table 2).Thus, the available data (Figure 1, Table 2) suggest that nucleotide excision repair activity is normally operating at near maximal efficiency and is increased to only a relatively small extent in platinum-resistant cell lines, while replicative bypass is limiting and appears to be increased to a far greater extent in the resistant cell lines. The data ((18,19,32), Table 1, Figure 2) also suggest that nucleotide excision repair discriminates very poorly between Pt-DNA adducts with different carrier ligands and that only the replicative bypass component of post-replicative repair reliably correlates with the carrier ligand specificity of platinum resistance.

The fact that the C13* cell line shows a significant increase in replicative bypass, but little difference in most parameters of repair with respect to 2008, has allowed us to explore the specificity of this process in some detail. The C13* cell line does not differ from 2008 in sensitivity to either UV or benzo(a)pyrene-7,8-diol-9,10-epoxide (BPDE) and we have found no difference in replicative bypass of UV or BPDE adducts in those cell lines (Table 3).

Table 3. Comparison of replicative bypass and cytoxicity in the 2008 and C13* cell lines[a]

DNA Damaging agent	Cell line	Replicative Bypass (D_0)[b]	Resistance (ID_{50})	
Cisplatin	2008	17.8	3.1 μM	
	C13*	81.0 (4.5)	37.1 μM	(12.0)
$PtCl_2$(dach)	2008	4.9	1.1 μM	
	C13*	10.0 (2.1)	3.4 μM	(3.1)
UV	2008	6.0	7.1 J/m^2	
	C13*	6.0 (1.0)	7.1 J/m^2	(1.0)
BPDE	2008	0.57	0.30 μM	
	C13*	0.65 (1.1)	0.31 μM	(1.0)

[a] from Mamenta *et al.* (25)

[b]D_0 refers to the number of adducts/10^5 base pairs required to inhibit DNA chain elongation by 63% using the steady-state replication assay.

Conversely, Boyer *et al.* (49) have shown that XP variant cells have a defect in replication past UV adducts, but do not differ in their ability to replicate past BPDE adducts compared to normal human fibroblasts. Taken together these data suggest at least 2 independent mechanisms of replicative bypass, one specific for UV adducts and other specific for Pt adducts. Neither system appears to afford significant bypass of BPDE adducts. The mechanism of this specificity is not known, but could involve recognition of the sequence specificity of the adducts, hydrophobicity of the adducts, or subtle differences in the conformation.

To further characterize the specificity of this bypass, we have examined the resistance of C13* to a variety of classical alkylating agents (Table 4). C13* is cross-resistant to melphalan, 4-OH-cyclophosphamide and partially cross-resistant to mechlorethamine. Preliminary data suggest that it is also cross-resistant to BCNU. Since there is little or no difference in nucleotide excision repair (12,24,25) and only a 2-fold increase in glutathione levels (50) in the C13* cell line compared to 2008, it seemed likely that replicative bypass contributed to this cross-resistance. We have examined that hypothesis in more detail for melphalan-treated cells. The inhibition of DNA synthesis by melphalan adducts was much less in the C13* cell line than in the parental 2008 cell line

Table 4. Cytotoxicity of alkylating agents in 2008 and C13*

Alkylating Agent	Cell Line	IC_{50}[a]
melphalan	2008	2.07μM
	C13*	28.0μM (13.5)
4-OH-cyclophosphamide	2008	2.03μM
	C13*	25.0μM (12.3)
mechlorethamine	2008	0.50μM
	C13*	2.27μM (4.5)

[a]Concentration of drug which, following a 1-hour exposure, caused a 50% inhibition of cell growth in the 3 day growth inhibition assay. Fold resistance is shown in parenthesis.

(Figure 4A). This differential effect of DNA adducts on DNA synthesis is very similar to that previously seen for Pt-DNA adducts in these two cell lines (Figure 4B).

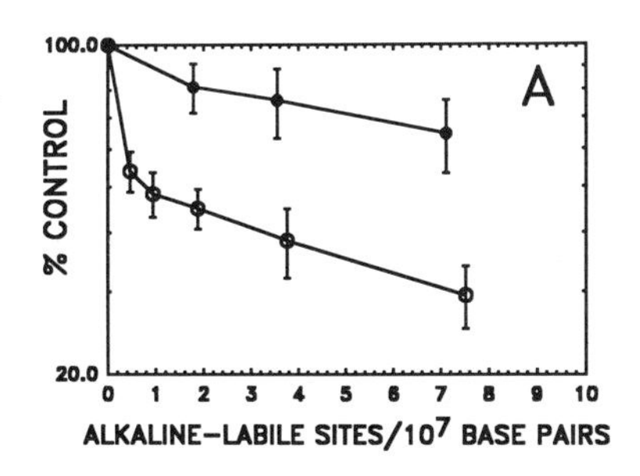

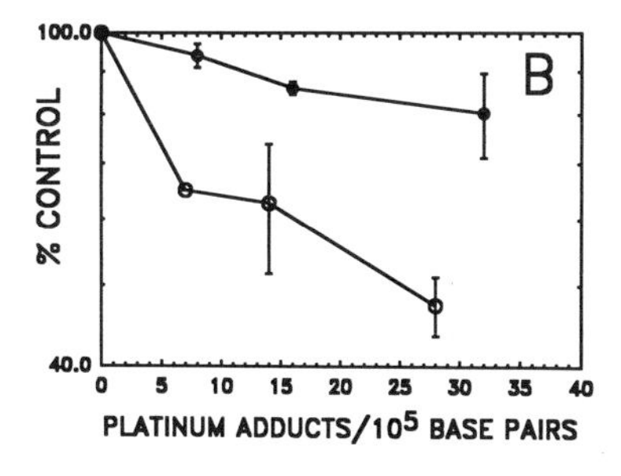

Figure 4. Inhibition of DNA synthesis by agents that form bulky DNA-adducts. C13* (● —— ●) or 2008 (0 —— 0) cells were treated for 15 minutes with various concentration of melphalan (A) or cisplatin (B). After a 45 minute chase in fresh culture medium, the cells were pulse labeled for 30 minutes with ^{3}H-thymidine to determine the rate of DNA synthesis. Melphalan-DNA adducts were determined by a neutral depurination and alkaline hydrolysis assay (51). Interstrand cross-links were measured by a rapid renaturation assay (52) and were a constant fraction of total adducts in both cell lines. Pt-DNA adducts were measured by atomic absorption (18).

Thus, the enhanced replicative bypass activity in the C13* cell line appears to be able to bypass Pt-DNA adducts and adducts formed by classical alkylating agents, but not adducts formed by UV or BPDE. What does this tell us about the specificity of this replicative bypass process? There appear to be two possibilities: 1) Alkylating agents and platinum complexes share the characteristic of forming diadducts at Pur-Pur sequences. Thus, the bypass activity that is enhanced in C13* cells might be specific for bypass of adducts at Pur-Pur sequences; 2) The alkylating agents and platinum complexes also share the characteristic of forming transient monoadducts which are slowly converted to the more stable diadducts (53). Moreover, at the time replicative bypass is assessed experimentally in these cell lines, the majority of adducts are still in the monoadduct form. While several experiments have shown that Pt-DNA monoadducts are less inhibitory for DNA chain elongation than diadducts (41,42), the elegant experiments of Holler *et al.* (54) demonstrate that monoadducts can inhibit DNA replication to a significant extent. Thus, the bypass activity which is enhanced in the C13* cell line also might be specific for the bypass of Pt-DNA monoadducts. Obviously, our current data cannot distinguish between these possibilities.

The molecular mechanism(s) of replicative bypass are ill-defined at present. However, there is experimental evidence for at least three mechanisms of replicative bypass in mammalian cells (40,55): 1) direct translesion replication, 2) template switching and 3) activation of alternative origins of replication near the site of damage, followed by gap filling. Based on models from *E. coli* (56,57), one might speculate that the mechanism of translesion replication would consists of damage recognition proteins (DRPs) and enzymes which increase the processivity of DNA polymerases. For example, PCNA has been shown to enhance the processivity of DNA polymerase δ (58) and to enhance its ability to synthesize DNA past pyrimidine dimers located in the template strand (59). However, the question of which DRPs might play a role in this process is less clear. Several DRPs have been identified with specificity for Pt-DNA adducts (60-63). In theory, such proteins could either enhance replicative bypass in resistant cell lines or interfere with replicative bypass in sensitive cell lines. However, the Pt-specific DRPs identified to date do not appear to be present at different levels in resistant or sensitive cell lines, or to be inducible by Pt treatment (60-63). We have recently shown that UV-DRP is Pt-inducible, that the extent of inducibility is greater in cisplatin-resistant cell lines, and that the inducibility correlates with replicative bypass rather than repair in those cell lines (64). However, no direct role of UV-DRP in replicative bypass or repair has been demonstrated at this time. We have recently initiated studies to characterize the molecular mechanism of this replicative bypass in more detail.

CONCLUDING REMARKS

The available data suggest that nucleotide excision repair makes an important contribution to resistance. There are currently over 70 reports in the literature of platinum-resistant cell lines with enhanced repair activity. However, the increased repair activity does not appear to be the only determinant of enhanced tolerance of Pt-DNA adducts in resistant cell lines. Not only does increased repair activity appear to make a relatively minor contribution to enhanced tolerance of Pt-DNA adducts in cell lines where both have been measured, but repair activity does not increase in proportion to tolerance of Pt-DNA adducts in cell lines with increasing platinum resistance. These data suggest that nucleotide excision repair activity may not be severely limiting, even in the platinum-sensitive cell lines. Furthermore, nucleotide excision repair activity does not appear to make any contribution to the carrier ligand specificity of resistance. Gene-specific repair may make a contribution to resistance, but recent data suggest that this contribution may need to be reassessed. Finally, our data suggest that the replicative bypass step of post-replication repair also makes a strong contribution to resistance. Increased replicative bypass and nucleotide excision repair activity together appear to account for most of the enhanced tolerance of Pt-DNA adducts in platinum-resistant cell lines. More importantly, replicative bypass is the only activity which correlates with the carrier ligand specificity of resistance at the DNA level. Thus, further characterization of the mechanisms and specificity of replicative bypass should greatly improve our understanding of platinum resistance. This could have important clinical applications. For example, dach-Pt complexes appear to be effective against some cisplatin resistant cell lines and human tumors, but not against others. Thus, a better understanding of the mechanisms leading to carrier-ligand specific replicative bypass might allow prediction of which tumors are likely to be responsive to dach-Pt complexes.

ACKNOWLEDGMENTS

This research was supported in part by U.S.PHS Grant CA 34082. We would like to thank Dr. Steven Wyrick for the preparation of the platinum compounds, Dr. Alan Eastman for the L1210 cell lines and Dr. Paul Andrews for the human ovarian carcinoma cell lines used in these experiments. The experiments in my laboratory were primarily conducted by Ed Mamenta, Heather Grady, Dr. Gregory Gibbons and Dr. Wendelyn Schmidt. The experiments on repair of ISC in the DHFR gene were performed by Dr. Lone Peterson in the laboratory of Dr. Vilhelm Bohr. I would also like to acknowledge the helpful comments and suggestions of colleges Dr. Aziz Sancar, Dr. William Kaufmann, Dr. Marila Cordeiro-Stone and Dr. David Holbrook.

REFERENCES

1. Andrews, P.A., and Howell, S.B., Cellular pharmacology of cisplatin: perspectives on mechanisms of acquired resistance. Cancer Cells, *2*: 35-43, 1990.
2. Johnson, S.W., Ozols, R.F., and Hamilton, T.C. Mechanisms of drug resistance in ovarian cancer. Cancer, *71*: 644-649, 1993.
3. Calsou, P. and Salles, B. Role of DNA repair in the mechanisms of cell resistance to alkylating agents and cisplatin. Cancer Chemother. Pharmacol., *32*: 85-89, 1993.
4. Sancar, A., and Tang, M.S. Nucleotide Excision Repair. Photochem. Photobiol., *57*: 905-921, 1993.
5. Sheibani, N., Jennerwein, M.M., and Eastman, A. DNA repair in cells sensitive and resistant to cis-diamminedichloroplatinum(II):host cell reactivation of damaged plasmid DNA. Biochemistry, *28*: 3120-3124, 1989.
6. Demarcq, C., Bunch, R.T., Creswell, D., and Eastman, A. The role of cell cycle progression in cisplatin-induced apoptosis in Chinese hamster ovary cells. Cell Growth Differ., *5*: 983-993, 1994.
7. Sancar, A. Mechanisms of DNA excision repair. Science, *266*: 1954-1956, 1994.
8. Dabholkar, M., Bostick-Bruton, F., Weber, C., Bohr, V.A., Egwuagu, C., and Reed, E. ERCC1 and ERCC2 expression in malignant tissues from ovarian cancer patients. J. Nat. Cancer Inst., *84*: 1512-1517, 1992.
9. Dabholkar, M., Vionnet, J., Bostick-Bruton, F., Yu, J.J., and Reed, E. Messenger RNA levels of XPAC and ERCC1 in ovarian cancer tissue correlate with response to platinum-based chemotherapy. J. Clin. Invest., *94*: 703-708, 1994.

10. Kratz, E.J., Andrews, P.A., and Howell, S.B. The effect of DNA polymerase inhibitors on the cytotoxicity of cisplatin in human ovarian carcinoma cells. Cancer Commun., *2*: 159-164, 1990.
11. Sheibani, N., and Eastman, A. Analysis of various mRNAs potentially involved in cisplatin resistance of murine leukemia L1210 cells. Cancer Lett., *52*: 179-185, 1990.
12. Zhen, W.P., Link, C.J., O'Connor, P.M., Reed, E., Parker, R., Howell, S.B., and Bohr, V.A. Increased gene-specific repair of cisplatin interstrand cross-links in cisplatin-resistant human ovarian cancer cell lines. Mol. Cell Biol, *12*: 3689-3698, 1992.
13. May, A., Nairn, R.S., Okumoto, D.S., Wassermann, K., Stevnsner, T., Jones, J.C., and Bohr, V.A. Repair of individual DNA strands in the hamster dihydrofolate reductase gene after treatment with ultraviolet light, alkylating agents, and cisplatin. J. Biol. Chem., *268*: 1650-1657, 1993.
14. Larminat, F., Zhen, W.P., and Bohr, V.A. Gene-specific DNA repair of interstrand cross-links induced by chemotherapeutic agents can be preferential. J. Biol. Chem., *268*: 2649-2654, 1993.
15. Larminat, F., and Bohr, V.A. Role of the human ERCC-1 gene in gene-specific repair of cisplatin-induced DNA damage. Nucleic. Acids. Res., *22*: 3005-3010, 1994.
16. Johnson, S.W., Perez, R.P., Godwin, A.K., Yeung, A.T., Handel, L.M., Ozols, R.F., and Hamilton, T.C. Role of platinum-DNA adduct formation and removal in cisplatin resistance in human ovarian cancer cell lines. Biochem. Pharmacol., *47*: 689-697, 1994.
17. Johnson, S.W., Swiggard, P.A., Handel, L.M., Brennan, J.M., Godwin, A.K., Ozols, R.F., and Hamilton, T.C. Relationship between platinum-DNA adduct formation and removal and cisplatin cytotoxicity in cisplatin-sensitive and -resistant human ovarian cancer cells. Cancer Res., *54*: 5911-5916, 1994.
18. Gibbons, G.R., Page, J.D., Mauldin, S.K., Husain, I., and Chaney, S.G. Role of carrier ligand in platinum resistance in L1210 cells. Cancer Res., *50*: 6497-6501, 1990.
19. Schmidt, W., and Chaney, S.G. Role of Carrier Ligand in Platinum Resistance of Human Carcinoma Cell Lines. Cancer Res., *53*: 799-805, 1993.
20. Rawlings, C.J., and Roberts, J.J. Walker rat carcinoma cells are exceptionally sensitive to cis-diamminedichloroplatinum(II) (cisplatin) and other difunctional agents but not defective in the removal of platinum-DNA adducts. Mutat. Res., *166*: 157-168, 1986.
21. Shellard, S.A., Hosking, L.K., and Hill, B.T. Anomolous relationship between cisplatin sensitivity and the formation and removal of platinum-DNA adducts in two human ovarian carcinoma cell lines *in vitro*. Cancer Res., *51*: 4557-4564, 1991.
22. Hill, B.T., Shellard, S.A., Hosking, L.K., Fichtinger-Schepman, A.M.J., and Bedford, P. Enhanced DNA repair and tolerance of DNA damage associated with resistance to cis-diamminedichloroplatinum(II) after in vitro exposure of a human teratoma cell line to fractionated X-irradiation. Int. J. Radiat. Oncol. Biol. Phys., *19*: 75-83, 1990.
23. Shellard, S.A., Fichtinger-Schepman, A.M.J., Lazo, J.S., and Hill, B.T. Evidence of differential cisplatin DNA adduct formation, removal and tolerance of DNA damage in three human lung carcinoma cell lines. Anti-Cancer. Drug, *4*: 491-500, 1993.
24. Jekunen, A.P., Hom, D.K., Alcarez, J.E., Eastman, A., and Howell, S.B. Cellular pharmacology of dichloro(ethylenediamine)platinum(II) in cisplatin-sensitive and -resistant human ovarian carcinoma cells. Cancer Res., *54*: 2680-2687, 1994.
25. Mamenta, E.L., Poma, E.E., Kaufmann, W.K., Delmastro, D.A., Grady, H.L., and Chaney, S.G. Enhanced replicative bypass of platinum-DNA adducts in cisplatin-resistant human ovarian carcinoma cell lines. Cancer Res., *54*: 3500-3505, 1994.
26. Eastman, A., and Schulte, N. Enhanced DNA repair as a mechanism of resistance to cis-diamminedichloroplatinum(II). Biochemistry, *27*: 4730-4734, 1988.
27. Wilkoff, L.J., Dulmadge, E.A., Trader, M.W., Harrison, S.D.,Jr., and Griswold, D.P. Evaluation of trans-tetrachloro-1,2-diaminocyclohexane platinum (IV) in murine leukemia L1210 resistant and sensitive to cis-diamminedichloroplatinum (II). Cancer Chemother. Pharmacol., *20*: 96-100, 1987.
28. Behrens, B.C., Hamilton, T.C., Masuda, H., Grotzinger, K.R., Whang-Peng, J., Louie, K.G., Knutsen, T., McKoy, W.M., Young, R.C., and Ozols, R.F. Characterization of a cis-diamminedichloroplatinum(II)-resistant human ovarian cancer cell line and its use in evaluation of platinum analogues. Cancer Res., *47*: 414-418, 1987.
29. Teicher, B.A., Holden, S.A., Herman, T.S., Stomayor, E.A., Khandekar, V., Rosbe, K.W., Brann, T.W., Korbut, T.T., and Frei, E. III Characteristics of five human tumor cell lines and sublines resistant to cis-diamminedichloroplatinum(II). Int. J. Cancer, *47*: 252-260, 1991.

30. Perez, R.P., O'Dwyer, P.J., Handel, L.M., Ozols, R.F., and Hamilton, T.C. Comparative cytotoxicity of CI-973, cisplatin, carboplatin and tetraplatin in human ovarian carcinoma cell lines. Int. J. Cancer, *48*: 265-269, 1991.
31. Hills, C.A., Kelland, L.R., Abel, G., Riracky, J., Wilson, A.P., and Harrap, K.R. Biological properties of ten human ovarian cell lines: calibration *in vitro* against four platinum complexes. Br. J. Cancer, *59*: 527-534, 1989.
32. Page, J.D., Husain, I., Sancar, A., and Chaney, S.G. Effect of the diaminocyclohexane carrier ligand on platinum adduct formation, repair, and lethality. Biochemistry, *29*: 1016-1024, 1990.
33. Kohn, K.W., and Grimek-Ewig, R.A. Alkaline elution analysis, a new approach to the study of DNA single-strand interuptions in cells. Cancer Res., *33*: 1849-1853, 1973.
34. Micetich, K., Zwelling, L.A., and Kohn, K.W. Quenching of DNA:platinum(II) monoadducts as a possible mechanism of resistance to cis-diamminedichloroplatinum(II) in L1210 cells. Cancer Res., *43*: 3609-3613, 1983.
35. Fram, R.J., Woda, B.A., Wilson, J.M., and Robichaud, N. Characterization of acquired resistance to cis-diamminedichloroplatinum(II) in BE human colon carcinoma cells. Cancer Res., *50*: 72-77, 1990.
36. Hospers, G.A.P., Mulder, N.H., de Jong, B., de Ley, L., Uges, D.R.A., Fichtinger-Schepman, A.M.J., Scheper, R.J., and de Vries, E.G.E. Characterization of a human small cell lung carcinoma cell line with acquired resistance to cis-diamminedichloroplatinum(II) in vitro. Cancer Res., *48*: 6803-6807, 1988.
37. Richon, V.M., Schulte, N., and Eastman, A. Multiple mechanisms of resistance to cis-diamminedichloroplatinum(II) in murine leukemia L1210 cells. Cancer Res., *47*: 2056-2061, 1987.
38. Strandberg, M.C., Bresnick, E., and Eastman, A. The significance of DNA cross linking to cis-diamminedichloroplatinum(II)-induced cytotoxicity in sensitive and resistant lines of murine leukemia L1210 cells. Chem. -Biol. Interact., *39*: 169-180, 1982.
39. Strandberg, M.C., Bresnick, E., and Eastman, A. DNA crosslinking induced by 1,2-diaminocyclohexanedichloroplatinum(II) in murine leukemia L1210 cells and comparison with other platinum analogs. Biochim. Biophys. Acta, *698*: 128-133, 1982.
40. Kaufmann, W.K. Pathways of human cell post-replication repair. Carcinogenesis, *10*: 1-11, 1989.
41. Hoffmann, J.-S., Johnson, N.P., and Villani, G. Conversion of monofunctional DNA adducts of cis-diamminedichloroplatinum(II) to bifunctional lesions. J. Biol. Chem., *264*: 15130-15135, 1989.
42. Comess, K.M., Burstyn, J.N., Essigmann, J.M., and Lippard, S.J. Replication inhibition and translesion synthesis on templates containing site-specifically placed cis-diamminedichloroplatinum(II) DNA adducts. Biochemistry, *31*: 3975-3990, 1992.
43. Gibbons, G.R., Kaufmann, W.K., and Chaney, S.G. Role of DNA replication in carrier-ligand-specific resistance to platinum compounds in L1210 cells. Carcinogenesis, *12*: 2253-2257, 1991.
44. Jennerwein, M.M., Eastman, A., and Khokhar, A.R. Characterization of adducts produced in DNA by isomeric 1,2-diaminocyclohexaneplatinum(II) complexes. Chem. -Biol. Interact., *70*: 39-49, 1989.
45. Fichtinger-Schepman, A.M.J., Vendrik, C.P.J., van Dijk-Knijnenburg, W.C.M., De Jong, W.H., van der Minnen, A.C.E., Claessen, A.M.E., van der Velde-Visser, S.D., de Groot, G., Wubs, K.L., Steerenberg, P.A., Schornagel, J.H., and Berends, F. Platinum concentrations and DNA adduct levels in tumors and organs of cisplatin-treated LOU/M rats innoculated with cisplatin-sensitive or resistant immunoglobulin M immunocytoma. Cancer Res., *49*: 2862-2867, 1989.
46. Salles, B., Butour, J.L., Lesca, C., and Macquet, J.P. cis-$Pt(NH_3)_2Cl_2$ and trans-$Pt(NH_3)_2Cl_2$ inhibit DNA synthesis in cultured L1210 leukemia cells. Biochem. Biophys,Res. Commun., *112*: 555-563, 1983.
47. Sorenson, C.M. and Eastman, A. Influence of cis-diamminedichloroplatinum(II) on DNA synthesis and cel l cycle progression in excision repair proficient and deficient chinese hamster ovary cells. Cancer Res., *48*: 6703-6707, 1988.
48. Just, G., and Holler, E. Platinum incorporation and differential effects of cis- and trans-diamminedichloroplatinum(II) on the growth of mouse leukemia P388/D1. Cancer Res., *49*: 7072-7077, 1989.
49. Boyer, J.C., Kaufmann, W.K., Brylawski, B.P., and Cordeiro-Stone, M. Defective post-replication repair in Xeroderma pigmentosum variant fibroblasts. Cancer Res., *50*: 2593-2598, 1990.
50. Zinkewich-Peotti, K., and Andrews, P.A. Loss of cis-diamminedichloroplatinum(II) resistance in human ovarian carcinoma cells selected for rhodamine-123 resistance. Cancer Res., *52*: 1902-1906, 1992.

51. Wassermann, K., Kohn, K.W., and Bohr, V.A. Heterogeneity of nitrogen-mustard-induced DNA damage and repair at the level of the gene in chinese hamster ovary cells. J. Biol. Chem., *265*: 13906-13913, 1990.
52. Matsuo, N., and Ross, P.M. Measurement of interstrand cross-link frequency and distance between interruptions in DNA exposed to 4,5',8-trimethylpsoralen and near-ultraviolet light. Biochemistry, *26*: 2001-2009, 1987.
53. Kohn, K.W. Molecular mechanisms of cross-linking by alkylating agents and platinum complexes. In: A.C. Sartorelli, J.S. Lazlo and J.R. Bertino (eds.), Molecular actions and targets for cancer chemotherapeutic agents, pp. 3-16, New York: Academic Press. 1981.
54. Holler, E., Bauer, R., and Bernges, F. Monofunctional DNA-platinum(II) adducts block frequently DNA polymerases. Nucleic. Acids Res., *20*: 2307-2312, 1992.
55. Naegeli, H. Roadblocks and detours during DNA replication: Mechanisms of mutagenesis in mammalian cells. Bioessays, *16*: 557-564, 1994.
56. Sweasy, J.B., Witkin, E.M., Sinha, N., and Roegner-Maniscalco, V. RecA protein of Escherichia coli has a third essential role in SOS mutator activity. J. Bacteriol., *172*: 3030-3036, 1990.
57. Frank, E.G., Hauser, J., Levine, A.S., and Woodgate, R. Targeting of the UmuD, UmuD', and MucA' mutagenesis proteins to DNA by RecA protein. Proc. Natl. Acad. Sci. USA, *90*: 8169-8173, 1993.
58. Hurwitz, J., Dean, F.B., Kwong, A.D., and Lee, S.-H. The in vitro replication of DNA containing the SV40 origen. J. Biol. Chem., *265*: 18043-18046, 1990.
59. O'Day, C.L., Burgers, P.M.J., and Taylor, J.S. PCNA-induced DNA synthesis past cis-syn and trans-syn-I thymine dimers by calf thymus DNA polymerase delta *in vitro*. Nucleic. Acids Res., *20*: 5403-5406, 1992.
60. Hughes, E.N., Engelsberg, B.N., and Billings, P.C. Purification of nuclear proteins that bind to cisplatin- damaged DNA - Identity with high mobility group protein-1 and protein-2. J. Biol Chem., *267*: 13520-13527, 1992.
61. Donahue, B.A., Augot, M., Bellon, S.F., Treiber, D.K., Tonry, J.H., Lippard, S.J., and Essigmann, J.M. Characterization of a DNA damage-recognition protein from mammalian cells that binds specifically to intrastrand d(GpG) and d(ApG) DNA adducts of the anticancer drug cisplatin. Biochemistry, *29*: 5872-5880, 1990.
62. Brown, S.J., Kellett, P.J., and Lippard, S.J. Ixr1, a yeast protein that binds to platinated DNA and confers sensitivity to cisplatin. Science, *261*: 603-605, 1993.
63. Andrews, P.A. and Jones, J.A. Characterization of binding proteins from ovarian carcinoma and kidney tubule cells that are specific for cisplatin modified DNA. Cancer Commun., *3*: 93-102, 1991.
64. Vaisman, A., and Chaney, S.G. Induction of UV-damage recognition protein by cisplatin treatment. Biochemistry, *34*: 105-114, 1995.

TRANSCRIPTION FACTOR DIFFERENCES IN CISPLATIN RESISTANT CELLS

John S. Lazo and Ya-Yun Yang

Department of Pharmacology and The Experimental Therapeutics Program, Pittsburgh Cancer Institute, University of Pittsburgh, Pittsburgh, PA 15261, USA

INTRODUCTION

Acquired resistance frequently develops in humans after the initial use of *cis*-diamminedichloroplatinum(II) (CP). Furthermore, both clinical and laboratory results suggest tumor cells with resistance to CP are often cross resistant to other structurally and mechanistically distinct agents. For example, the human squamous cell carcinoma SCC25/CP, which has 12-fold acquired resistance to CP, is also approximately 7-fold resistant to methotrexate, 5-fold resistant to melphalan and 3-fold resistant to cyclophosphamide (Kelley et al., 1988). Studies with other cells suggest CP resistance can be associated with cross resistance to x-radiation in human tumor cell lines (Schwartz et al., 1988). Thus, brief or repeated exposure to CP can lead to cross resistance to other cancer therapeutics but the mechanism for the drug cross-resistance is not known. Moreover, the generation of anticancer drug cross-resistance is not unique to CP and can be seen with other agents. Because many anticancer agents are genotoxic, altered gene expression might occur after drug treatment. Indeed, increased gene expression is frequently seen in cells with acquired resistance to anticancer drugs. The focus of the work described in this manuscript has been to examine the mechanistic basis for alter gene expression in a human cell line with acquired resistance to CP.

MT EXPRESSION AND CP RESISTANCE

All higher eukaryotes have the potential to synthesize multiple isoforms of the low molecular weight zinc-binding proteins metallothioneins (MTs). There is little evidence, however, that any of these MT isoforms, with the exception of neuronal specific MT III, have different biological functions. Putative factors in CP resistance are decreased drug accumulation, increased DNA repair, elevated glutathione levels and overexpression of MT (Yang et al. 1994). The relative importance of each of these factors has been the subject of considerable controversy and in many cell lines several factors may be responsible for the resistant phenotype. Numerous studies support a role for MT in CP resistance. For example, MT levels are elevated in some tumors with acquired resistance to MT (Lazo and Pitt, In Press). Cells with acquired resistance to heavy metals, such as cadmium, which induces the transcription of MT genes, are also cross resistant to CP (Lazo and Pitt, In Press). Transient induction of MT with dexamethasone or cytokines, such as interleukin-1β, can cause transient resistance to

Platinum and Other Metal Coordination Compounds in Cancer Chemotherapy 2
Edited by H.M. Pinedo and J.H. Schornagel, Plenum Press, New York, 1996

CP (Basu and Lazo, 1991, Kondo et al., 1994). Elevated MT mRNA and protein content has been seen in some human tumor samples. The levels of MT protein have been correlated with cellular responsiveness to CP by some investigators (Bahnson et al., 1994) but not others (Murphy et al., 1991). The differences, however, may reflect the different types of tumors sampled, the relatively small sample sizes, or the failure to account for differences in MT isoform composition or subcellular localization. CP can covalently interact with MT both *in vitro* and *in vivo* (Lazo and Pitt, In Press). For example, Petering and coworkers have precisely defined the stoichometery and kinetics of covalent interactions between CP and MT *in vitro* (Lemkuil et al., 1994). Although covalent interactions between CP and MT have been reported with cells resistant to CP and overexpressing MT, in our opinion, it remains to be determined whether CP directly interacts with MT within tumors with physiologically relevant levels of MT. Nevertheless, transfer and expression of genes encoding human MT IIa or mouse MT I demonstrate resistance to several electrophilic agents, including CP (Kelley et al., 1988).

Recently mice have been generated that lack both MT I and II expression due to homologous recombination with disrupted MT I and II genes (Michalska and Choo, 1993). Mouse embryo fibroblasts have been isolated from these mice and characterized; they have a similar morphology when grown in 20% fetal bovine serum containing medium and have a similar cell cycle distribution (Lazo et al., 1995). As anticipated they completely lack any basal or inducible MT expression and are approximately 3-4 fold more sensitive to cadmium compared to the wild-type cells. These MT null cells are approximately 5-fold more sensitive to CP than the wild-type mouse cells (Kondo et al., In Press). Thus, MT clearly can protect cells from the toxic actions of CP.

To determine the isotype of human MT that might be involved in CP resistance, we examined the isotype expression in three pairs of CP sensitive and resistant human cells with acquired drug resistance (Table 1) (see Yang et al., 1994 for additional details). No significant basal expression of human MT Ia, Ib, If or Ig was detected in any cells, suggesting overexpression of these isoforms was not commonly associated with the CP-resistant phenotype. The universal overexpression of human MT IIa and the relatively high overexpression of this isotype in SCC25/CP compared to the parental SCC25 cells focused our attention on this cell pair and the MT IIa isotype. The SCC25/CP cells were selected by exposure to increasing concentrations of CP and showed cross resistance to methotrexate and melphalan. Using nuclear run-on studies, we observed a 3-fold increase in MT IIa transcription rate (Yang et al., 1994). No significant differences in the degradation of MT IIa mRNA has been noted by us between SCC25/CP and SCC25 cells (Y.-Y. Yang, Ph.D. dissertation, 1994). These results suggest alterations in transcription rates might be responsible for elevated levels of human MT IIa in SCC25/CP cells.

Table 1. Basal level of human MT isoform mRNA in CP-resistant cells

	Fold Increase			
Cell Lines[a]	MT Ie	MT If	MT Ig	MT IIa
SCC25/CP	5.4 ± 0.3[b]	NE	NE	9.2 ± 0.5[b]
H69/CP	NE[c]	ND	ND[d]	2.0 ± 0.2[b]
SW2/CP	NE	NE	ND	3.1 ± 0.6[b]

[a] Levels relative to the parental values.

[b] Each value is the average of two independent experiments with the range.

[c] NE, not estimated because no basal levels were detected in the parental cells, only in the CP-resistant cells.

[d] ND, not detectable in either drug sensitive or drug resistant cells.

TRANSCRIPTIONAL CONTROL OF HUMAN MT IIa EXPRESSION

Each of the seven known human MT promoters have been sequenced and appear to have some common functional domains but also have unique sequences, which presumably defines their ligand responsiveness and tissue specificity. The 5′-flanking region of human MT IIa has been characterized extensively and a number of enhancer and promoter regions have been defined. The 5′-enhancer and promoter region of the human MT IIa gene can be functionally subdivided into two categories: the basal regulatory elements and the inducible elements, such as the metal regulatory and glucocorticoid regulatory elements. Distinct basal recognition sequences including TATA-box, GC-box, proximal and distal basal regulatory element sequences have been identified. Sequences involved in MT induciblity include Sp1, AP-1 and AP-2 sites (Haslinger and Karin, 1985; Scholer et al., 1986; Lee et al., 1987). These elements are used by cytokines, such as interleukin Iβ, heavy metals, such as cadmium or zinc, hypoxia, or drugs, such as dexamethasone, to induce MT IIa. Thus, transcriptional control of human MT IIa is complex. Interestingly, these previously characterized elements all appear to be positive, that is they increase the expression of MT rather than decrease it. A schematic of the known regulatory regions that have been identified on the human MT IIa are shown in Figure 1.

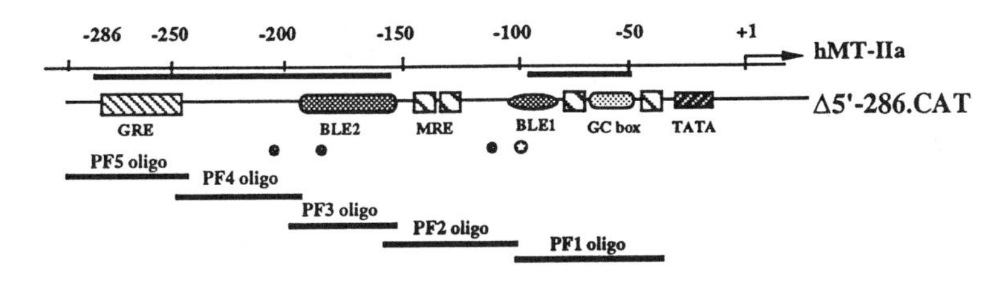

Figure 1. Schematic of the regulatory region of human MT IIa.

TRANSCRIPTIONAL CONTROL OF HUMAN MT IIa EXPRESSION IN CP RESISTANT CELLS

A correlation between DNA methylation of the MT promoter region and MT expression has previously been observed (Compere and Palmiter, 1981). Lieberman et al. (1983) correlated induction of murine MT by UV irradiation with the extent of DNA hypomethylation. The MT Ib gene, which is normally highly methylated in HeLa cells and not expressed, can be induced after treatment of cells with the demethylating agent 5′-azacytidine (Heguy et al., 1986). Thus, we used an MT IIa-specific probe and 5′-azacytidine to determine if hypermethylation in the SCC25 cells was responsible for the lower level of MT expression. Treatment of SCC25 cells with 16 μM azacytidine for 72 h produced a marked increase in MT IIa mRNA levels (Yang et al., 1994). In contrast we saw little increase in the MT IIa expression of SCC25/CP cells. Attempts to identify differences in the status of the promoter in CP-sensitive and -resistant cells using methylation-sensitive enzymes were not successful, although not all potential CpG methylation sites could be examined by these methods (Yang et al., 1994). Thus, we cannot exclude the possibility that methylation of human MT IIa promoter CpG site is responsible for the reduced MT expression in SCC25. Alternatively, the overexpression could be caused by demethylation of a regulatory region of a *trans* factor that affects MT IIa expression.

We have used a series of deletion mutations of the human MT IIa promoter linked to the bacterial chloramphenicol acetyltransferase (CAT) gene, which have been described (Scholer et al., 1986), to examine the promoter regions responsible for the increased transcription of MT IIa. Comparing the CAT activity profiles of SCC25 and SCC25/CP, we have found that multiple *cis*-acting elements were important in the region between -286 and -51. In particular, deletions from -286 to -160 produced a 50% reduction of CAT activity in SCC25/CP cells while a 90% reduction was seen in SCC25 cells. We then synthesized a series of MT IIa promoter fragments of 60 base pairs and probes nuclear extracts from both SCC25 and SCC25/CP cells using a competitive gel mobility shift assay. These results indicate that SCC25/CP have increased *trans*-acting factors binding to the -100 to -40, -195 to -151 and -245 to -190 region of the human MT IIa promoter.

INCREASED NUCLEAR PROTEIN BINDING TO HUMAN IIA PROMOTER IN CP RESISTANT CELLS

To examine further the potential *trans*-acting factors responsible for the increased transcription of MT IIa, we used synthetic oliognucleotides that bind to known transcription factors and competitive gel mobility shift assays. We found significant increases in Sp1 and AP-2 DNA binding activity in CP resistant SCC25/CP cells relative to CP sensitivity SCC25 cells. The identity of these factors was confirmed using antibodies specific for each factor and supershift assays. We do not know if other CP resistant cells exhibit a similar difference in Sp1 or AP-2 activity. Initial studies suggest the amount of Sp1 is not different in nuclear extracts of CP-sensitive and resistant cells so it is possible the increase in activity is due to altered posttranslational modifications or other accessory proteins.

POTENTIAL SIGNIFICANCE OF ENHANCED TRANSCRIPTION FACTORS IN DRUG-RESISTANT CELLS

Multiple mechanisms have been proposed for both acquired and intrinsic CP resistance in human tumor cells. Another common characteristic of acquired CP resistance has been the observation that cross resistance exist to anticancer drugs with different mechanistic and structural features. Altered transcription factor activity is an attractive mechanistic vehicle that could explain the diversity of resistance processes found in a single drug resistant population and the complex cross-resistant phenotype. Several drug resistant genes appear to be controlled in part by Sp1 or AP-2 activity. These include mdr1 (Cornwell et al., 1993), the excision repair gene XPBC/ERCC-3 (Ma et al.,1991), poly(ADP-ribose) polymerase (Yokoyama et al., 1990), and ERCC1 (Lee et al., 1993). It seem likely that other drug resistant genes will be responsive to these transcription factors, which are ubiquitous and multigenomic in their targets.

SUMMARY

In summary, we have found that expression of MT can be an important factor in determining the sensitivity of cells to anticancer drugs, most notably CP. For example, MT null cells are more sensitive to CP than wild type cells (Kondo et al., In Press). This supports other work using both pharmacologic and genetic methods to overexpress MT and reduce sensitivity to anticancer drugs (Kelley et al., 1988). We have also found the overexpression of human MT IIa in CP resistant SCC25/CP cells was followed by transcriptional activation of the promoter region of this gene. This correlated well with increased Sp1 and AP-2 DNA binding activity. We do not know the origins of the increased Sp1 and AP-2 activity currently but initial studies indicate some increase in the AP-2 amount in SCC25/CP cell nuclear extracts. We believe the increase expression of *trans* activating factor activity after treatment with genotoxic substances, such as CP, could altered the cellular phenotype and produce increased resistance to other antineoplastic agents. This might lead to multiple drug resistance commonly seen in the clinical situation. We believe additional studies should be directed to examining the transcriptional factors that are altered during the development of drug resistance. The resulting information should provide the bases for understanding better the regulatory mechanisms for the appearance of drug resistance phenotypes during cancer chemotherapy.

REFERENCES

Bahnson, R.R., Becich, M., Ernstoff, M.S., Sandlow, J., Cohen, M., and Williams, R.D., 1974, Absence of immunohistochemical metallothionein staining in bladder tumors specimens predicts response to neoadjuvant cisplatin, methotrexate and vinblastine chemotherapy, *J. Urol.* 152:2272-2275.

Basu, A. and Lazo, J.S., 1991, Suppression of dexamethasone-induced metallothionein expression and *cis*-diamminedichloroplatinum (II) resistance by v-mos, *Cancer Res.* 51:893-896.

Compere S.J. and Palmiter R.D., 1981, DNA methylation controls the inducibility of the mouse metallothionein-I gene in lymphoid cells, *Cell* 25:233-240.

Cornwell, M.M. and Smith, D.E., 1993, SP1 activates the MDR1 promoter through one of two distinct G-rich regions that modulate promoter activity, *J. Biol. Chem.* 268:19505-19511.

Haslinger, A. and Karin, M., 1985, Upstream promoter element of the human metallothionein-IIA gene can act like an enhancer element, *Proc. Natl. Acad. Sci., USA* 82:8572-8576.

Heguy, A., West, A., Richards, R.I., and Karin, M., 1986, Structure and tissue-specific expression of the human metallothionein Iβ gene, *Mol. Cell Biol.* 6:288149-2157.

Kelley, S.L., Basu, A., Teicher, B.A., Hacker, M.P., Hamer, D.H. and Lazo, J.S., 1988, Overexpression of metallothionein confers resistance to anticancer drugs, *Science* 241:1813-1815.

Kondo, Y., Kuo, S.-M. and Lazo, J.S., 1994, Interleukin 1β mediated metallothionein induction and cytoprotection against cadmium and *cis*-diamminedichloroplatinum, *J. Pharmacol. Exp. Therap.* 270:1313-1318.

Kondo, Y., Woo, E.S., Michalska, A.E., Choo, K.H.A. and Lazo, J.S., In Press, Metallothionein null cells have increased sensitivity to anticancer drugs, *Cancer Res.*

Lazo, J.S., Kondo, Y. Dellapiazza, D., Michalska, A.E., Choo, K.H.A., and Pitt, B.R., 1995, Enhanced sensitivity to oxidative stress in cultured embryonic cells from transgenic mice deficient in metallothionein I and II genes, *J. Biol. Chem.* 270:5506-5510.

Lazo, J.S. and Pitt, B.R., In Press, Metallothionein and Cell Death, *Ann. Rev. Pharmacol and Toxicol.*

Lee, W. Mitchell, P. and Tjian, R, 1987, Activation of transcription by two factors that bind promoter and enhancer sequence of the human metallothionein gene and SV40, *Nature* 325:368-372.

Lee, K.B., Parker, R.J., Bohr, V., Cornelison, T. and Reed, E., 1993, Cisplatin sensitivity/resistance to UV repair-deficient Chinese hamster ovary cells of complementation groups 1 and 3, *Carcinogenesis* 14:2177-2180.

Lemkuil, D.C., Nettesheim, D., Shaw, C.F., III and Petering, D.H., 1994, Reaction of Cd^7

metallothionein with *cis*-diamminedichloroplatinum(II), *J. Biol. Chem.* 269:24792-24797.

Lieberman, M.W., Beacher, L.R., and Palmiter, R.D., 1993, Ultraviolet radiation-induced metallothionein-I gene activation is associated with extensive DNA demethylation, *Cell* 35:207-214.

Ma, L., Weeda, G., Jochemsen, A.G., Bootsma, D., Hoeijmakers, J.H. and Eb, A.J.v.d. , 1991, Molecular and functional analysis of the XPBC/ERCC promoter: transcription activity is dependent on the integrity of an Sp-1 binding site, *Nucleic Acids Res.* 20:217-2241.

Michalska, A. E.and Choo,K.H.A., 1993, Targeting and germ-line transmission of a null mutation at the metallothionein I and II loci in mice, *Proc. Natl. Acad. Sci. USA* 90:8088-8092.

Murphy D.A., McGrown, A.T., Crowther, D., Manered, A. and Fox, B.W., 1991, Metallothionein levels in ovarian tumors before and after chemotherapy. *Br. J. Cancer* 63:711-714.

Scholer, H., Haslinger, A., Heguy, A., Holtgreve, H., and Karin, M., 1986, In vivo competition between metallothionein regulatory elements and the SV40 enhancer. *Science* 232:76-80.

Schwartz, J.L., Rotmensch, J., Beckett, M.A., Jaffe, D.R., Toohill, M., Giovanazzi, S.M., McIntosh, J. and Weichselbaum, R. R, 1988, X-ray and cis-diamminedichloroplatinum(II) is cross-resistant in human tumor cell lines, *Cancer Res.* 48:5133-5135.

Yang, Y.-Y., Woo, E.S., Reese, C.E., Bahnson, R.R., Saijo, N. and Lazo, J.S., 1994, Human metallothionein isoform gene expression in cisplatin-sensitive and resistant cells, *Mol. Pharm.* 45:453-460.

Yang, Y.-Y., Ph.D. Dissertation, 1994, Regulation of human metallothionein gene expression in cisplatin sensitive and resistant cells, University Microfilms, Inc.

Yokoyama, Y., Kawamoto, T., Mitsuuchi, Y., Kurosaki, T., Toda, K., Ushiro, H., Terashima, M., Summoto, H., Kuribayashi, I., Yamamoto, Y., Maeda, T., Ikeda, H., Sagara, Y. and Shizuta, Y., 1991, Human poly(ADP-ribose) polymerase gene: cloning of the promoter region, *Eur. J. Biochem.* 194:521-526.

INDEX